Statistical Inference via Data Science

A ModernDive into R and the Tidyverse

Chapman & Hall/CRC
The R Series

Series Editors

John M. Chambers, Department of Statistics, Stanford University, California, USA
Torsten Hothorn, Division of Biostatistics, University of Zurich, Switzerland
Duncan Temple Lang, Department of Statistics, University of California, Davis, USA
Hadley Wickham, RStudio, Boston, Massachusetts, USA

Recently Published Titles

For more information about this series, please visit: https://www.crcpress.com/go/the-r-series

Statistical Inference via Data Science

A ModernDive into R and the Tidyverse

Chester Ismay
Albert Y. Kim

CRC Press
Taylor & Francis Group
Boca Raton London New York

CRC Press is an imprint of the
Taylor & Francis Group, an **informa** business

A CHAPMAN & HALL BOOK

CRC Press
Taylor & Francis Group
6000 Broken Sound Parkway NW, Suite 300
Boca Raton, FL 33487-2742

© 2020 by Taylor & Francis Group, LLC
CRC Press is an imprint of Taylor & Francis Group, an Informa business

No claim to original U.S. Government works

International Standard Book Number-13: 978-0-367-40982-1 (paperback)
International Standard Book Number-13: 978-0-367-40987-6 (hardback)

Library of Congress Cataloging-in-Publication Data

Names: Ismay, Chester, author. | Kim, Albert Young-Sun, 1980- author.
Title: Statistical inference via data science : a ModernDive, into R and
the tidyverse / Chester Ismay, Albert Y. Kim.
Description: Boca Raton : Taylor and Francis, 2019. | Series: Chapman &
hall/crc the r series | Includes bibliographical references and index. |
Summary: "Statistical Inference via Data Science: A ModernDive into R
and the Tidyverse provides a pathway for learning about statistical
inference using data science tools widely used in industry, academia,
and government. It introduces the tidyverse suite of R packages,
including the ggplot2 package for data visualization, and the dplyr
package for data wrangling. After equipping readers with just enough of
these data science tools to perform effective exploratory data analyses,
the book covers traditional introductory statistics topics like
confidence intervals, hypothesis testing, and multiple regression
modeling, while focusing on visualization throughout"-- Provided by publisher.
Identifiers: LCCN 2019042572 (print) | LCCN 2019042573 (ebook) | ISBN 9780367409821
(paperback) | ISBN 9780367409876 (hardback) | ISBN 9780367409913 (ebook)
Subjects: LCSH: Statistics--Data processing. | Quantitative research. | R
(Computer program language)
Classification: LCC QA276.45.R3 I86 2019 (print) |LCC QA276.45.R3
(ebook) | DDC 519.5/4--dc23
LC record available at https://lccn.loc.gov/2019042572
LC ebook record available at https://lccn.loc.gov/2019042573

Visit the Taylor & Francis Web site at
http://www.taylorandfrancis.com

and the CRC Press Web site at
http://www.crcpress.com

Chester: To Karolyn, who has always been patient and continued to support me to write and finish edits on this book even sometimes when that meant working through a weekend. Your continued pushes for me to think about new ways to teach novices about R have meant the world to me.

Albert: 엄마와 아빠: 나 한번도 표현 못해도 그냥 다 고마워요. 행복하게 살수 있는 지금 모습이 모두가 다 당신때문이예요. To Ginna: Thanks for tolerating my playing of "Nothing In This World Will Ever Break My Heart Again" on repeat while I finished this book. I love you.

Contents

4 Data Importing and "Tidy" Data 99

II Data Modeling with `moderndive` 119

5 Basic Regression 121

6 Multiple Regression 161

III   Statistical Inference with `infer`   193

Foreword

These are exciting times in statistics and data science education. (I am predicting this statement will continue to be true regardless of whether you are reading this foreword in 2020 or 2050.) But (isn't there always a but?), as a statistics educator, it can also feel a bit overwhelming to stay on top of all the new statistical, technological, and pedagogical innovations. I find myself constantly asking, "Am I teaching my students the correct content, with the relevant software, and in the most effective way?". Before I make all of us feel lost at sea, let me point out how great a life raft I have found in *ModernDive*. In a sea of intro stats textbooks, *ModernDive* floats to the top of my list, and let me tell you why. (Note my use of *ModernDive* here refers to the book in its shortened title version. This also matches up nicely with the neat hex sticker[1] Drs. Ismay and Kim created for the cover of *ModernDive*, too.)

Why I ♥ ModernDive

* Provides students experience with the whole data analysis 🪈 line.
* Incorporates contemporary, user-friendly R packages directly into the text.
* Emphasizes models that prepare students for our multivariate 🌍.

My favorite aspect of *ModernDive*, if I must pick a favorite, is that students gain experience with the whole data analysis pipeline (see Figure 2). In particular, *ModernDive* is one of the few intro stats textbooks that teaches students how to wrangle data. And, while data cleaning may not be as groovy as model building, it's often a prerequisite step! The world is full of messy data and *ModernDive* equips students to transform their data via the `dplyr` package.

---

[1] https://moderndive.com/images/logos/hex_blue_text.png

Speaking of `dplyr`, students of *ModernDive* are exposed to the `tidyverse` suite of R packages. Designed with a common structure, `tidyverse` functions are written to be easy to learn and use. And, since most intro stats students are programming newbies, *ModernDive* carefully walks the students through each new function it presents and provides frequent reinforcement through the many *Learning checks* dispersed throughout the chapters.

Overall, *ModernDive* includes wise choices for the placement of topics. Starting with data visualization, *ModernDive* gets students building `ggplot2` graphs early on and then continues to reinforce important concepts graphically throughout the book. After moving through data wrangling and data importing, modeling plays a prominent role, with two chapters devoted to building regression models and a later chapter on inference for regression. Lastly, statistical inference is presented through a computational lens with calculations done via the `infer` package.

I first met Drs. Ismay and Kim while attending their workshop at the 2017 US Conference on Teaching Statistics[2]. They pushed us as participants to put data first and to use computers, instead of math, as the engine for statistical inference. That experience helped me modernize my own intro stats course and introduced me to two really forward-thinking statistics and data science educators. It has been exciting to see *ModernDive* develop and grow into such a wonderful, timely textbook. I hope you have decided to dive on in!

Kelly S. McConville, Reed College

---

[2]https://www.causeweb.org/cause/uscots/uscots17/workshop/3

Preface

Help! I'm new to R and RStudio and I need to learn about them! However, I'm completely new to coding! What do I do?

If you're asking yourself this question, then you've come to the right place! Start with the "Introduction for students" section.

- *Are you an instructor hoping to use this book in your courses? We recommend you first read the "Introduction for students" section first. Then, read the "Introduction for instructors" section for more information on how to teach with this book.*
- *Are you looking to connect with and contribute to ModernDive? Then, read the "Connect and contribute" section for information on how.*
- *Are you curious about the publishing of this book? Then, read the "About this book" section for more information on the open-source technology, in particular R Markdown and the bookdown package.*

Introduction for students

This book assumes no prerequisites: no algebra, no calculus, and no prior programming/coding experience. This is intended to be a gentle introduction to the practice of analyzing data and answering questions using data the way data scientists, statisticians, data journalists, and other researchers would.

We present a map of your upcoming journey in Figure 1.

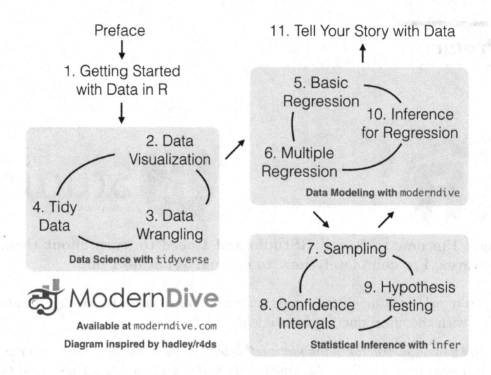

FIGURE 1: *ModernDive* flowchart.

You'll first get started with data in Chapter 1 where you'll learn about the difference between R and RStudio, start coding in R, install and load your first R packages, and explore your first dataset: all domestic departure `flights` from a New York City airport in 2013. Then you'll cover the following three portions of this book (Parts 2 and 4 are combined into a single portion):

1. Data science with `tidyverse`. You'll assemble your data science toolbox using `tidyverse` packages. In particular, you'll
    - Ch.2: Visualize data using the `ggplot2` package.
    - Ch.3: Wrangle data using the `dplyr` package.
    - Ch.4: Learn about the concept of "tidy" data as a standardized data input and output format for all packages in the `tidyverse`. Furthermore, you'll learn how to import spreadsheet files into R using the `readr` package.
2. Data modeling with `moderndive`. Using these data science tools and helper functions from the `moderndive` package, you'll fit your first data models. In particular, you'll

- Ch.5: Discover basic regression models with only one explanatory variable.
- Ch.6: Examine multiple regression models with more than one explanatory variable.

3. Statistical inference with `infer`. Once again using your newly acquired data science tools, you'll unpack statistical inference using the `infer` package. In particular, you'll:

- Ch.7: Learn about the role that sampling variability plays in statistical inference and the role that sample size plays in this sampling variability.
- Ch.8: Construct confidence intervals using bootstrapping.
- Ch.9: Conduct hypothesis tests using permutation.

4. Data modeling with `moderndive` (revisited): Armed with your understanding of statistical inference, you'll revisit and review the models you'll construct in Ch.5 and Ch.6. In particular, you'll:

- Ch.10: Interpret confidence intervals and hypothesis tests in a regression setting.

We'll end with a discussion on what it means to "tell your story with data" in Chapter 11 by presenting example case studies.[3]

What we hope you will learn from this book

We hope that by the end of this book, you'll have learned how to:

1. Use R and the `tidyverse` suite of R *packages* for data science.
2. Fit your first *models* to data, using a method known as *linear regression*.
3. Perform *statistical inference* using *sampling, confidence intervals.* and *hypothesis tests.*
4. *Tell your story with data* using these tools.

What do we mean by data stories? We mean any analysis involving data that engages the reader in answering questions with careful visuals and thoughtful discussion. Further discussions on data stories can be found in the blog post "Tell a Meaningful Story With Data."[4]

---

[3]Note that you'll see different versions of the word "ModernDive" in this book: (1) `moderndive` refers to the R package. (2) *ModernDive* is an abbreviated version of *Statistical Inference via Data Science: A ModernDive into R and the Tidyverse*. It's essentially a nickname we gave the book. (3) ModernDive (without italics) corresponds to both the book and the corresponding R package together as an entity.

[4]https://www.thinkwithgoogle.com/marketing-resources/data-measurement/tell-meaningful-stories-with-data/

Over the course of this book, you will develop your "data science toolbox," equipping yourself with tools such as data visualization, data formatting, data wrangling, and data modeling using regression.

In particular, this book will lean heavily on data visualization. In today's world, we are bombarded with graphics that attempt to convey ideas. We will explore what makes a good graphic and what the standard ways are used to convey relationships within data. In general, we'll use visualization as a way of building almost all of the ideas in this book.

To impart the statistical lessons of this book, we have intentionally minimized the number of mathematical formulas used. Instead, you'll develop a conceptual understanding of statistics using data visualization and computer simulations. We hope this is a more intuitive experience than the way statistics has traditionally been taught in the past and how it is commonly perceived.

Finally, you'll learn the importance of literate programming. By this we mean you'll learn how to write code that is useful not just for a computer to execute, but also for readers to understand exactly what your analysis is doing and how you did it. This is part of a greater effort to encourage reproducible research (see the "Reproducible research" subsection in this Preface for more details). Hal Abelson coined the phrase that we will follow throughout this book:

Programs must be written for people to read, and only incidentally for machines to execute.

We understand that there may be challenging moments as you learn to program. Both of us continue to struggle and find ourselves often using web searches to find answers and reach out to colleagues for help. In the long run though, we all can solve problems faster and more elegantly via programming. We wrote this book as our way to help you get started and you should know that there is a huge community of R users that are happy to help everyone along as well. This community exists in particular on the internet on various forums and websites such as stackoverflow.com[5].

Data/science pipeline

You may think of statistics as just being a bunch of numbers. We commonly hear the phrase "statistician" when listening to broadcasts of sporting events.

---

[5]https://stackoverflow.com/

Statistics (in particular, data analysis), in addition to describing numbers like with baseball batting averages, plays a vital role in all of the sciences.

You'll commonly hear the phrase "statistically significant" thrown around in the media. You'll see articles that say, "Science now shows that chocolate is good for you." Underpinning these claims is data analysis. By the end of this book, you'll be able to better understand whether these claims should be trusted or whether we should be wary. Inside data analysis are many sub-fields that we will discuss throughout this book (though not necessarily in this order):

- data collection
- data wrangling
- data visualization
- data modeling
- inference
- correlation and regression
- interpretation of results
- data communication/storytelling

These sub-fields are summarized in what Grolemund and Wickham have previously termed the "data/science pipeline"[6] in Figure 2.

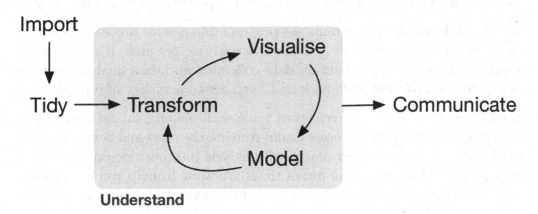

FIGURE 2: Data/science pipeline.

We will begin by digging into the grey **Understand** portion of the cycle with data visualization, then with a discussion on what is meant by tidy data and data wrangling, and then conclude by talking about interpreting and discussing the results of our models via **Communication**. These steps are vital to any statistical analysis. But, why should you care about statistics?

---

[6]http://r4ds.had.co.nz/explore-intro.html

There's a reason that many fields require a statistics course. Scientific knowledge grows through an understanding of statistical significance and data analysis. You needn't be intimidated by statistics. It's not the beast that it used to be and, paired with computation, you'll see how reproducible research in the sciences particularly increases scientific knowledge.

Reproducible research

The most important tool is the *mindset*, when starting, that the end product will be reproducible. – Keith Baggerly

---

Another goal of this book is to help readers understand the importance of reproducible analyses. The hope is to get readers into the habit of making their analyses reproducible from the very beginning. This means we'll be trying to help you build new habits. This will take practice and be difficult at times. You'll see just why it is so important for you to keep track of your code and document it well to help yourself later and any potential collaborators as well.

Copying and pasting results from one program into a word processor is not an ideal way to conduct efficient and effective scientific research. It's much more important for time to be spent on data collection and data analysis and not on copying and pasting plots back and forth across a variety of programs.

In traditional analyses, if an error was made with the original data, we'd need to step through the entire process again: recreate the plots and copy-and-paste all of the new plots and our statistical analysis into our document. This is error prone and a frustrating use of time. We want to help you to get away from this tedious activity so that we can spend more time doing science.

We are talking about *computational* reproducibility. - Yihui Xie

---

Reproducibility means a lot of things in terms of different scientific fields. Are experiments conducted in a way that another researcher could follow the steps and get similar results? In this book, we will focus on what is known as **computational reproducibility**. This refers to being able to pass all of one's data analysis, datasets, and conclusions to someone else and have them get exactly the same results on their machine. This allows for time to be spent interpreting results and considering assumptions instead of the more error prone way of starting from scratch or following a list of steps that may be different from machine to machine.

Final note for students

At this point, if you are interested in instructor perspectives on this book, ways to contribute and collaborate, or the technical details of this book's construction and publishing, then continue with the rest of the chapter. Otherwise, let's get started with R and RStudio in Chapter 1!

Introduction for instructors

Resources

Here are some resources to help you use *ModernDive*:

1. We've included review questions posed as *Learning checks*. You can find all the solutions to all *Learning checks* in Appendix D of the online version of the book at https://moderndive.com/D-appendixD.html.
2. Dr. Jenny Smetzer and Albert Y. Kim have written a series of labs and problem sets. You can find them at https://moderndive.com/labs.
3. You can see the webpages for two courses that use *ModernDive*:
   - Smith College "SDS192 Introduction to Data Science": https://rudeboybert.github.io/SDS192/.
   - Smith College "SDS220 Introduction to Probability and Statistics": https://rudeboybert.github.io/SDS220/.

Why did we write this book?

This book is inspired by

- *Mathematical Statistics with Resampling and R* (Chihara and Hesterberg, 2011)
- *OpenIntro: Intro Stat with Randomization and Simulation* (Diez et al., 2014)
- *R for Data Science* (Grolemund and Wickham, 2017)

The first book, designed for upper-level undergraduates and graduate students, provides an excellent resource on how to use resampling to impart statistical concepts like sampling distributions using computation instead of large-sample approximations and other mathematical formulas. The last two books are free options for learning about introductory statistics and data science, providing an alternative to the many traditionally expensive introductory statistics textbooks.

When looking over the introductory statistics textbooks that currently exist, we found there wasn't one that incorporated many newly developed R packages directly into the text, in particular the many packages included in the `tidyverse`[7] set of packages, such as `ggplot2`, `dplyr`, `tidyr`, and `readr` that will be the focus of this book's first part on "Data Science with `tidyverse`."

Additionally, there wasn't an open-source and easily reproducible textbook available that exposed new learners to all four of the learning goals we listed in the "Introduction for students" subsection. We wanted to write a book that could develop theory via computational techniques and help novices master the R language in doing so.

Who is this book for?

This book is intended for instructors of traditional introductory statistics classes using RStudio, who would like to inject more data science topics into their syllabus. RStudio can be used in either the server version or the desktop version. (This is discussed further in Subsection 1.1.1.) We assume that students taking the class will have no prior algebra, no calculus, nor programming/coding experience.

Here are some principles and beliefs we kept in mind while writing this text. If you agree with them, this is the book for you.

1. **Blur the lines between lecture and lab**
   - With increased availability and accessibility of laptops and open-source non-proprietary statistical software, the strict dichotomy between lab and lecture can be loosened.
   - It's much harder for students to understand the importance of using software if they only use it once a week or less. They forget the syntax in much the same way someone learning a foreign language forgets the grammar rules. Frequent reinforcement is key.
2. **Focus on the entire data/science research pipeline**
   - We believe that the entirety of Grolemund and Wickham's data/science pipeline[8] as seen in Figure 2 should be taught.

---

[7] http://tidyverse.org/
[8] http://r4ds.had.co.nz/introduction.html

- We heed George Cobb's call to "minimize prerequisites to research"[9]: students should be answering questions with data as soon as possible.

3. **It's all about the data**
   - We leverage R packages for rich, real, and realistic datasets that at the same time are easy-to-load into R, such as the `nycflights13` and `fivethirtyeight` packages.
   - We believe that data visualization is a "gateway drug" for statistics[10] and that the grammar of graphics as implemented in the `ggplot2` package is the best way to impart such lessons. However, we often hear: "You can't teach `ggplot2` for data visualization in intro stats!" We, like David Robinson[11], are much more optimistic and have found our students have been largely successful in learning it.
   - `dplyr` has made data wrangling much more accessible[12] to novices, and hence much more interesting datasets can be explored.

4. **Use simulation/resampling to introduce statistical inference, not probability/mathematical formulas**
   - Instead of using formulas, large-sample approximations, and probability tables, we teach statistical concepts using simulation-based inference.
   - This allows for a de-emphasis of traditional probability topics, freeing up room in the syllabus for other topics. Bridges to these mathematical concepts are given as well to help with relation of these traditional topics with more modern approaches.

5. **Don't fence off students from the computation pool, throw them in!**
   - Computing skills are essential to working with data in the 21st century. Given this fact, we feel that to shield students from computing is to ultimately do them a disservice.
   - We are not teaching a course on coding/programming per se, but rather just enough of the computational and algorithmic thinking necessary for data analysis.

6. **Complete reproducibility and customizability**
   - We are frustrated when textbooks give examples, but not the source code and the data itself. We give you the source code for all examples as well as the whole book! While we have made choices

---

[9] https://arxiv.org/abs/1507.05346
[10] http://escholarship.org/uc/item/84v3774z
[11] http://varianceexplained.org/r/teach_ggplot2_to_beginners/
[12] http://chance.amstat.org/2015/04/setting-the-stage/

to occasionally hide the code that produces more complicated figures, reviewing the book's GitHub repository will provide you with all the code (see below).

- Ultimately the best textbook is one you've written yourself. You know best your audience, their background, and their priorities. You know best your own style and the types of examples and problems you like best. Customization is the ultimate end. We encourage you to take what we've provided and make it work for your own needs. For more about how to make this book your own, see "About this book" later in this Preface.

Connect and contribute

If you would like to connect with ModernDive, check out the following links:

- If you would like to receive periodic updates about ModernDive (roughly every 6 months), please sign up for our mailing list[13].
- Contact   Albert   at   `albert.ys.kim@gmail.com`   and   Chester   at `chester.ismay@gmail.com`.
- We're on Twitter at `https://twitter.com/ModernDive`.

If you would like to contribute to *ModernDive*, there are many ways! We would love your help and feedback to make this book as great as possible! For example, if you find any errors, typos, or areas for improvement, then please email us or post an issue on our GitHub issues[14] page. If you are familiar with GitHub and would like to contribute, see the "About this book" section.

Acknowledgements

The authors would like to thank Nina Sonneborn[15], Dr. Alison Hill[16], Kristin Bott[17], Dr. Jenny Smetzer[18], and the participants of our 2017[19] and 2019[20]

---

[13]`http://eepurl.com/cBkItf`
[14]`https://github.com/moderndive/moderndive_book/issues`
[15]`https://github.com/nsonneborn`
[16]`https://alison.rbind.io/`
[17]`https://twitter.com/rhobott?lang=en`
[18]`https://www.smith.edu/academics/faculty/jennifer-smetzer`
[19]`https://www.causeweb.org/cause/uscots/uscots17/workshop/3`
[20]`https://www.causeweb.org/cause/uscots/uscots19/workshop/4`

USCOTS workshops for their feedback and suggestions. We'd also like to thank Dr. Andrew Heiss[21] for contributing nearly all of Subsection 1.2.3 on "Errors, warnings, and messages," Evgeni Chasnovski[22] for creating the new `geom_parallel_slopes()` extension to the `ggplot2` package for plotting parallel slopes models, and Starry Zhou[23] for her many edits to the book. A special thanks goes to Dr. Yana Weinstein, cognitive psychological scientist and co-founder of The Learning Scientists[24], for her extensive feedback.

We were both honored to have Dr. Kelly S. McConville[25] write the Foreword of the book. Dr. McConville is a pioneer in statistics education and was a source of great inspiration to both of us as we continued to update the book to get it to its current form. Thanks additionally to the continued contributions by members of the community[26] to the book on GitHub and to the many individuals that have recommended this book to others. We are so very appreciative of all of you!

Lastly, a very special shout out to any student who has ever taken a class with us at either Pacific University, Reed College, Middlebury College, Amherst College, or Smith College. We couldn't have made this book without you!

About this book

This book was written using RStudio's bookdown[27] package by Yihui Xie (Xie, 2019). This package simplifies the publishing of books by having all content written in R Markdown[28]. The bookdown/R Markdown source code for all versions of ModernDive is available on GitHub:

- **Latest online version** The most up-to-date release:
  - Version 1.0.0 released on November 25, 2019 (source code[29])
  - Available at `https://moderndive.com/`
- **Print version** The CRC Press print version of *ModernDive* corresponds to Version 1.0.0.

---

[21] `https://twitter.com/andrewheiss`
[22] `https://github.com/echasnovski`
[23] `https://github.com/Starryz`
[24] `http://www.learningscientists.org/yana-weinstein/`
[25] `https://www.reed.edu/faculty-profiles/profiles/mcconville-kelly.html`
[26] `https://github.com/moderndive/ModernDive_book/graphs/contributors`
[27] `https://bookdown.org/`
[28] `http://rmarkdown.rstudio.com/html_document_format.html`
[29] `https://github.com/moderndive/moderndive_book/releases/tag/v1.0.0`

- **Development online version** The working copy of the next version which is currently being edited:
    - Preview of development version is available at `https://moderndive.netlify.com/`.
    - Source code: Available on ModernDive's GitHub repository page at `https://github.com/moderndive/moderndive_book`.
- **Previous online versions** Older versions that may be out of date:
    - Version 0.6.1[30] released on August 28, 2019 (source code[31])
    - Version 0.6.0[32] released on August 7, 2019 (source code[33])
    - Version 0.5.0[34] released on February 24, 2019 (source code[35])
    - Version 0.4.0[36] released on July 21, 2018 (source code[37])
    - Version 0.3.0[38] released on February 3, 2018 (source code[39])
    - Version 0.2.0[40] released on August 2, 2017 (source code[41])
    - Version 0.1.3[42] released on February 9, 2017 (source code[43])
    - Version 0.1.2[44] released on January 22, 2017 (source code[45])

Could this be a new paradigm for textbooks? Instead of the traditional model of textbook companies publishing updated *editions* of the textbook every few years, we apply a software design influenced model of publishing more easily updated *versions*. We can then leverage open-source communities of instructors and developers for ideas, tools, resources, and feedback. As such, we welcome your GitHub pull requests.

Finally, since this book is under a Creative Commons Attribution - NonCommercial - ShareAlike 4.0 license[46], feel free to modify the book as you wish for your own non-commercial needs, but please list the authors at the top of `index.Rmd` as: "Chester Ismay, Albert Y. Kim, and YOU!"

---

[30]`https://moderndive.com/previous_versions/v0.6.1/index.html`
[31]`https://github.com/moderndive/ModernDive_book/releases/tag/v0.6.1`
[32]`https://moderndive.com/previous_versions/v0.6.0/index.html`
[33]`https://github.com/moderndive/moderndive_book/releases/tag/v0.6.0`
[34]`https://moderndive.com/previous_versions/v0.5.0/index.html`
[35]`https://github.com/moderndive/moderndive_book/releases/tag/v0.5.0`
[36]`https://moderndive.com/previous_versions/v0.4.0/index.html`
[37]`https://github.com/moderndive/moderndive_book/releases/tag/v0.4.0`
[38]`https://moderndive.com/previous_versions/v0.3.0/index.html`
[39]`https://github.com/moderndive/moderndive_book/releases/tag/v0.3.0`
[40]`https://moderndive.com/previous_versions/v0.2.0/index.html`
[41]`https://github.com/moderndive/moderndive_book/releases/tag/v0.2.0`
[42]`https://moderndive.com/previous_versions/v0.1.3/index.html`
[43]`https://github.com/moderndive/moderndive_book/releases/tag/v0.1.3`
[44]`https://moderndive.com/previous_versions/v0.1.2/index.html`
[45]`https://github.com/moderndive/moderndive_book/releases/tag/v0.1.2`
[46]`https://creativecommons.org/licenses/by-nc-sa/4.0/`

About the authors

Chester Ismay	Albert Y. Kim

Chester Ismay is a Data Science Evangelist at DataRobot in Portland, OR, USA. In this role, he leads data science, machine learning, and data engineering in-person workshops for DataRobot University. He completed his PhD in statistics from Arizona State University in 2013. He has previously worked in a variety of roles including as an actuary at Scottsdale Insurance Company (now Nationwide E&S/Specialty), as a freelance data science consultant, and at Ripon College, Reed College, and Pacific University. In addition to his work for *ModernDive*, he also contributed as initial developer of the infer[47] R package and is author and maintainer of the thesisdown[48] R package.

- Email: chester.ismay@gmail.com
- Webpage: https://chester.rbind.io/
- Twitter: old_man_chester[49]
- GitHub: https://github.com/ismayc

---

[47]https://cran.r-project.org/package=infer
[48]https://github.com/ismayc/thesisdown
[49]https://twitter.com/old_man_chester

Albert Y. Kim is an Assistant Professor of Statistical & Data Sciences at Smith College in Northampton, MA, USA. He completed his PhD in statistics at the University of Washington in 2011. Previously he worked in the Search Ads Metrics Team at Google Inc. as well as at Reed, Middlebury, and Amherst Colleges. In addition to his work for *ModernDive*, he is a co-author of the resampledata[50] and SpatialEpi[51] R packages.

- Email: albert.ys.kim@gmail.com
- Webpage: https://rudeboybert.rbind.io/
- Twitter: rudeboybert[52]
- GitHub: https://github.com/rudeboybert

Both Drs. Ismay and Kim, along with Jennifer Chunn[53], are co-authors of the fivethirtyeight[54] package of code and datasets published by the data journalism website FiveThirtyEight.com[55].

---

[50]https://cran.r-project.org/package=resampledata
[51]https://cran.r-project.org/package=SpatialEpi
[52]https://twitter.com/rudeboybert
[53]https://github.com/jchunn
[54]https://fivethirtyeight-r.netlify.com/
[55]https://fivethirtyeight.com/

1

Getting Started with Data in R

Before we can start exploring data in R, there are some key concepts to understand first:

1. What are R and RStudio?
2. How do I code in R?
3. What are R packages?

We'll introduce these concepts in the upcoming Sections 1.1-1.3. If you are already somewhat familiar with these concepts, feel free to skip to Section 1.4 where we'll introduce our first dataset: all domestic flights departing one of the three main New York City (NYC) airports in 2013. This is a dataset we will explore in depth for much of the rest of this book.

1.1 What are R and RStudio?

Throughout this book, we will assume that you are using R via RStudio. First time users often confuse the two. At its simplest, R is like a car's engine while RStudio is like a car's dashboard as illustrated in Figure 1.1.

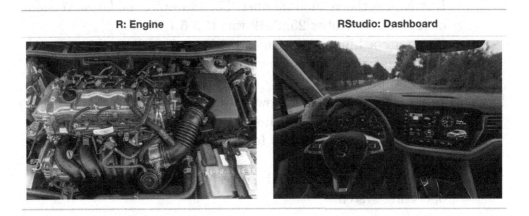

R: Engine   **RStudio: Dashboard**

FIGURE 1.1: Analogy of difference between R and RStudio.

More precisely, R is a programming language that runs computations, while RStudio is an *integrated development environment (IDE)* that provides an interface by adding many convenient features and tools. So just as the way of having access to a speedometer, rearview mirrors, and a navigation system makes driving much easier, using RStudio's interface makes using R much easier as well.

1.1.1 Installing R and RStudio

Note about RStudio Server or RStudio Cloud: If your instructor has provided you with a link and access to RStudio Server or RStudio Cloud, then you can skip this section. We do recommend after a few months of working on RStudio Server/Cloud that you return to these instructions to install this software on your own computer though.

You will first need to download and install both R and RStudio (Desktop version) on your computer. It is important that you install R first and then install RStudio.

1. **You must do this first:** Download and install R by going to `https://cloud.r-project.org/`.
   - If you are a Windows user: Click on "Download R for Windows", then click on "base", then click on the Download link.
   - If you are macOS user: Click on "Download R for (Mac) OS X", then under "Latest release:" click on R-X.X.X.pkg, where R-X.X.X is the version number. For example, the latest version of R as of November 25, 2019 was R-3.6.1.
   - If you are a Linux user: Click on "Download R for Linux" and choose your distribution for more information on installing R for your setup.
2. **You must do this second:** Download and install RStudio at `https://www.rstudio.com/products/rstudio/download/`.
   - Scroll down to "Installers for Supported Platforms" near the bottom of the page.
   - Click on the download link corresponding to your computer's operating system.

1.1.2 Using R via RStudio

Recall our car analogy from earlier. Much as we don't drive a car by interacting directly with the engine but rather by interacting with elements on the car's dashboard, we won't be using R directly but rather we will use RStudio's interface. After you install R and RStudio on your computer, you'll have two new *programs* (also called *applications*) you can open. We'll always work in RStudio and not in the R application. Figure 1.2 shows what icon you should be clicking on your computer.

FIGURE 1.2: Icons of R versus RStudio on your computer.

After you open RStudio, you should see something similar to Figure 1.3. (Note that slight differences might exist if the RStudio interface is updated after 2019 to not be this by default.)

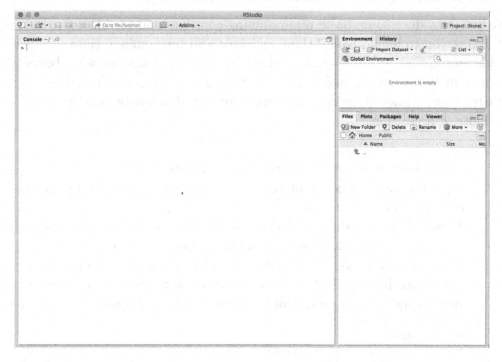

FIGURE 1.3: RStudio interface to R.

Note the three *panes* which are three panels dividing the screen: the *console pane*, the *files pane*, and the *environment pane*. Over the course of this chapter, you'll come to learn what purpose each of these panes serves.

1.2 How do I code in R?

Now that you're set up with R and RStudio, you are probably asking yourself, "OK. Now how do I use R?". The first thing to note is that unlike other statistical software programs like Excel, SPSS, or Minitab that provide point-and-click[1] interfaces, R is an interpreted language[2]. This means you have to type in commands written in *R code*. In other words, you have to code/program in R. Note that we'll use the terms "coding" and "programming" interchangeably in this book.

While it is not required to be a seasoned coder/computer programmer to use R, there is still a set of basic programming concepts that new R users need to understand. Consequently, while this book is not a book on programming, you will still learn just enough of these basic programming concepts needed to explore and analyze data effectively.

1.2.1 Basic programming concepts and terminology

We now introduce some basic programming concepts and terminology. Instead of asking you to memorize all these concepts and terminology right now, we'll guide you so that you'll "learn by doing." To help you learn, we will always use a different font to distinguish regular text from `computer_code`. The best way to master these topics is, in our opinions, through deliberate practice[3] with R and lots of repetition.

- Basics:
    - *Console pane*: where you enter in commands.
    - *Running code*: the act of telling R to perform an act by giving it commands in the console.
    - *Objects*: where values are saved in R. We'll show you how to *assign* values to objects and how to display the contents of objects.
    - *Data types*: integers, doubles/numerics, logicals, and characters. Integers are values like -1, 0, 2, 4092. Doubles or numerics are a larger set of values containing both the integers but also fractions and decimal values

---

[1]https://en.wikipedia.org/wiki/Point_and_click
[2]https://en.wikipedia.org/wiki/Interpreted_language
[3]https://jamesclear.com/deliberate-practice-theory

like -24.932 and 0.8. Logicals are either TRUE or FALSE while characters are text such as "cabbage", "Hamilton", "The Wire is the greatest TV show ever", and "This ramen is delicious." Note that characters are often denoted with the quotation marks around them.

- *Vectors*: a series of values. These are created using the c() function, where c() stands for "combine" or "concatenate." For example, c(6, 11, 13, 31, 90, 92) creates a six element series of positive integer values .   .
- *Factors*: *categorical data* are commonly represented in R as factors. Categorical data can also be represented as *strings*. We'll study this difference as we progress through the book.
- *Data frames*: rectangular spreadsheets. They are representations of datasets in R where the rows correspond to *observations* and the columns correspond to *variables* that describe the observations. We'll cover data frames later in Section 1.4.
- *Conditionals*:
  - Testing for equality in R using == (and not =, which is typically used for assignment). For example, 2 + 1 == 3 compares 2 + 1 to 3 and is correct R code, while 2 + 1 = 3 will return an error.
  - Boolean algebra: TRUE/FALSE statements and mathematical operators such as < (less than), <= (less than or equal), and != (not equal to). For example, 4 + 2 >= 3 will return TRUE, but 3 + 5 <= 1 will return FALSE.
  - Logical operators: & representing "and" as well as | representing "or." For example, (2 + 1 == 3) & (2 + 1 == 4) returns FALSE since both clauses are not TRUE (only the first clause is TRUE). On the other hand, (2 + 1 == 3) | (2 + 1 == 4) returns TRUE since at least one of the two clauses is TRUE.
- *Functions*, also called *commands*: Functions perform tasks in R. They take in inputs called *arguments* and return outputs. You can either manually specify a function's arguments or use the function's *default values*.
  - For example, the function seq() in R generates a sequence of numbers. If you just run seq() it will return the value 1. That doesn't seem very useful! This is because the default arguments are set as seq(from = 1, to = 1). Thus, if you don't pass in different values for from and to to change this behavior, R just assumes all you want is the number 1. You can change the argument values by updating the values after the = sign. If we try out seq(from = 2, to = 5) we get the result 2 3 4 5 that we might expect.
  - We'll work with functions a lot throughout this book and you'll get lots of practice in understanding their behaviors. To further assist you in understanding when a function is mentioned in the book, we'll also include the () after them as we did with seq() above.

This list is by no means an exhaustive list of all the programming concepts and terminology needed to become a savvy R user; such a list would be so large it wouldn't be very useful, especially for novices. Rather, we feel this is a minimally viable list of programming concepts and terminology you need to know before getting started. We feel that you can learn the rest as you go. Remember that your mastery of all of these concepts and terminology will build as you practice more and more.

1.2.2 Errors, warnings, and messages

One thing that intimidates new R and RStudio users is how it reports *errors*, *warnings*, and *messages*. R reports errors, warnings, and messages in a glaring red font, which makes it seem like it is scolding you. However, seeing red text in the console is not always bad.

R will show red text in the console pane in three different situations:

- **Errors**: When the red text is a legitimate error, it will be prefaced with "Error in..." and will try to explain what went wrong. Generally when there's an error, the code will not run. For example, we'll see in Subsection 1.3.3 if you see `Error in ggplot(...) : could not find function "ggplot"`, it means that the `ggplot()` function is not accessible because the package that contains the function (`ggplot2`) was not loaded with `library(ggplot2)`. Thus you cannot use the `ggplot()` function without the `ggplot2` package being loaded first.
- **Warnings**: When the red text is a warning, it will be prefaced with "Warning:" and R will try to explain why there's a warning. Generally your code will still work, but with some caveats. For example, you will see in Chapter 2 if you create a scatterplot based on a dataset where two of the rows of data have missing entries that would be needed to create points in the scatterplot, you will see this warning: `Warning: Removed 2 rows containing missing values (geom_point)`. R will still produce the scatterplot with all the remaining non-missing values, but it is warning you that two of the points aren't there.
- **Messages**: When the red text doesn't start with either "Error" or "Warning", it's *just a friendly message*. You'll see these messages when you load *R packages* in the upcoming Subsection 1.3.2 or when you read data saved in spreadsheet files with the `read_csv()` function as you'll see in Chapter 4. These are helpful diagnostic messages and they don't stop your code from working. Additionally, you'll see these messages when you install packages too using `install.packages()` as discussed in Subsection 1.3.1.

Remember, when you see red text in the console, *don't panic*. It doesn't necessarily mean anything is wrong. Rather:

- If the text starts with "Error", figure out what's causing it. Think of errors as a red traffic light: something is wrong!
- If the text starts with "Warning", figure out if it's something to worry about. For instance, if you get a warning about missing values in a scatterplot and you know there are missing values, you're fine. If that's surprising, look at your data and see what's missing. Think of warnings as a yellow traffic light: everything is working fine, but watch out/pay attention.
- Otherwise, the text is just a message. Read it, wave back at R, and thank it for talking to you. Think of messages as a green traffic light: everything is working fine and keep on going!

1.2.3   Tips on learning to code

Learning to code/program is quite similar to learning a foreign language. It can be daunting and frustrating at first. Such frustrations are common and it is normal to feel discouraged as you learn. However, just as with learning a foreign language, if you put in the effort and are not afraid to make mistakes, anybody can learn and improve.

Here are a few useful tips to keep in mind as you learn to program:

- **Remember that computers are not actually that smart**: You may think your computer or smartphone is "smart," but really people spent a lot of time and energy designing them to appear "smart." In reality, you have to tell a computer everything it needs to do. Furthermore, the instructions you give your computer can't have any mistakes in them, nor can they be ambiguous in any way.
- **Take the "copy, paste, and tweak" approach**: Especially when you learn your first programming language or you need to understand particularly complicated code, it is often much easier to take existing code that you know works and modify it to suit your ends. This is as opposed to trying to type out the code from scratch. We call this the *"copy, paste, and tweak"* approach. So early on, we suggest not trying to write code from memory, but rather take existing examples we have provided you, then copy, paste, and tweak them to suit your goals. After you start feeling more confident, you can slowly move away from this approach and write code from scratch. Think of the "copy, paste, and tweak" approach as training wheels for a child learning to ride a bike. After getting comfortable, they won't need them anymore.
- **The best way to learn to code is by doing**: Rather than learning to code for its own sake, we find that learning to code goes much smoother when you have a goal in mind or when you are working on a particular

project, like analyzing data that you are interested in and that is important
to you.

- **Practice is key**: Just as the only method to improve your foreign language
  skills is through lots of practice and speaking, the only method to improving
  your coding skills is through lots of practice. Don't worry, however, we'll give
  you plenty of opportunities to do so!

1.3   What are R packages?

Another point of confusion with many new R users is the idea of an R package.
R packages extend the functionality of R by providing additional functions,
data, and documentation. They are written by a worldwide community of R
users and can be downloaded for free from the internet.

For example, among the many packages we will use in this book are the `ggplot2`
package (Wickham et al., 2019a) for data visualization in Chapter 2, the
`dplyr` package (Wickham et al., 2019b) for data wrangling in Chapter 3, the
`moderndive` package (Kim and Ismay, 2019) that accompanies this book, and
the `infer` package (Bray et al., 2019) for "tidy" and transparent statistical
inference in Chapters 8, 9, and 10.

A good analogy for R packages is they are like apps you can download onto a
mobile phone:

FIGURE 1.4: Analogy of R versus R packages.

So R is like a new mobile phone: while it has a certain amount of features
when you use it for the first time, it doesn't have everything. R packages are
like the apps you can download onto your phone from Apple's App Store or
Android's Google Play.

Let's continue this analogy by considering the Instagram app for editing and sharing pictures. Say you have purchased a new phone and you would like to share a photo you have just taken with friends on Instagram. You need to:

1. *Install the app*: Since your phone is new and does not include the Instagram app, you need to download the app from either the App Store or Google Play. You do this once and you're set for the time being. You might need to do this again in the future when there is an update to the app.
2. *Open the app*: After you've installed Instagram, you need to open it.

Once Instagram is open on your phone, you can then proceed to share your photo with your friends and family. The process is very similar for using an R package. You need to:

1. *Install the package*: This is like installing an app on your phone. Most packages are not installed by default when you install R and RStudio. Thus if you want to use a package for the first time, you need to install it first. Once you've installed a package, you likely won't install it again unless you want to update it to a newer version.
2. *"Load" the package*: "Loading" a package is like opening an app on your phone. Packages are not "loaded" by default when you start RStudio on your computer; you need to "load" each package you want to use every time you start RStudio.

Let's perform these two steps for the `ggplot2` package for data visualization.

1.3.1 Package installation

Note about RStudio Server or RStudio Cloud: If your instructor has provided you with a link and access to RStudio Server or RStudio Cloud, you might not need to install packages, as they might be preinstalled for you by your instructor. That being said, it is still a good idea to know this process for later on when you are not using RStudio Server or Cloud, but rather RStudio Desktop on your own computer.

There are two ways to install an R package: an easy way and a more advanced way. Let's install the `ggplot2` package the easy way first as shown in Figure 1.5. In the Files pane of RStudio:

a) Click on the "Packages" tab.
b) Click on "Install" next to Update.
c) Type the name of the package under "Packages (separate multiple with space or comma):" In this case, type `ggplot2`.
d) Click "Install."

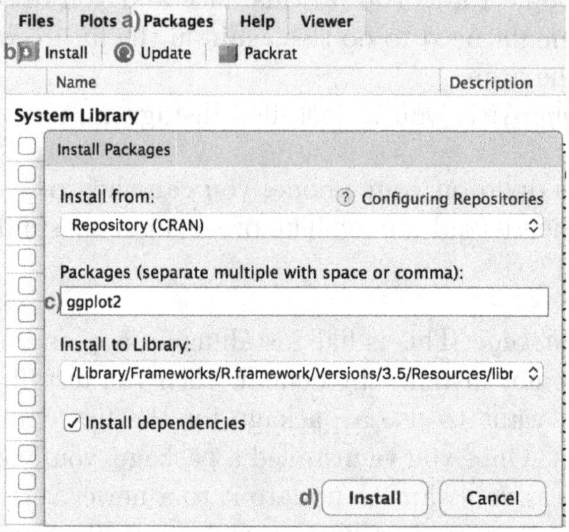

FIGURE 1.5: Installing packages in R the easy way.

An alternative but slightly less convenient way to install a package is by typing `install.packages("ggplot2")` in the console pane of RStudio and pressing Return/Enter on your keyboard. Note you must include the quotation marks around the name of the package.

Much like an app on your phone, you only have to install a package once. However, if you want to update a previously installed package to a newer version, you need to reinstall it by repeating the earlier steps.

Learning check

(LC1.1) Repeat the earlier installation steps, but for the `dplyr`, `nycflights13`, and `knitr` packages. This will install the earlier mentioned `dplyr` package for data wrangling, the `nycflights13` package containing data on all domestic flights leaving a NYC airport in 2013, and the `knitr` package for generating easy-to-read tables in R. We'll use these packages in the next section.

Note that if you'd like your output on your computer to match up exactly with the output presented throughout the book, you may want to use the exact versions of the packages that we used. You can find a full listing of these packages and their versions in Appendix B. This likely won't be relevant for novices, but we included it for reproducibility reasons.

1.3.2 Package loading

Recall that after you've installed a package, you need to "load it." In other words, you need to "open it." We do this by using the `library()` command.

For example, to load the `ggplot2` package, run the following code in the console pane. What do we mean by "run the following code"? Either type or copy-and-paste the following code into the console pane and then hit the Enter key.

```
library(ggplot2)
```

If after running the earlier code, a blinking cursor returns next to the > "prompt" sign, it means you were successful and the `ggplot2` package is now loaded and ready to use. If, however, you get a red "error message" that reads ...

```
Error in library(ggplot2) : there is no package called 'ggplot2'
```

... it means that you didn't successfully install it. This is an example of an "error message" we discussed in Subsection 1.2.2. If you get this error message, go back to Subsection 1.3.1 on R package installation and make sure to install the `ggplot2` package before proceeding.

Learning check

(LC1.2) "Load" the `dplyr`, `nycflights13`, and `knitr` packages as well by repeating the earlier steps.

1.3.3 Package use

One very common mistake new R users make when wanting to use particular packages is they forget to "load" them first by using the `library()` command we just saw. Remember: *you have to load each package you want to use every time you start RStudio.* If you don't first "load" a package, but attempt to use one of its features, you'll see an error message similar to:

```
Error: could not find function
```

This is a different error message than the one you just saw on a package not having been installed yet. R is telling you that you are trying to use a function in a package that has not yet been "loaded." R doesn't know where to find the function you are using. Almost all new users forget to do this when starting out, and it is a little annoying to get used to doing it. However, you'll remember with practice and after some time it will become second nature for you.

1.4 Explore your first datasets

Let's put everything we've learned so far into practice and start exploring some real data! Data comes to us in a variety of formats, from pictures to text to numbers. Throughout this book, we'll focus on datasets that are saved in "spreadsheet"-type format. This is probably the most common way data are collected and saved in many fields. Remember from Subsection 1.2.1 that these "spreadsheet"-type datasets are called *data frames* in R. We'll focus on working with data saved as data frames throughout this book.

Let's first load all the packages needed for this chapter, assuming you've already installed them. Read Section 1.3 for information on how to install and load R packages if you haven't already.

```
library(nycflights13)
library(dplyr)
library(knitr)
```

At the beginning of all subsequent chapters in this book, we'll always have a list of packages that you should have installed and loaded in order to work with that chapter's R code.

1.4.1 nycflights13 package

Many of us have flown on airplanes or know someone who has. Air travel has become an ever-present aspect of many people's lives. If you look at the Departures flight information board at an airport, you will frequently see that some flights are delayed for a variety of reasons. Are there ways that we can understand the reasons that cause flight delays?

We'd all like to arrive at our destinations on time whenever possible. (Unless you secretly love hanging out at airports. If you are one of these people, pretend for a moment that you are very much anticipating being at your final destination.) Throughout this book, we're going to analyze data related to all domestic flights departing from one of New York City's three main airports in 2013: Newark Liberty International (EWR), John F. Kennedy International (JFK), and LaGuardia Airport (LGA). We'll access this data using the `nycflights13` R package, which contains five datasets saved in five data frames:

- `flights`: Information on all 336,776 flights.
- `airlines`: A table matching airline names and their two-letter International Air Transport Association (IATA) airline codes (also known as carrier codes) for 16 airline companies. For example, "DL" is the two-letter code for Delta.
- `planes`: Information about each of the 3,322 physical aircraft used.
- `weather`: Hourly meteorological data for each of the three NYC airports. This data frame has 26,115 rows, roughly corresponding to the $365 \times 24 \times 3 = 26,280$ possible hourly measurements one can observe at three locations over the course of a year.
- `airports`: Names, codes, and locations of the 1,458 domestic destinations.

1.4.2 `flights` data frame

We'll begin by exploring the `flights` data frame and get an idea of its structure. Run the following code in your console, either by typing it or by cutting-and-pasting it. It displays the contents of the `flights` data frame in your console. Note that depending on the size of your monitor, the output may vary slightly.

```
flights
```

```
# A tibble: 336,776 x 19
     year month   day dep_time sched_dep_time dep_delay arr_time
    <int> <int> <int>    <int>          <int>     <dbl>    <int>
 1   2013     1     1      517            515         2      830
 2   2013     1     1      533            529         4      850
 3   2013     1     1      542            540         2      923
 4   2013     1     1      544            545        -1     1004
 5   2013     1     1      554            600        -6      812
 6   2013     1     1      554            558        -4      740
 7   2013     1     1      555            600        -5      913
 8   2013     1     1      557            600        -3      709
 9   2013     1     1      557            600        -3      838
10   2013     1     1      558            600        -2      753
# ... with 336,766 more rows, and 12 more variables: sched_arr_time <int>,
```

```
#   arr_delay <dbl>, carrier <chr>, flight <int>, tailnum <chr>,
#   origin <chr>, dest <chr>, air_time <dbl>, distance <dbl>, hour <dbl>,
#   minute <dbl>, time_hour <dttm>
```

Let's unpack this output:

- `A tibble: 336,776 x 19`: A `tibble` is a specific kind of data frame in R. This particular data frame has
  - `336,776` rows corresponding to different *observations*. Here, each observation is a flight.
  - 19 columns corresponding to 19 *variables* describing each observation.
- `year`, `month`, `day`, `dep_time`, `sched_dep_time`, `dep_delay`, and `arr_time` are the different columns, in other words, the different variables of this dataset.
- We then have a preview of the first 10 rows of observations corresponding to the first 10 flights. R is only showing the first 10 rows, because if it showed all `336,776` rows, it would overwhelm your screen.
- `... with 336,766 more rows, and 11 more variables:` indicating to us that 336,766 more rows of data and 11 more variables could not fit in this screen.

Unfortunately, this output does not allow us to explore the data very well, but it does give a nice preview. Let's look at some different ways to explore data frames.

1.4.3 Exploring data frames

There are many ways to get a feel for the data contained in a data frame such as `flights`. We present three functions that take as their "argument" (their input) the data frame in question. We also include a fourth method for exploring one particular column of a data frame:

1. Using the `View()` function, which brings up RStudio's built-in data viewer.
2. Using the `glimpse()` function, which is included in the `dplyr` package.
3. Using the `kable()` function, which is included in the `knitr` package.
4. Using the `$` "extraction operator," which is used to view a single variable/column in a data frame.

1. `View()`:

Run `View(flights)` in your console in RStudio, either by typing it or cutting-and-pasting it into the console pane. Explore this data frame in the resulting pop up viewer. You should get into the habit of viewing any data frames you encounter. Note the uppercase `V` in `View()`. R is case-sensitive, so you'll get an error message if you run `view(flights)` instead of `View(flights)`.

Learning check

(LC1.3) What does any *ONE* row in this `flights` dataset refer to?

- A. Data on an airline
- B. Data on a flight
- C. Data on an airport
- D. Data on multiple flights

By running `View(flights)`, we can explore the different *variables* listed in the columns. Observe that there are many different types of variables. Some of the variables like `distance`, `day`, and `arr_delay` are what we will call *quantitative* variables. These variables are numerical in nature. Other variables here are *categorical*.

Note that if you look in the leftmost column of the `View(flights)` output, you will see a column of numbers. These are the row numbers of the dataset. If you glance across a row with the same number, say row 5, you can get an idea of what each row is representing. This will allow you to identify what object is being described in a given row by taking note of the values of the columns in that specific row. This is often called the *observational unit*. The observational unit in this example is an individual flight departing from New York City in 2013. You can identify the observational unit by determining what "thing" is being measured or described by each of the variables. We'll talk more about observational units in Subsection 1.4.4 on *identification* and *measurement* variables.

2. `glimpse()`:

The second way we'll cover to explore a data frame is using the `glimpse()` function included in the `dplyr` package. Thus, you can only use the `glimpse()` function after you've loaded the `dplyr` package by running `library(dplyr)`. This function provides us with an alternative perspective for exploring a data frame than the `View()` function:

```
glimpse(flights)
```

```
Observations: 336,776
Variables: 19
$ year        <int> 2013, 2013, 2013, 2013, 2013, 2013, 2013, 2013,...
$ month       <int> 1, 1, 1, 1, 1, 1, 1, 1, 1, 1, 1, 1, 1, 1, 1, 1,...
```

```
$ day            <int> 1, 1, 1, 1, 1, 1, 1, 1, 1, 1, 1, 1, 1, 1, 1, 1,...
$ dep_time       <int> 517, 533, 542, 544, 554, 554, 555, 557, 557, 55...
$ sched_dep_time <int> 515, 529, 540, 545, 600, 558, 600, 600, 600, 60...
$ dep_delay      <dbl> 2, 4, 2, -1, -6, -4, -5, -3, -3, -2, -2, -2, -2...
$ arr_time       <int> 830, 850, 923, 1004, 812, 740, 913, 709, 838, 7...
$ sched_arr_time <int> 819, 830, 850, 1022, 837, 728, 854, 723, 846, 7...
$ arr_delay      <dbl> 11, 20, 33, -18, -25, 12, 19, -14, -8, 8, -2, -...
$ carrier        <chr> "UA", "UA", "AA", "B6", "DL", "UA", "B6", "EV",...
$ flight         <int> 1545, 1714, 1141, 725, 461, 1696, 507, 5708, 79...
$ tailnum        <chr> "N14228", "N24211", "N619AA", "N804JB", "N668DN...
$ origin         <chr> "EWR", "LGA", "JFK", "JFK", "LGA", "EWR", "EWR"...
$ dest           <chr> "IAH", "IAH", "MIA", "BQN", "ATL", "ORD", "FLL"...
$ air_time       <dbl> 227, 227, 160, 183, 116, 150, 158, 53, 140, 138...
$ distance       <dbl> 1400, 1416, 1089, 1576, 762, 719, 1065, 229, 94...
$ hour           <dbl> 5, 5, 5, 5, 6, 5, 6, 6, 6, 6, 6, 6, 6, 6, 6, 5,...
$ minute         <dbl> 15, 29, 40, 45, 0, 58, 0, 0, 0, 0, 0, 0, 0, 0, ...
$ time_hour      <dttm> 2013-01-01 05:00:00, 2013-01-01 05:00:00, 2013...
```

Observe that `glimpse()` will give you the first few entries of each variable in a row after the variable name. In addition, the *data type* (see Subsection 1.2.1) of the variable is given immediately after each variable's name inside < >. Here, `int` and `dbl` refer to "integer" and "double", which are computer coding terminology for quantitative/numerical variables. "Doubles" take up twice the size to store on a computer compared to integers.

In contrast, `chr` refers to "character", which is computer terminology for text data. In most forms, text data, such as the `carrier` or `origin` of a flight, are categorical variables. The `time_hour` variable is another data type: `dttm`. These types of variables represent date and time combinations. However, we won't work with dates and times in this book; we leave this topic for other data science books like *Introduction to Data Science* by Tiffany-Anne Timbers, Melissa Lee, and Trevor Campbell[4] or *R for Data Science*[5] (Grolemund and Wickham, 2017).

Learning check

(LC1.4) What are some other examples in this dataset of *categorical* variables? What makes them different than *quantitative* variables?

---

[4]https://ubc-dsci.github.io/introduction-to-datascience/
[5]https://r4ds.had.co.nz/dates-and-times.html

3. kable():

The final way to explore the entirety of a data frame is using the kable() function from the knitr package. Let's explore the different carrier codes for all the airlines in our dataset two ways. Run both of these lines of code in the console:

```
airlines
kable(airlines)
```

At first glance, it may not appear that there is much difference in the outputs. However, when using tools for producing reproducible reports such as R Markdown[6], the latter code produces output that is much more legible and reader-friendly. You'll see us use this reader-friendly style in many places in the book when we want to print a data frame as a nice table.

4. $ operator

Lastly, the $ operator allows us to extract and then explore a single variable within a data frame. For example, run the following in your console

```
airlines$name
```

We used the $ operator to extract only the name variable and return it as a vector of length 16. We'll only be occasionally exploring data frames using the $ operator, instead favoring the View() and glimpse() functions.

1.4.4 Identification and measurement variables

There is a subtle difference between the kinds of variables that you will encounter in data frames. There are *identification variables* and *measurement variables*. For example, let's explore the airports data frame by showing the output of glimpse(airports):

```
glimpse(airports)
```

```
Observations: 1,458
Variables: 8
$ faa    <chr> "04G", "06A", "06C", "06N", "09J", "0A9", "0G6", "0G7", ...
$ name   <chr> "Lansdowne Airport", "Moton Field Municipal Airport", "S...
$ lat    <dbl> 41.1, 32.5, 42.0, 41.4, 31.1, 36.4, 41.5, 42.9, 39.8, 48...
```

---

[6]http://rmarkdown.rstudio.com/lesson-1.html

```
$ lon    <dbl> -80.6, -85.7, -88.1, -74.4, -81.4, -82.2, -84.5, -76.8, ...
$ alt    <dbl> 1044, 264, 801, 523, 11, 1593, 730, 492, 1000, 108, 409,...
$ tz     <dbl> -5, -6, -6, -5, -5, -5, -5, -5, -5, -8, -5, -6, -5, -5, ...
$ dst    <chr> "A", "A", "A", "A", "A", "A", "A", "A", "U", "A", "A", "...
$ tzone <chr> "America/New_York", "America/Chicago", "America/Chicago"...
```

The variables faa and name are what we will call *identification variables*, variables that uniquely identify each observational unit. In this case, the identification variables uniquely identify airports. Such variables are mainly used in practice to uniquely identify each row in a data frame. faa gives the unique code provided by the FAA for that airport, while the name variable gives the longer official name of the airport. The remaining variables (lat, lon, alt, tz, dst, tzone) are often called *measurement* or *characteristic* variables: variables that describe properties of each observational unit. For example, lat and long describe the latitude and longitude of each airport.

Furthermore, sometimes a single variable might not be enough to uniquely identify each observational unit: combinations of variables might be needed. While it is not an absolute rule, for organizational purposes it is considered good practice to have your identification variables in the leftmost columns of your data frame.

Learning check

(LC1.5) What properties of each airport do the variables lat, lon, alt, tz, dst, and tzone describe in the airports data frame? Take your best guess.

(LC1.6) Provide the names of variables in a data frame with at least three variables where one of them is an identification variable and the other two are not. Further, create your own tidy data frame that matches these conditions.

1.4.5 Help files

Another nice feature of R are help files, which provide documentation for various functions and datasets. You can bring up help files by adding a ? before the name of a function or data frame and then run this in the console. You will then be presented with a page showing the corresponding documentation if it exists. For example, let's look at the help file for the flights data frame.

```
?flights
```

The help file should pop up in the Help pane of RStudio. If you have questions about a function or data frame included in an R package, you should get in the habit of consulting the help file right away.

Learning check

(LC1.7) Look at the help file for the `airports` data frame. Revise your earlier guesses about what the variables `lat`, `lon`, `alt`, `tz`, `dst`, and `tzone` each describe.

1.5 Conclusion

We've given you what we feel is a minimally viable set of tools to explore data in R. Does this chapter contain everything you need to know? Absolutely not. To try to include everything in this chapter would make the chapter so large it wouldn't be useful! As we said earlier, the best way to add to your toolbox is to get into RStudio and run and write code as much as possible.

1.5.1 Additional resources

Solutions to all *Learning checks* can be found online in Appendix D[7].

If you are new to the world of coding, R, and RStudio and feel you could benefit from a more detailed introduction, we suggest you check out the short book, *Getting Used to R, RStudio, and R Markdown*[8] (Ismay and Kennedy, 2016). It includes screencast recordings that you can follow along and pause as you learn. This book also contains an introduction to R Markdown, a tool used for reproducible research in R.

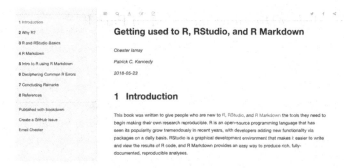

FIGURE 1.6: Preview of *Getting Used to R, RStudio, and R Markdown.*

---

[7]https://moderndive.com/D-appendixD.html

[8]https://rbasics.netlify.com/

1.5.2   What's to come?

We're now going to start the "Data Science with `tidyverse`" portion of this book in Chapter 2 as shown in Figure 1.7 with what we feel is the most important tool in a data scientist's toolbox: data visualization. We'll continue to explore the data included in the `nycflights13` package using the `ggplot2` package for data visualization. You'll see that data visualization is a powerful tool to add to your toolbox for data exploration that provides additional insight to what the `View()` and `glimpse()` functions can provide.

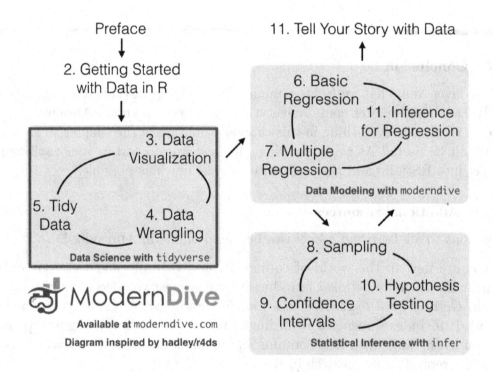

FIGURE 1.7: *ModernDive* flowchart - on to Part I!

Part I

Data Science with tidyverse

2

Data Visualization

We begin the development of your data science toolbox with data visualization. By visualizing data, we gain valuable insights we couldn't initially obtain from just looking at the raw data values. We'll use the `ggplot2` package, as it provides an easy way to customize your plots. `ggplot2` is rooted in the data visualization theory known as *the grammar of graphics* (Wilkinson, 2005), developed by Leland Wilkinson.

At their most basic, graphics/plots/charts (we use these terms interchangeably in this book) provide a nice way to explore the patterns in data, such as the presence of *outliers*, *distributions* of individual variables, and *relationships* between groups of variables. Graphics are designed to emphasize the findings and insights you want your audience to understand. This does, however, require a balancing act. On the one hand, you want to highlight as many interesting findings as possible. On the other hand, you don't want to include so much information that it overwhelms your audience.

As we will see, plots also help us to identify patterns and outliers in our data. We'll see that a common extension of these ideas is to compare the *distribution* of one numerical variable, such as what are the center and spread of the values, as we go across the levels of a different categorical variable.

Needed packages

Let's load all the packages needed for this chapter (this assumes you've already installed them). Read Section 1.3 for information on how to install and load R packages.

```
library(nycflights13)
library(ggplot2)
library(dplyr)
```

2.1   The grammar of graphics

We start with a discussion of a theoretical framework for data visualization known as "the grammar of graphics." This framework serves as the foundation for the `ggplot2` package which we'll use extensively in this chapter. Think of how we construct and form sentences in English by combining different elements, like nouns, verbs, articles, subjects, objects, etc. We can't just combine these elements in any arbitrary order; we must do so following a set of rules known as a linguistic grammar. Similarly to a linguistic grammar, "the grammar of graphics" defines a set of rules for constructing *statistical graphics* by combining different types of *layers*. This grammar was created by Leland Wilkinson (Wilkinson, 2005) and has been implemented in a variety of data visualization software platforms like R, but also Plotly[1] and Tableau[2].

2.1.1   Components of the grammar

In short, the grammar tells us that:

A statistical graphic is a `mapping` of `data` variables to `aesthetic` attributes of `geometric` objects.

Specifically, we can break a graphic into the following three essential components:

1. `data`: the dataset containing the variables of interest.
2. `geom`: the geometric object in question. This refers to the type of object we can observe in a plot. For example: points, lines, and bars.
3. `aes`: aesthetic attributes of the geometric object. For example, x/y position, color, shape, and size. Aesthetic attributes are *mapped* to variables in the dataset.

You might be wondering why we wrote the terms `data`, `geom`, and `aes` in a computer code type font. We'll see very shortly that we'll specify the elements of the grammar in R using these terms. However, let's first break down the grammar with an example.

---
[1]`https://plot.ly/`
[2]`https://www.tableau.com/`

2.1.2 Gapminder data

In February 2006, a Swedish physician and data advocate named Hans Rosling gave a TED talk titled "The best stats you've ever seen"[3] where he presented global economic, health, and development data from the website gapminder.org[4]. For example, for data on 142 countries in 2007, let's consider only a few countries in Table 2.1 as a peak into the data.

TABLE 2.1: Gapminder 2007 Data: First 3 of 142 countries

Country	Continent	Life Expectancy	Population	GDP per Capita
Afghanistan	Asia	43.8	31889923	975
Albania	Europe	76.4	3600523	5937
Algeria	Africa	72.3	33333216	6223

Each row in this table corresponds to a country in 2007. For each row, we have 5 columns:

1. **Country**: Name of country.
2. **Continent**: Which of the five continents the country is part of. Note that "Americas" includes countries in both North and South America and that Antarctica is excluded.
3. **Life Expectancy**: Life expectancy in years.
4. **Population**: Number of people living in the country.
5. **GDP per Capita**: Gross domestic product (in US dollars).

Now consider Figure 2.1, which plots this for all 142 of the data's countries.

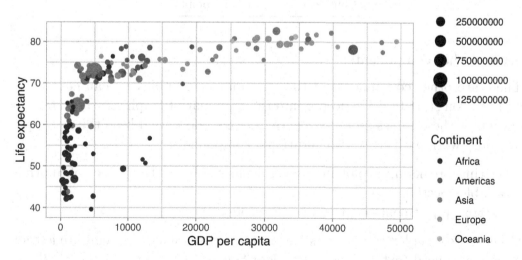

FIGURE 2.1: Life expectancy over GDP per capita in 2007.

---

[3]https://www.ted.com/talks/hans_rosling_shows_the_best_stats_you_ve_ever_seen
[4]http://www.gapminder.org/tools/#_locale_id=en;&chart-type=bubbles

Let's view this plot through the grammar of graphics:

1. The `data` variable **GDP per Capita** gets mapped to the x-position aesthetic of the points.
2. The `data` variable **Life Expectancy** gets mapped to the y-position aesthetic of the points.
3. The `data` variable **Population** gets mapped to the `size` aesthetic of the points.
4. The `data` variable **Continent** gets mapped to the `color` aesthetic of the points.

We'll see shortly that `data` corresponds to the particular data frame where our data is saved and that "data variables" correspond to particular columns in the data frame. Furthermore, the type of `geometric` object considered in this plot are points. That being said, while in this example we are considering points, graphics are not limited to just points. We can also use lines, bars, and other geometric objects.

Let's summarize the three essential components of the grammar in Table 2.2.

TABLE 2.2: Summary of the grammar of graphics for this plot

data variable	aes	geom
GDP per Capita	x	point
Life Expectancy	y	point
Population	size	point
Continent	color	point

2.1.3 Other components

There are other components of the grammar of graphics we can control as well. As you start to delve deeper into the grammar of graphics, you'll start to encounter these topics more frequently. In this book, we'll keep things simple and only work with these two additional components:

- `faceting` breaks up a plot into several plots split by the values of another variable (Section 2.6)
- `position` adjustments for barplots (Section 2.8)

Other more complex components like `scales` and `coordinate` systems are left for a more advanced text such as *R for Data Science*[5] (Grolemund and Wickham, 2017). Generally speaking, the grammar of graphics allows for a high degree of

---

[5]http://r4ds.had.co.nz/data-visualisation.html#aesthetic-mappings

customization of plots and also a consistent framework for easily updating and modifying them.

2.1.4 ggplot2 package

In this book, we will use the `ggplot2` package for data visualization, which is an implementation of the grammar of graphics for R (Wickham et al., 2019a). As we noted earlier, a lot of the previous section was written in a computer code type font. This is because the various components of the grammar of graphics are specified in the `ggplot()` function included in the `ggplot2` package. For the purposes of this book, we'll always provide the `ggplot()` function with the following arguments (i.e., inputs) at a minimum:

- The data frame where the variables exist: the `data` argument.
- The mapping of the variables to aesthetic attributes: the `mapping` argument which specifies the `aesthetic` attributes involved.

After we've specified these components, we then add *layers* to the plot using the + sign. The most essential layer to add to a plot is the layer that specifies which type of `geometric` object we want the plot to involve: points, lines, bars, and others. Other layers we can add to a plot include the plot title, axes labels, visual themes for the plots, and facets (which we'll see in Section 2.6).

Let's now put the theory of the grammar of graphics into practice.

2.2   Five named graphs - the 5NG

In order to keep things simple in this book, we will only focus on five different types of graphics, each with a commonly given name. We term these "five named graphs" or in abbreviated form, the **5NG**:

1. scatterplots
2. linegraphs
3. boxplots
4. histograms
5. barplots

We'll also present some variations of these plots, but with this basic repertoire of five graphics in your toolbox, you can visualize a wide array of different variable types. Note that certain plots are only appropriate for categorical variables, while others are only appropriate for numerical variables.

2.3 5NG#1: Scatterplots

The simplest of the 5NG are *scatterplots*, also called *bivariate plots*. They allow you to visualize the *relationship* between two numerical variables. While you may already be familiar with scatterplots, let's view them through the lens of the grammar of graphics we presented in Section 2.1. Specifically, we will visualize the relationship between the following two numerical variables in the flights data frame included in the nycflights13 package:

1. dep_delay: departure delay on the horizontal "x" axis and
2. arr_delay: arrival delay on the vertical "y" axis

for Alaska Airlines flights leaving NYC in 2013. This requires paring down the data from all 336,776 flights that left NYC in 2013, to only the 714 *Alaska Airlines* flights that left NYC in 2013. We do this so our scatterplot will involve a manageable 714 points, and not an overwhelmingly large number like 336,776. To achieve this, we'll take the flights data frame, filter the rows so that only the 714 rows corresponding to Alaska Airlines flights are kept, and save this in a new data frame called alaska_flights using the <- *assignment* operator:

```
alaska_flights <- flights %>%
  filter(carrier == "AS")
```

For now, we suggest you don't worry if you don't fully understand this code. We'll see later in Chapter 3 on data wrangling that this code uses the dplyr package for data wrangling to achieve our goal: it takes the flights data frame and filters it to only return the rows where carrier is equal to "AS", Alaska Airlines' carrier code. Recall from Section 1.2 that testing for equality is specified with == and not =. Convince yourself that this code achieves what it is supposed to by exploring the resulting data frame by running View(alaska_flights). You'll see that it has 714 rows, consisting of only 714 Alaska Airlines flights.

Learning check

(LC2.1) Take a look at both the flights and alaska_flights data frames by running View(flights) and View(alaska_flights). In what respect do these data frames differ? For example, think about the number of rows in each dataset.

2.3.1   Scatterplots via `geom_point`

Let's now go over the code that will create the desired scatterplot, while keeping in mind the grammar of graphics framework we introduced in Section 2.1. Let's take a look at the code and break it down piece-by-piece.

Note: The printed version of this book uses `theme_light()` instead of the default `theme_grey()` for the plots created with `ggplot2` throughout the book. Bars and points are also converted to greyscale using `scale_color_grey()` and `scale_fill_grey()`. This helps with readability of the plots in the printed copy. As you follow along and run the code yourself, your plots will have a grey background instead of the white background in the printed book. Also, your plots will have colors beyond the greyscale versions provided in this printing.

```
ggplot(data = alaska_flights, mapping = aes(x = dep_delay, y = arr_delay)) +
  geom_point()
```

Within the `ggplot()` function, we specify two of the components of the grammar of graphics as arguments (i.e., inputs):

1. The `data` as the `alaska_flights` data frame via `data = alaska_flights`.
2. The aesthetic `mapping` by setting `mapping = aes(x = dep_delay, y = arr_delay)`. Specifically, the variable `dep_delay` maps to the x position aesthetic, while the variable `arr_delay` maps to the y position.

We then add a layer to the `ggplot()` function call using the `+` sign. The added layer in question specifies the third component of the grammar: the `geometric` object. In this case, the geometric object is set to be points by specifying `geom_point()`. After running these two lines of code in your console, you'll notice two outputs: a warning message and the graphic shown in Figure 2.2.

```
Warning: Removed 5 rows containing missing values (geom_point).
```

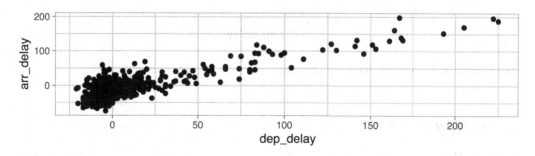

FIGURE 2.2: Arrival delays versus departure delays for Alaska Airlines flights from NYC in 2013.

Let's first unpack the graphic in Figure 2.2. Observe that a *positive relationship* exists between dep_delay and arr_delay: as departure delays increase, arrival delays tend to also increase. Observe also the large mass of points clustered near (0, 0), the point indicating flights that neither departed nor arrived late.

Let's turn our attention to the warning message. R is alerting us to the fact that five rows were ignored due to them being missing. For these 5 rows, either the value for dep_delay or arr_delay or both were missing (recorded in R as NA), and thus these rows were ignored in our plot.

Before we continue, let's make a few more observations about this code that created the scatterplot. Note that the + sign comes at the end of lines, and not at the beginning. You'll get an error in R if you put it at the beginning of a line. When adding layers to a plot, you are encouraged to start a new line after the + (by pressing the Return/Enter button on your keyboard) so that the code for each layer is on a new line. As we add more and more layers to plots, you'll see this will greatly improve the legibility of your code.

To stress the importance of adding the layer specifying the geometric object, consider Figure 2.3 where no layers are added. Because the geometric object was not specified, we have a blank plot which is not very useful!

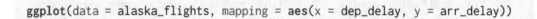

```
ggplot(data = alaska_flights, mapping = aes(x = dep_delay, y = arr_delay))
```

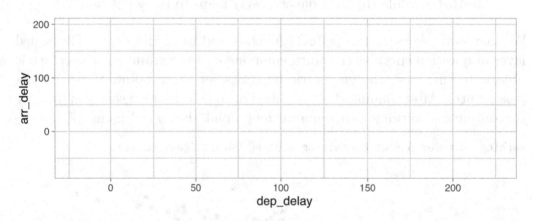

FIGURE 2.3: A plot with no layers.

Learning check

(LC2.2) What are some practical reasons why dep_delay and arr_delay have a positive relationship?

(LC2.3) What variables in the `weather` data frame would you expect to have a negative correlation (i.e., a negative relationship) with `dep_delay`? Why? Remember that we are focusing on numerical variables here. Hint: Explore the `weather` dataset by using the `View()` function.

(LC2.4) Why do you believe there is a cluster of points near (0, 0)? What does (0, 0) correspond to in terms of the Alaska Air flights?

(LC2.5) What are some other features of the plot that stand out to you?

(LC2.6) Create a new scatterplot using different variables in the `alaska_flights` data frame by modifying the example given.

2.3.2 Overplotting

The large mass of points near (0, 0) in Figure 2.2 can cause some confusion since it is hard to tell the true number of points that are plotted. This is the result of a phenomenon called *overplotting*. As one may guess, this corresponds to points being plotted on top of each other over and over again. When overplotting occurs, it is difficult to know the number of points being plotted. There are two methods to address the issue of overplotting. Either by

1. Adjusting the transparency of the points or
2. Adding a little random "jitter", or random "nudges", to each of the points.

Method 1: Changing the transparency

The first way of addressing overplotting is to change the transparency/opacity of the points by setting the `alpha` argument in `geom_point()`. We can change the `alpha` argument to be any value between `0` and `1`, where `0` sets the points to be 100% transparent and `1` sets the points to be 100% opaque. By default, `alpha` is set to `1`. In other words, if we don't explicitly set an `alpha` value, R will use `alpha = 1`.

Note how the following code is identical to the code in Section 2.3 that created the scatterplot with overplotting, but with `alpha = 0.2` added to the `geom_point()` function:

```
ggplot(data = alaska_flights, mapping = aes(x = dep_delay, y = arr_delay)) +
  geom_point(alpha = 0.2)
```

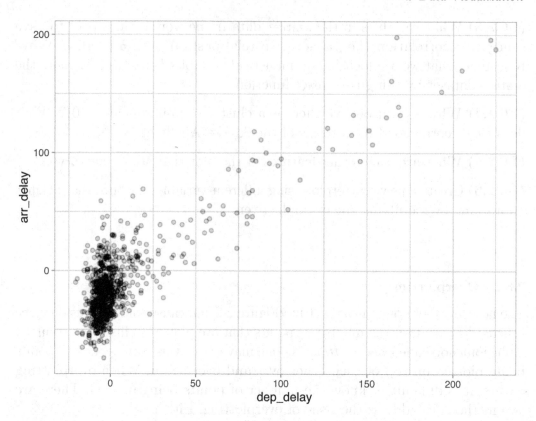

FIGURE 2.4: Arrival vs. departure delays scatterplot with alpha = 0.2.

The key feature to note in Figure 2.4 is that the transparency of the points is cumulative: areas with a high-degree of overplotting are darker, whereas areas with a lower degree are less dark. Note furthermore that there is no `aes()` surrounding `alpha = 0.2`. This is because we are not mapping a variable to an aesthetic attribute, but rather merely changing the default setting of `alpha`. In fact, you'll receive an error if you try to change the second line to read `geom_point(aes(alpha = 0.2))`.

Method 2: Jittering the points

The second way of addressing overplotting is by *jittering* all the points. This means giving each point a small "nudge" in a random direction. You can think of "jittering" as shaking the points around a bit on the plot. Let's illustrate using a simple example first. Say we have a data frame with 4 identical rows of x and y values: (0,0), (0,0), (0,0), and (0,0). In Figure 2.5, we present both the regular scatterplot of these 4 points (on the left) and its jittered counterpart (on the right).

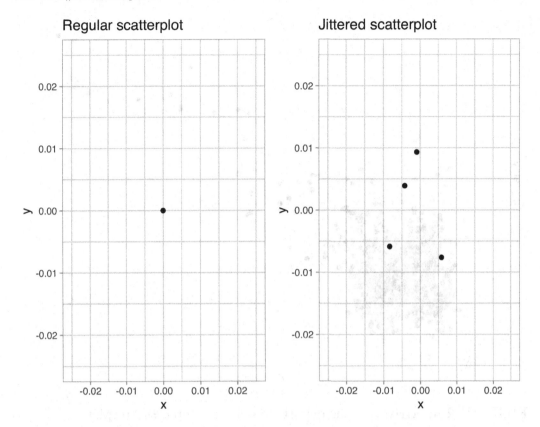

FIGURE 2.5: Regular and jittered scatterplot.

In the left-hand regular scatterplot, observe that the 4 points are superimposed on top of each other. While we know there are 4 values being plotted, this fact might not be apparent to others. In the right-hand jittered scatterplot, it is now plainly evident that this plot involves four points since each point is given a random "nudge."

Keep in mind, however, that jittering is strictly a visualization tool; even after creating a jittered scatterplot, the original values saved in the data frame remain unchanged.

To create a jittered scatterplot, instead of using geom_point(), we use geom_jitter(). Observe how the following code is very similar to the code that created the scatterplot with overplotting in Subsection 2.3.1, but with geom_point() replaced with geom_jitter().

```
ggplot(data = alaska_flights, mapping = aes(x = dep_delay, y = arr_delay)) +
   geom_jitter(width = 30, height = 30)
```

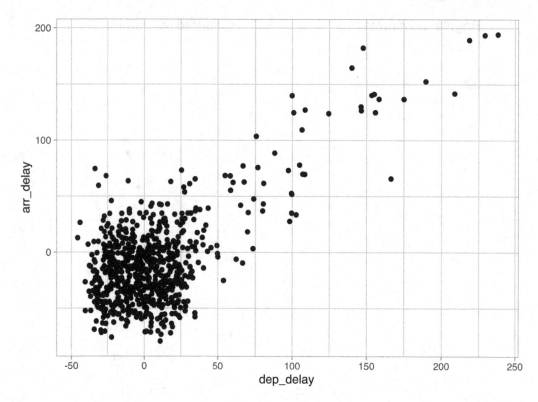

FIGURE 2.6: Arrival versus departure delays jittered scatterplot.

In order to specify how much jitter to add, we adjusted the `width` and `height` arguments to `geom_jitter()`. This corresponds to how hard you'd like to shake the plot in horizontal x-axis units and vertical y-axis units, respectively. In this case, both axes are in minutes. How much jitter should we add using the `width` and `height` arguments? On the one hand, it is important to add just enough jitter to break any overlap in points, but on the other hand, not so much that we completely alter the original pattern in points.

As can be seen in the resulting Figure 2.6, in this case jittering doesn't really provide much new insight. In this particular case, it can be argued that changing the transparency of the points by setting `alpha` proved more effective. When would it be better to use a jittered scatterplot? When would it be better to alter the points' transparency? There is no single right answer that applies to all situations. You need to make a subjective choice and own that choice. At the very least when confronted with overplotting, however, we suggest you make both types of plots and see which one better emphasizes the point you are trying to make.

Learning check

(LC2.7) Why is setting the `alpha` argument value useful with scatterplots? What further information does it give you that a regular scatterplot cannot?

(LC2.8) After viewing Figure 2.4, give an approximate range of arrival delays and departure delays that occur most frequently. How has that region changed compared to when you observed the same plot without `alpha = 0.2` set in Figure 2.2?

2.3.3 Summary

Scatterplots display the relationship between two numerical variables. They are among the most commonly used plots because they can provide an immediate way to see the trend in one numerical variable versus another. However, if you try to create a scatterplot where either one of the two variables is not numerical, you might get strange results. Be careful!

With medium to large datasets, you may need to play around with the different modifications to scatterplots we saw such as changing the transparency/opacity of the points or by jittering the points. This tweaking is often a fun part of data visualization, since you'll have the chance to see different relationships emerge as you tinker with your plots.

2.4 5NG#2: Linegraphs

The next of the five named graphs are linegraphs. Linegraphs show the relationship between two numerical variables when the variable on the x-axis, also called the *explanatory* variable, is of a sequential nature. In other words, there is an inherent ordering to the variable.

The most common examples of linegraphs have some notion of time on the x-axis: hours, days, weeks, years, etc. Since time is sequential, we connect consecutive observations of the variable on the y-axis with a line. Linegraphs that have some notion of time on the x-axis are also called *time series* plots. Let's illustrate linegraphs using another dataset in the `nycflights13` package: the `weather` data frame.

Let's explore the weather data frame by running View(weather) and glimpse(weather). Furthermore let's read the associated help file by running ?weather to bring up the help file.

Observe that there is a variable called temp of hourly temperature recordings in Fahrenheit at weather stations near all three major airports in New York City: Newark (origin code EWR), John F. Kennedy International (JFK), and LaGuardia (LGA). However, instead of considering hourly temperatures for all days in 2013 for all three airports, for simplicity let's only consider hourly temperatures at Newark airport for the first 15 days in January.

Recall in Section 2.3, we used the filter() function to only choose the subset of rows of flights corresponding to Alaska Airlines flights. We similarly use filter() here, but by using the & operator we only choose the subset of rows of weather where the origin is "EWR", the month is January, **and** the day is between 1 and 15. Recall we performed a similar task in Section 2.3 when creating the alaska_flights data frame of only Alaska Airlines flights, a topic we'll explore more in Chapter 3 on data wrangling.

```
early_january_weather <- weather %>%
  filter(origin == "EWR" & month == 1 & day <= 15)
```

Learning check

(LC2.9) Take a look at both the weather and early_january_weather data frames by running View(weather) and View(early_january_weather). In what respect do these data frames differ?

(LC2.10) View() the flights data frame again. Why does the time_hour variable uniquely identify the hour of the measurement, whereas the hour variable does not?

2.4.1 Linegraphs via geom_line

Let's create a time series plot of the hourly temperatures saved in the early_january_weather data frame by using geom_line() to create a linegraph, instead of using geom_point() like we used previously to create scatterplots:

```
ggplot(data = early_january_weather,
       mapping = aes(x = time_hour, y = temp)) +
  geom_line()
```

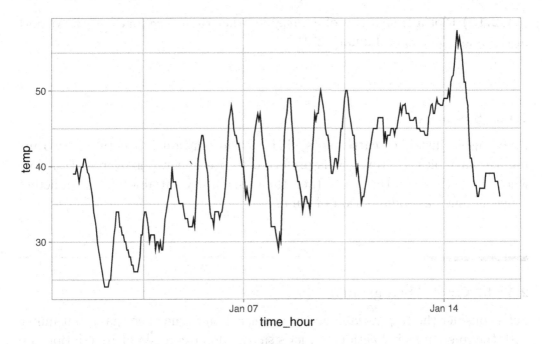

FIGURE 2.7: Hourly temperature in Newark for January 1-15, 2013.

Much as with the `ggplot()` code that created the scatterplot of departure and arrival delays for Alaska Airlines flights in Figure 2.2, let's break down this code piece-by-piece in terms of the grammar of graphics:

Within the `ggplot()` function call, we specify two of the components of the grammar of graphics as arguments:

1. The `data` to be the `early_january_weather` data frame by setting `data =` `early_january_weather`.
2. The aesthetic `mapping` by setting `mapping = aes(x = time_hour, y =` `temp)`. Specifically, the variable `time_hour` maps to the x position aesthetic, while the variable `temp` maps to the y position aesthetic.

We add a layer to the `ggplot()` function call using the `+` sign. The layer in question specifies the third component of the grammar: the `geometric` object in question. In this case, the geometric object is a `line` set by specifying `geom_line()`.

Learning check

(LC2.11) Why should linegraphs be avoided when there is not a clear ordering of the horizontal axis?

(LC2.12) Why are linegraphs frequently used when time is the explanatory variable on the x-axis?

(LC2.13) Plot a time series of a variable other than `temp` for Newark Airport in the first 15 days of January 2013.

2.4.2 Summary

Linegraphs, just like scatterplots, display the relationship between two numerical variables. However, it is preferred to use linegraphs over scatterplots when the variable on the x-axis (i.e., the explanatory variable) has an inherent ordering, such as some notion of time.

2.5 5NG#3: Histograms

Let's consider the `temp` variable in the `weather` data frame once again, but unlike with the linegraphs in Section 2.4, let's say we don't care about its relationship with time, but rather we only care about how the values of `temp` *distribute*. In other words:

1. What are the smallest and largest values?
2. What is the "center" or "most typical" value?
3. How do the values spread out?
4. What are frequent and infrequent values?

One way to visualize this *distribution* of this single variable `temp` is to plot them on a horizontal line as we do in Figure 2.8:

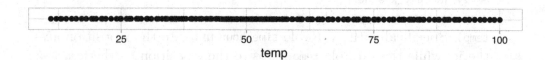

FIGURE 2.8: Plot of hourly temperature recordings from NYC in 2013.

This gives us a general idea of how the values of `temp` distribute: observe that temperatures vary from around 11°F (-11°C) up to 100°F (38°C). Furthermore, there appear to be more recorded temperatures between 40°F and 60°F than outside this range. However, because of the high degree of overplotting in the points, it's hard to get a sense of exactly how many values are between say 50°F and 55°F.

What is commonly produced instead of Figure 2.8 is known as a *histogram*. A histogram is a plot that visualizes the *distribution* of a numerical value as follows:

1. We first cut up the x-axis into a series of *bins*, where each bin represents a range of values.
2. For each bin, we count the number of observations that fall in the range corresponding to that bin.
3. Then for each bin, we draw a bar whose height marks the corresponding count.

Let's drill-down on an example of a histogram, shown in Figure 2.9.

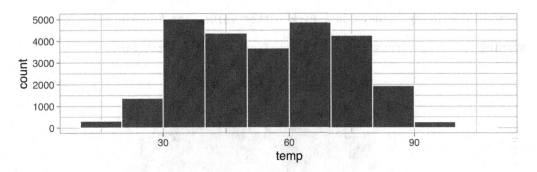

FIGURE 2.9: Example histogram.

Let's focus only on temperatures between 30°F (-1°C) and 60°F (15°C) for now. Observe that there are three bins of equal width between 30°F and 60°F. Thus we have three bins of width 10°F each: one bin for the 30-40°F range, another bin for the 40-50°F range, and another bin for the 50-60°F range. Since:

1. The bin for the 30-40°F range has a height of around 5000. In other words, around 5000 of the hourly temperature recordings are between 30°F and 40°F.
2. The bin for the 40-50°F range has a height of around 4300. In other words, around 4300 of the hourly temperature recordings are between 40°F and 50°F.
3. The bin for the 50-60°F range has a height of around 3500. In other words, around 3500 of the hourly temperature recordings are between 50°F and 60°F.

All nine bins spanning 10°F to 100°F on the x-axis have this interpretation.

2.5.1 Histograms via `geom_histogram`

Let's now present the `ggplot()` code to plot your first histogram! Unlike with scatterplots and linegraphs, there is now only one variable being mapped in `aes()`: the single numerical variable `temp`. The y-aesthetic of a histogram, the count of the observations in each bin, gets computed for you automatically. Furthermore, the geometric object layer is now a `geom_histogram()`. After running the following code, you'll see the histogram in Figure 2.10 as well as warning messages. We'll discuss the warning messages first.

```
ggplot(data = weather, mapping = aes(x = temp)) +
  geom_histogram()
```

```
`stat_bin()` using `bins = 30`. Pick better value with `binwidth`.
```

```
Warning: Removed 1 rows containing non-finite values (stat_bin).
```

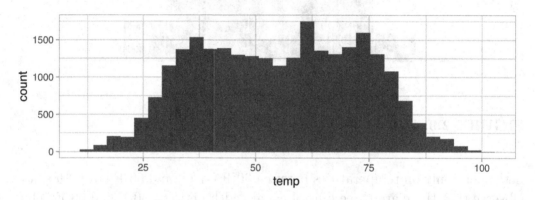

FIGURE 2.10: Histogram of hourly temperatures at three NYC airports.

The first message is telling us that the histogram was constructed using `bins = 30` for 30 equally spaced bins. This is known in computer programming as a default value; unless you override this default number of bins with a number you specify, R will choose 30 by default. We'll see in the next section how to change the number of bins to another value than the default.

The second message is telling us something similar to the warning message we received when we ran the code to create a scatterplot of departure and arrival delays for Alaska Airlines flights in Figure 2.2: that because one row has a missing `NA` value for `temp`, it was omitted from the histogram. R is just giving us a friendly heads up that this was the case.

Now let's unpack the resulting histogram in Figure 2.10. Observe that values less than 25°F as well as values above 80°F are rather rare. However, because of the large number of bins, it's hard to get a sense for which range of temperatures is spanned by each bin; everything is one giant amorphous blob. So let's add white vertical borders demarcating the bins by adding a `color = "white"` argument to `geom_histogram()` and ignore the warning about setting the number of bins to a better value:

```
ggplot(data = weather, mapping = aes(x = temp)) +
  geom_histogram(color = "white")
```

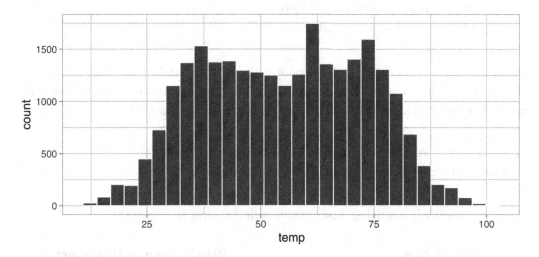

FIGURE 2.11: Histogram of hourly temperatures at three NYC airports with white borders.

We now have an easier time associating ranges of temperatures to each of the bins in Figure 2.11. We can also vary the color of the bars by setting the `fill` argument. For example, you can set the bin colors to be "blue steel" by setting `fill = "steelblue"`:

```
ggplot(data = weather, mapping = aes(x = temp)) +
  geom_histogram(color = "white", fill = "steelblue")
```

If you're curious, run `colors()` to see all 657 possible choice of colors in R!

2.5.2 Adjusting the bins

Observe in Figure 2.11 that in the 50-75°F range there appear to be roughly 8 bins. Thus each bin has width 25 divided by 8, or 3.125°F, which is not a very easily interpretable range to work with. Let's improve this by adjusting the number of bins in our histogram in one of two ways:

1. By adjusting the number of bins via the `bins` argument to `geom_histogram()`.
2. By adjusting the width of the bins via the `binwidth` argument to `geom_histogram()`.

Using the first method, we have the power to specify how many bins we would like to cut the x-axis up in. As mentioned in the previous section, the default number of bins is 30. We can override this default, to say 40 bins, as follows:

```
ggplot(data = weather, mapping = aes(x = temp)) +
  geom_histogram(bins = 40, color = "white")
```

Using the second method, instead of specifying the number of bins, we specify the width of the bins by using the `binwidth` argument in the `geom_histogram()` layer. For example, let's set the width of each bin to be 10°F.

```
ggplot(data = weather, mapping = aes(x = temp)) +
  geom_histogram(binwidth = 10, color = "white")
```

We compare both resulting histograms side-by-side in Figure 2.12.

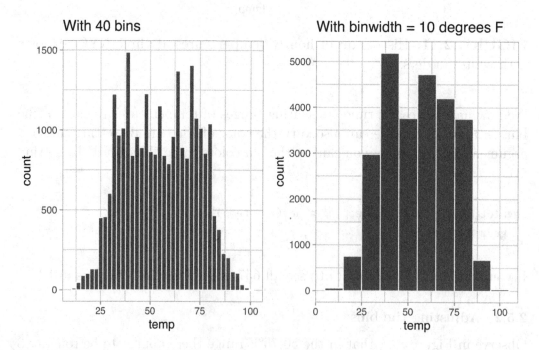

FIGURE 2.12: Setting histogram bins in two ways.

Learning check

(LC2.14) What does changing the number of bins from 30 to 40 tell us about the distribution of temperatures?

(LC2.15) Would you classify the distribution of temperatures as symmetric or skewed in one direction or another?

(LC2.16) What would you guess is the "center" value in this distribution? Why did you make that choice?

(LC2.17) Is this data spread out greatly from the center or is it close? Why?

2.5.3 Summary

Histograms, unlike scatterplots and linegraphs, present information on only a single numerical variable. Specifically, they are visualizations of the distribution of the numerical variable in question.

2.6 Facets

Before continuing with the next of the 5NG, let's briefly introduce a new concept called *faceting*. Faceting is used when we'd like to split a particular visualization by the values of another variable. This will create multiple copies of the same type of plot with matching x and y axes, but whose content will differ.

For example, suppose we were interested in looking at how the histogram of hourly temperature recordings at the three NYC airports we saw in Figure 2.9 differed in each month. We could "split" this histogram by the 12 possible months in a given year. In other words, we would plot histograms of temp for each month separately. We do this by adding facet_wrap(~ month) layer. Note the ~ is a "tilde" and can generally be found on the key next to the "1" key on US keyboards. The tilde is required and you'll receive the error Error in as.quoted(facets) : object 'month' not found if you don't include it here.

```
ggplot(data = weather, mapping = aes(x = temp)) +
  geom_histogram(binwidth = 5, color = "white") +
  facet_wrap(~ month)
```

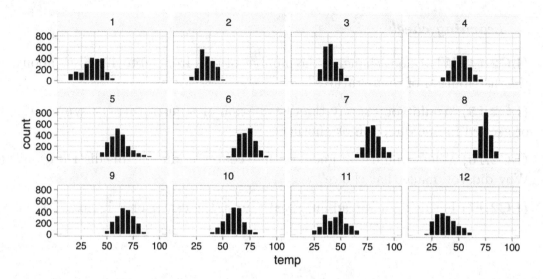

FIGURE 2.13: Faceted histogram of hourly temperatures by month.

We can also specify the number of rows and columns in the grid by using the nrow and ncol arguments inside of facet_wrap(). For example, say we would like our faceted histogram to have 4 rows instead of 3. We simply add an nrow = 4 argument to facet_wrap(~ month)

```
ggplot(data = weather, mapping = aes(x = temp)) +
  geom_histogram(binwidth = 5, color = "white") +
  facet_wrap(~ month, nrow = 4)
```

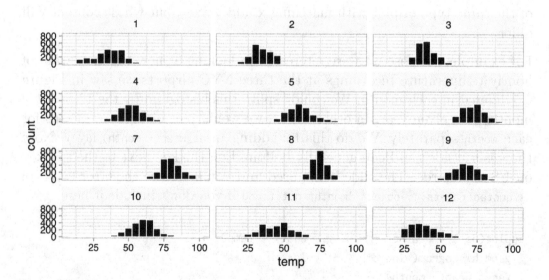

FIGURE 2.14: Faceted histogram with 4 instead of 3 rows.

Observe in both Figures 2.13 and 2.14 that as we might expect in the Northern Hemisphere, temperatures tend to be higher in the summer months, while they tend to be lower in the winter.

Learning check

(LC2.18) What other things do you notice about this faceted plot? How does a faceted plot help us see relationships between two variables?

(LC2.19) What do the numbers 1-12 correspond to in the plot? What about 25, 50, 75, 100?

(LC2.20) For which types of datasets would faceted plots not work well in comparing relationships between variables? Give an example describing the nature of these variables and other important characteristics.

(LC2.21) Does the temp variable in the weather dataset have a lot of variability? Why do you say that?

2.7 5NG#4: Boxplots

While faceted histograms are one type of visualization used to compare the distribution of a numerical variable split by the values of another variable, another type of visualization that achieves this same goal is a *side-by-side boxplot*. A boxplot is constructed from the information provided in the *five-number summary* of a numerical variable (see Appendix A.1).

To keep things simple for now, let's only consider the 2141 hourly temperature recordings for the month of November, each represented as a jittered point in Figure 2.15.

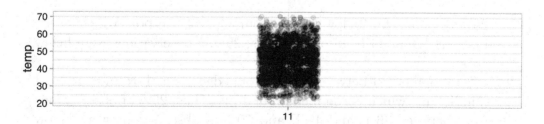

FIGURE 2.15: November temperatures represented as jittered points.

These 2141 observations have the following *five-number summary*:

1. Minimum: 21°F
2. First quartile (25th percentile): 36°F
3. Median (second quartile, 50th percentile): 45°F
4. Third quartile (75th percentile): 52°F
5. Maximum: 71°F

In the leftmost plot of Figure 2.16, let's mark these 5 values with dashed horizontal lines on top of the 2141 points. In the middle plot of Figure 2.16 let's add the *boxplot*. In the rightmost plot of Figure 2.16, let's remove the points and the dashed horizontal lines for clarity's sake.

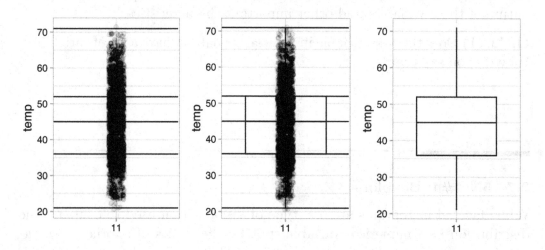

FIGURE 2.16: Building up a boxplot of November temperatures.

What the boxplot does is visually summarize the 2141 points by cutting the 2141 temperature recordings into *quartiles* at the dashed lines, where each quartile contains roughly 2141 ÷ 4 ≈ 535 observations. Thus

1. 25% of points fall below the bottom edge of the box, which is the first quartile of 36°F. In other words, 25% of observations were below 36°F.
2. 25% of points fall between the bottom edge of the box and the solid middle line, which is the median of 45°F. Thus, 25% of observations were between 36°F and 45°F and 50% of observations were below 45°F.
3. 25% of points fall between the solid middle line and the top edge of the box, which is the third quartile of 52°F. It follows that 25% of

observations were between 45°F and 52°F and 75% of observations were below 52°F.

4. 25% of points fall above the top edge of the box. In other words, 25% of observations were above 52°F.

5. The middle 50% of points lie within the *interquartile range (IQR)* between the first and third quartile. Thus, the IQR for this example is 52 - 36 = 16°F. The interquartile range is a measure of a numerical variable's *spread*.

Furthermore, in the rightmost plot of Figure 2.16, we see the *whiskers* of the boxplot. The whiskers stick out from either end of the box all the way to the minimum and maximum observed temperatures of 21°F and 71°F, respectively. However, the whiskers don't always extend to the smallest and largest observed values as they do here. They in fact extend no more than 1.5 × the interquartile range from either end of the box. In this case of the November temperatures, no more than 1.5 × 16°F = 24°F from either end of the box. Any observed values outside this range get marked with points called *outliers*, which we'll see in the next section.

2.7.1 Boxplots via `geom_boxplot`

Let's now create a side-by-side boxplot of hourly temperatures split by the 12 months as we did previously with the faceted histograms. We do this by mapping the `month` variable to the x-position aesthetic, the `temp` variable to the y-position aesthetic, and by adding a `geom_boxplot()` layer:

```
ggplot(data = weather, mapping = aes(x = month, y = temp)) +
  geom_boxplot()
```

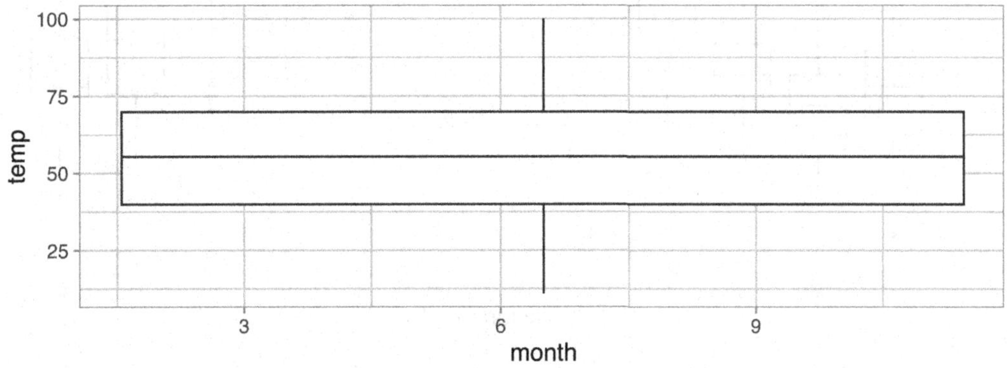

FIGURE 2.17: Invalid boxplot specification.

```
Warning messages:
1: Continuous x aesthetic -- did you forget aes(group=...)?
2: Removed 1 rows containing non-finite values (stat_boxplot).
```

Observe in Figure 2.17 that this plot does not provide information about temperature separated by month. The first warning message clues us in as to why. It is telling us that we have a "continuous", or numerical variable, on the x-position aesthetic. Boxplots, however, require a categorical variable to be mapped to the x-position aesthetic. The second warning message is identical to the warning message when plotting a histogram of hourly temperatures: that one of the values was recorded as NA missing.

We can convert the numerical variable month into a factor categorical variable by using the factor() function. So after applying factor(month), month goes from having numerical values just the 1, 2, ..., and 12 to having an associated ordering. With this ordering, ggplot() now knows how to work with this variable to produce the needed plot.

```
ggplot(data = weather, mapping = aes(x = factor(month), y = temp)) +
    geom_boxplot()
```

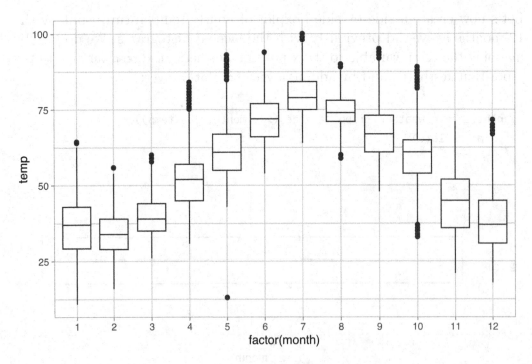

FIGURE 2.18: Side-by-side boxplot of temperature split by month.

The resulting Figure 2.18 shows 12 separate "box and whiskers" plots similar to the rightmost plot of Figure 2.16 of only November temperatures. Thus the different boxplots are shown "side-by-side."

- The "box" portions of the visualization represent the 1st quartile, the median (the 2nd quartile), and the 3rd quartile.
- The height of each box (the value of the 3rd quartile minus the value of the 1st quartile) is the interquartile range (IQR). It is a measure of the spread of the middle 50% of values, with longer boxes indicating more variability.
- The "whisker" portions of these plots extend out from the bottoms and tops of the boxes and represent points less than the 25th percentile and greater than the 75th percentiles, respectively. They're set to extend out no more than $1.5 \times IQR$ units away from either end of the boxes. We say "no more than" because the ends of the whiskers have to correspond to observed temperatures. The length of these whiskers show how the data outside the middle 50% of values vary, with longer whiskers indicating more variability.
- The dots representing values falling outside the whiskers are called *outliers*. These can be thought of as anomalous ("out-of-the-ordinary") values.

It is important to keep in mind that the definition of an outlier is somewhat arbitrary and not absolute. In this case, they are defined by the length of the whiskers, which are no more than $1.5 \times IQR$ units long for each boxplot. Looking at this side-by-side plot we can see, as expected, that summer months (6 through 8) have higher median temperatures as evidenced by the higher solid lines in the middle of the boxes. We can easily compare temperatures across months by drawing imaginary horizontal lines across the plot. Furthermore, the heights of the 12 boxes as quantified by the interquartile ranges are informative too; they tell us about variability, or spread, of temperatures recorded in a given month.

Learning check

(LC2.22) What does the dot at the bottom of the plot for May correspond to? Explain what might have occurred in May to produce this point.

(LC2.23) Which months have the highest variability in temperature? What reasons can you give for this?

(LC2.24) We looked at the distribution of the numerical variable `temp` split by the numerical variable `month` that we converted using the `factor()` function in order to make a side-by-side boxplot. Why would a boxplot of `temp` split by the numerical variable `pressure` similarly converted to a categorical variable using the `factor()` not be informative?

(LC2.25) Boxplots provide a simple way to identify outliers. Why may outliers be easier to identify when looking at a boxplot instead of a faceted histogram?

2.7.2 Summary

Side-by-side boxplots provide us with a way to compare the distribution of a numerical variable across multiple values of another variable. One can see where the median falls across the different groups by comparing the solid lines in the center of the boxes.

To study the spread of a numerical variable within one of the boxes, look at both the length of the box and also how far the whiskers extend from either end of the box. Outliers are even more easily identified when looking at a boxplot than when looking at a histogram as they are marked with distinct points.

2.8 5NG#5: Barplots

Both histograms and boxplots are tools to visualize the distribution of numerical variables. Another commonly desired task is to visualize the distribution of a categorical variable. This is a simpler task, as we are simply counting different categories within a categorical variable, also known as the *levels* of the categorical variable. Often the best way to visualize these different counts, also known as *frequencies*, is with barplots (also called barcharts).

One complication, however, is how your data is represented. Is the categorical variable of interest "pre-counted" or not? For example, run the following code that manually creates two data frames representing a collection of fruit: 3 apples and 2 oranges.

```
fruits <- tibble(
  fruit = c("apple", "apple", "orange", "apple", "orange")
)
fruits_counted <- tibble(
  fruit = c("apple", "orange"),
  number = c(3, 2)
)
```

We see both the `fruits` and `fruits_counted` data frames represent the same collection of fruit. Whereas `fruits` just lists the fruit individually...

```
# A tibble: 5 x 1
  fruit
  <chr>
1 apple
2 apple
3 orange
4 apple
5 orange
```

... `fruits_counted` has a variable `count` which represent the "pre-counted" values of each fruit.

```
# A tibble: 2 x 2
  fruit   number
  <chr>   <dbl>
1 apple       3
2 orange      2
```

Depending on how your categorical data is represented, you'll need to add a different geometric layer type to your `ggplot()` to create a barplot, as we now explore.

2.8.1 Barplots via `geom_bar` or `geom_col`

Let's generate barplots using these two different representations of the same basket of fruit: 3 apples and 2 oranges. Using the `fruits` data frame where all 5 fruits are listed individually in 5 rows, we map the `fruit` variable to the x-position aesthetic and add a `geom_bar()` layer:

```
ggplot(data = fruits, mapping = aes(x = fruit)) +
  geom_bar()
```

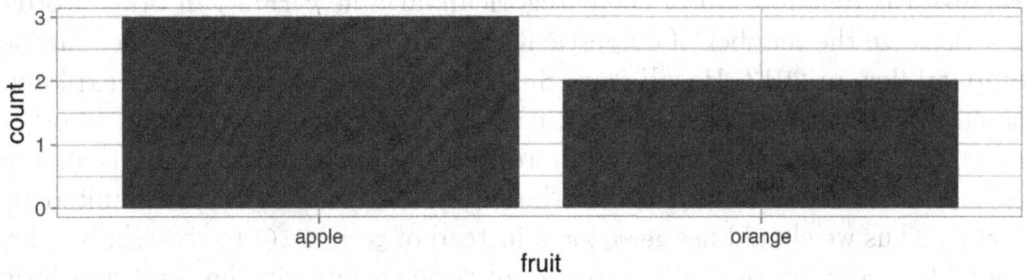

FIGURE 2.19: Barplot when counts are not pre-counted.

However, using the `fruits_counted` data frame where the fruits have been "pre-counted", we once again map the `fruit` variable to the x-position aesthetic, but here we also map the `count` variable to the y-position aesthetic, and add a `geom_col()` layer instead.

```
ggplot(data = fruits_counted, mapping = aes(x = fruit, y = number)) +
    geom_col()
```

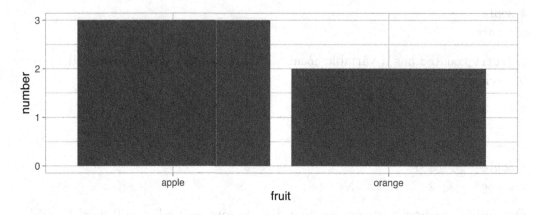

FIGURE 2.20: Barplot when counts are pre-counted.

Compare the barplots in Figures 2.19 and 2.20. They are identical because they reflect counts of the same five fruits. However, depending on how our categorical data is represented, either "pre-counted" or not, we must add a different `geom` layer. When the categorical variable whose distribution you want to visualize

- Is *not* pre-counted in your data frame, we use `geom_bar()`.
- Is pre-counted in your data frame, we use `geom_col()` with the y-position aesthetic mapped to the variable that has the counts.

Let's now go back to the `flights` data frame in the `nycflights13` package and visualize the distribution of the categorical variable `carrier`. In other words, let's visualize the number of domestic flights out of New York City each airline company flew in 2013. Recall from Subsection 1.4.3 when you first explored the `flights` data frame, you saw that each row corresponds to a flight. In other words, the `flights` data frame is more like the `fruits` data frame than the `fruits_counted` data frame because the flights have not been pre-counted by `carrier`. Thus we should use `geom_bar()` instead of `geom_col()` to create a barplot. Much like a `geom_histogram()`, there is only one variable in the `aes()` aesthetic mapping: the variable `carrier` gets mapped to the x-position. As a difference

though, histograms have bars that touch whereas bar graphs have white space
between the bars going from left to right.

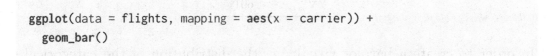

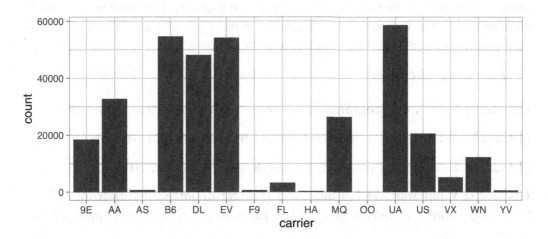

FIGURE 2.21: Number of flights departing NYC in 2013 by airline using
geom_bar().

Observe in Figure 2.21 that United Airlines (UA), JetBlue Airways (B6), and
ExpressJet Airlines (EV) had the most flights depart NYC in 2013. If you don't
know which airlines correspond to which carrier codes, then run View(airlines)
to see a directory of airlines. For example, B6 is JetBlue Airways. Alternatively,
say you had a data frame where the number of flights for each carrier was
pre-counted as in Table 2.3.

TABLE 2.3: Number of flights pre-counted for each carrier

carrier	number
9E	18460
AA	32729
AS	714
B6	54635
DL	48110
EV	54173
F9	685
FL	3260
HA	342
MQ	26397
OO	32
UA	58665

US	20536
VX	5162
WN	12275
YV	601

In order to create a barplot visualizing the distribution of the categorical variable carrier in this case, we would now use geom_col() instead of geom_bar(), with an additional y = number in the aesthetic mapping on top of the x = carrier. The resulting barplot would be identical to Figure 2.21.

Learning check

(LC2.26) Why are histograms inappropriate for categorical variables?

(LC2.27) What is the difference between histograms and barplots?

(LC2.28) How many Envoy Air flights departed NYC in 2013?

(LC2.29) What was the 7th highest airline for departed flights from NYC in 2013? How could we better present the table to get this answer quickly?

2.8.2 Must avoid pie charts!

One of the most common plots used to visualize the distribution of categorical data is the pie chart. While they may seem harmless enough, pie charts actually present a problem in that humans are unable to judge angles well. As Naomi Robbins describes in her book, *Creating More Effective Graphs* (Robbins, 2013), we overestimate angles greater than 90 degrees and we underestimate angles less than 90 degrees. In other words, it is difficult for us to determine the relative size of one piece of the pie compared to another.

Let's examine the same data used in our previous barplot of the number of flights departing NYC by airline in Figure 2.21, but this time we will use a pie chart in Figure 2.22. Try to answer the following questions:

- How much larger is the portion of the pie for ExpressJet Airlines (EV) compared to US Airways (US)?
- What is the third largest carrier in terms of departing flights?
- How many carriers have fewer flights than United Airlines (UA)?

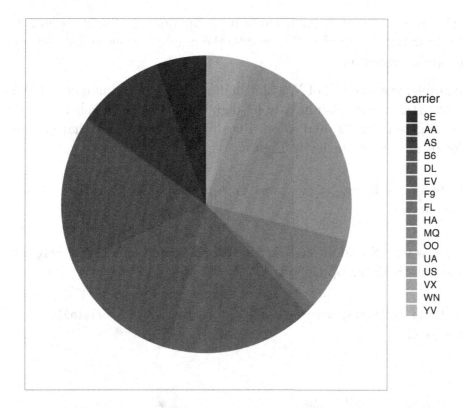

FIGURE 2.22: The dreaded pie chart.

While it is quite difficult to answer these questions when looking at the pie chart in Figure 2.22, we can much more easily answer these questions using the barchart in Figure 2.21. This is true since barplots present the information in a way such that comparisons between categories can be made with single horizontal lines, whereas pie charts present the information in a way such that comparisons must be made by comparing angles.

Learning check

(LC2.30) Why should pie charts be avoided and replaced by barplots?

(LC2.31) Why do you think people continue to use pie charts?

2.8.3 Two categorical variables

Barplots are a very common way to visualize the frequency of different categories, or levels, of a single categorical variable. Another use of barplots is to visualize the *joint* distribution of two categorical variables at the same time.

Let's examine the *joint* distribution of outgoing domestic flights from NYC by carrier as well as origin. In other words, the number of flights for each carrier and origin combination.

For example, the number of WestJet flights from JFK, the number of WestJet flights from LGA, the number of WestJet flights from EWR, the number of American Airlines flights from JFK, and so on. Recall the ggplot() code that created the barplot of carrier frequency in Figure 2.21:

```
ggplot(data = flights, mapping = aes(x = carrier)) +
  geom_bar()
```

We can now map the additional variable origin by adding a fill = origin inside the aes() aesthetic mapping.

```
ggplot(data = flights, mapping = aes(x = carrier, fill = origin)) +
  geom_bar()
```

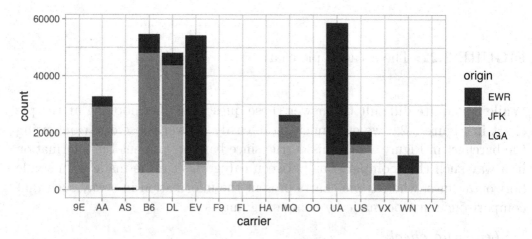

FIGURE 2.23: Stacked barplot of flight amount by carrier and origin.

Figure 2.23 is an example of a *stacked barplot*. While simple to make, in certain aspects it is not ideal. For example, it is difficult to compare the heights of the different colors between the bars, corresponding to comparing the number of flights from each origin airport between the carriers.

Before we continue, let's address some common points of confusion among new R users. First, the fill aesthetic corresponds to the color used to fill the bars, while the color aesthetic corresponds to the color of the outline of the bars. This is identical to how we added color to our histogram in Subsection

2.5.1: we set the outline of the bars to white by setting color = "white" and the colors of the bars to blue steel by setting fill = "steelblue". Observe in Figure 2.24 that mapping origin to color and not fill yields grey bars with different colored outlines.

```
ggplot(data = flights, mapping = aes(x = carrier, color = origin)) +
  geom_bar()
```

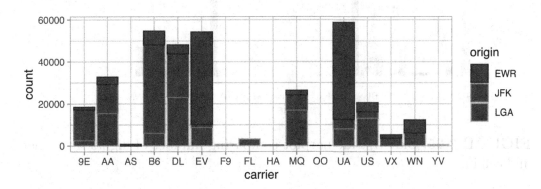

FIGURE 2.24: Stacked barplot with color aesthetic used instead of fill.

Second, note that fill is another aesthetic mapping much like x-position; thus we were careful to include it within the parentheses of the aes() mapping. The following code, where the fill aesthetic is specified outside the aes() mapping will yield an error. This is a fairly common error that new ggplot users make:

```
ggplot(data = flights, mapping = aes(x = carrier), fill = origin) +
  geom_bar()
```

An alternative to stacked barplots are *side-by-side barplots*, also known as *dodged barplots*, as seen in Figure 2.25. The code to create a side-by-side barplot is identical to the code to create a stacked barplot, but with a position = "dodge" argument added to geom_bar(). In other words, we are overriding the default barplot type, which is a *stacked* barplot, and specifying it to be a side-by-side barplot instead.

```
ggplot(data = flights, mapping = aes(x = carrier, fill = origin)) +
  geom_bar(position = "dodge")
```

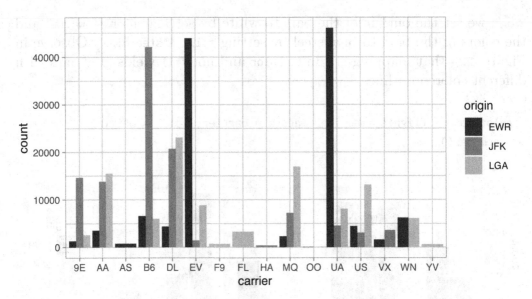

FIGURE 2.25: Side-by-side barplot comparing number of flights by carrier and origin.

Note the width of the bars for AS, F9, FL, HA and YV is different than the others. We can make one tweak to the position argument to get them to be the same size in terms of width as the other bars by using the more robust position_dodge() function.

```
ggplot(data = flights, mapping = aes(x = carrier, fill = origin)) +
  geom_bar(position = position_dodge(preserve = "single"))
```

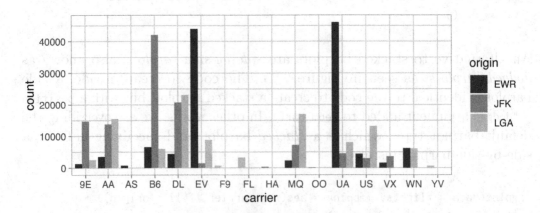

FIGURE 2.26: Side-by-side barplot comparing number of flights by carrier and origin (with formatting tweak).

Lastly, another type of barplot is a *faceted barplot*. Recall in Section 2.6 we visualized the distribution of hourly temperatures at the 3 NYC airports *split* by month using facets. We apply the same principle to our barplot visualizing the frequency of `carrier` split by `origin`: instead of mapping `origin` to `fill` we include it as the variable to create small multiples of the plot across the levels of `origin`.

```
ggplot(data = flights, mapping = aes(x = carrier)) +
  geom_bar() +
  facet_wrap(~ origin, ncol = 1)
```

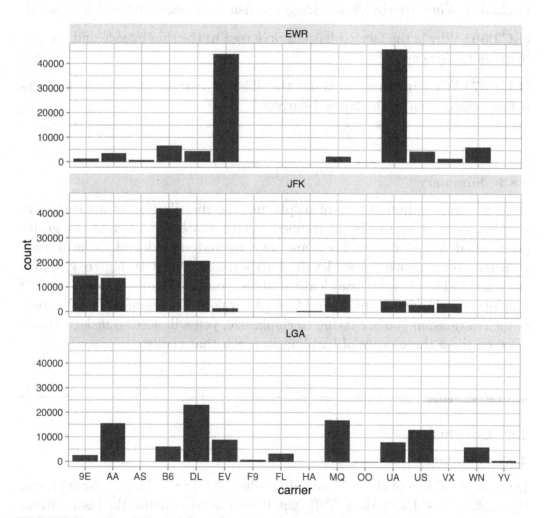

FIGURE 2.27: Faceted barplot comparing the number of flights by carrier and origin.

Learning check

(LC2.32) What kinds of questions are not easily answered by looking at Figure 2.23?

(LC2.33) What can you say, if anything, about the relationship between airline and airport in NYC in 2013 in regards to the number of departing flights?

(LC2.34) Why might the side-by-side barplot be preferable to a stacked barplot in this case?

(LC2.35) What are the disadvantages of using a dodged barplot, in general?

(LC2.36) Why is the faceted barplot preferred to the side-by-side and stacked barplots in this case?

(LC2.37) What information about the different carriers at different airports is more easily seen in the faceted barplot?

2.8.4 Summary

Barplots are a common way of displaying the distribution of a categorical variable, or in other words the frequency with which the different categories (also called *levels*) occur. They are easy to understand and make it easy to make comparisons across levels. Furthermore, when trying to visualize the relationship of two categorical variables, you have many options: stacked barplots, side-by-side barplots, and faceted barplots. Depending on what aspect of the relationship you are trying to emphasize, you will need to make a choice between these three types of barplots and own that choice.

2.9 Conclusion

2.9.1 Summary table

Let's recap all five of the five named graphs (5NG) in Table 2.4 summarizing their differences. Using these 5NG, you'll be able to visualize the distributions and relationships of variables contained in a wide array of datasets. This will be even more the case as we start to map more variables to more of each geometric object's aesthetic attribute options, further unlocking the awesome power of the ggplot2 package.

TABLE 2.4: Summary of Five Named Graphs

	Named graph	Shows	Geometric object	Notes
1	Scatterplot	Relationship between 2 numerical variables	geom_point()	
2	Linegraph	Relationship between 2 numerical variables	geom_line()	Used when there is a sequential order to x-variable, e.g., time
3	Histogram	Distribution of 1 numerical variable	geom_histogram()	Facetted histograms show the distribution of 1 numerical variable split by the values of another variable
4	Boxplot	Distribution of 1 numerical variable split by the values of another variable	geom_boxplot()	
5	Barplot	Distribution of 1 categorical variable	geom_bar() when counts are not pre-counted, geom_col() when counts are pre-counted	Stacked, side-by-side, and faceted barplots show the joint distribution of 2 categorical variables

2.9.2   Function argument specification

Let's go over some important points about specifying the arguments (i.e., inputs) to functions. Run the following two segments of code:

```
# Segment 1:
ggplot(data = flights, mapping = aes(x = carrier)) +
  geom_bar()

# Segment 2:
ggplot(flights, aes(x = carrier)) +
  geom_bar()
```

You'll notice that both code segments create the same barplot, even though in the second segment we omitted the `data =` and `mapping =` code argument names. This is because the `ggplot()` function by default assumes that the `data` argument comes first and the `mapping` argument comes second. As long as you specify the data frame in question first and the `aes()` mapping second, you can omit the explicit statement of the argument names `data =` and `mapping =`.

Going forward for the rest of this book, all `ggplot()` code will be like the second segment: with the `data =` and `mapping =` explicit naming of the argument omitted with the default ordering of arguments respected. We'll do this for brevity's sake; it's common to see this style when reviewing other R users' code.

2.9.3   Additional resources

Solutions to all *Learning checks* can be found online in Appendix D[6].

An R script file of all R code used in this chapter is available at `https://www.moderndive.com/scripts/02-visualization.R`.

If you want to further unlock the power of the `ggplot2` package for data visualization, we suggest that you check out RStudio's "Data Visualization with ggplot2" cheatsheet. This cheatsheet summarizes much more than what we've discussed in this chapter. In particular, it presents many more than the 5 geometric objects we covered in this chapter while providing quick and easy to read visual descriptions. For all the geometric objects, it also lists all the possible aesthetic attributes one can tweak. In the current version of RStudio in late 2019, you can access this cheatsheet by going to the RStudio Menu Bar -> Help -> Cheatsheets -> "Data Visualization with ggplot2."

2.9.4   What's to come

Recall in Figure 2.2 in Section 2.3 we visualized the relationship between departure delay and arrival delay for Alaska Airlines flights. This necessitated paring down the `flights` data frame to a new data frame `alaska_flights` consisting of only `carrier == AS` flights first:

```
alaska_flights <- flights %>%
  filter(carrier == "AS")

ggplot(data = alaska_flights, mapping = aes(x = dep_delay, y = arr_delay)) +
  geom_point()
```

Furthermore recall in Figure 2.7 in Section 2.4 we visualized hourly temperature recordings at Newark airport only for the first 15 days of January 2013. This necessitated paring down the `weather` data frame to a new data frame

---

[6]`https://moderndive.com/D-appendixD.html`

`early_january_weather` consisting of hourly temperature recordings only for `origin == "EWR"`, `month == 1`, and day less than or equal to 15 first:

```
early_january_weather <- weather %>%
  filter(origin == "EWR" & month == 1 & day <= 15)

ggplot(data = early_january_weather, mapping = aes(x = time_hour, y = temp))
+ geom_line()
```

These two code segments were a preview of Chapter 3 on data wrangling using the `dplyr` package. Data wrangling is the process of transforming and modifying existing data with the intent of making it more appropriate for analysis purposes. For example, these two code segments used the `filter()` function to create new data frames (`alaska_flights` and `early_january_weather`) by choosing only a subset of rows of existing data frames (`flights` and `weather`). In the next chapter, we'll formally introduce the `filter()` and other data wrangling functions as well as the *pipe operator* `%>%` which allows you to combine multiple data wrangling actions into a single sequential *chain* of actions. On to Chapter 3 on data wrangling!

3

Data Wrangling

So far in our journey, we've seen how to look at data saved in data frames using the `glimpse()` and `View()` functions in Chapter 1, and how to create data visualizations using the `ggplot2` package in Chapter 2. In particular we studied what we term the "five named graphs" (5NG):

1. scatterplots via `geom_point()`
2. linegraphs via `geom_line()`
3. boxplots via `geom_boxplot()`
4. histograms via `geom_histogram()`
5. barplots via `geom_bar()` or `geom_col()`

We created these visualizations using the grammar of graphics, which maps variables in a data frame to the aesthetic attributes of one of the 5 geometric objects. We can also control other aesthetic attributes of the geometric objects such as the size and color as seen in the Gapminder data example in Figure 2.1.

Recall however that for two of our visualizations, we first needed to transform/modify existing data frames a little. For example, recall the scatterplot in Figure 2.2 of departure and arrival delays *only* for Alaska Airlines flights. In order to create this visualization, we first needed to pare down the `flights` data frame to an `alaska_flights` data frame consisting of only `carrier == "AS"` flights. Thus, `alaska_flights` will have fewer rows than `flights`. We did this using the `filter()` function:

```
alaska_flights <- flights %>%
  filter(carrier == "AS")
```

In this chapter, we'll extend this example and we'll introduce a series of functions from the `dplyr` package for data wrangling that will allow you to take a data frame and "wrangle" it (transform it) to suit your needs. Such functions include:

1. `filter()` a data frame's existing rows to only pick out a subset of them. For example, the `alaska_flights` data frame.

2. `summarize()` one or more of its columns/variables with a *summary statistic*. Examples of summary statistics include the median and interquartile range of temperatures as we saw in Section 2.7 on boxplots.

3. `group_by()` its rows. In other words, assign different rows to be part of the same *group*. We can then combine `group_by()` with `summarize()` to report summary statistics for each group *separately*. For example, say you don't want a single overall average departure delay `dep_delay` for all three `origin` airports combined, but rather three separate average departure delays, one computed for each of the three `origin` airports.

4. `mutate()` its existing columns/variables to create new ones. For example, convert hourly temperature recordings from degrees Fahrenheit to degrees Celsius.

5. `arrange()` its rows. For example, sort the rows of `weather` in ascending or descending order of `temp`.

6. `join()` it with another data frame by matching along a "key" variable. In other words, merge these two data frames together.

Notice how we used `computer_code` font to describe the actions we want to take on our data frames. This is because the `dplyr` package for data wrangling has intuitively verb-named functions that are easy to remember.

There is a further benefit to learning to use the `dplyr` package for data wrangling: its similarity to the database querying language SQL[1] (pronounced "sequel" or spelled out as "S", "Q", "L"). SQL (which stands for "Structured Query Language") is used to manage large databases quickly and efficiently and is widely used by many institutions with a lot of data. While SQL is a topic left for a book or a course on database management, keep in mind that once you learn `dplyr`, you can learn SQL easily. We'll talk more about their similarities in Subsection 3.7.4.

Needed packages

Let's load all the packages needed for this chapter (this assumes you've already installed them). If needed, read Section 1.3 for information on how to install and load R packages.

---

[1]`https://en.wikipedia.org/wiki/SQL`

```
library(dplyr)
library(ggplot2)
library(nycflights13)
```

3.1 The pipe operator: %>%

Before we start data wrangling, let's first introduce a nifty tool that gets loaded with the dplyr package: the pipe operator %>%. The pipe operator allows us to combine multiple operations in R into a single sequential *chain* of actions.

Let's start with a hypothetical example. Say you would like to perform a hypothetical sequence of operations on a hypothetical data frame x using hypothetical functions f(), g(), and h():

1. Take x *then*
2. Use x as an input to a function f() *then*
3. Use the output of f(x) as an input to a function g() *then*
4. Use the output of g(f(x)) as an input to a function h()

One way to achieve this sequence of operations is by using nesting parentheses as follows:

```
h(g(f(x)))
```

This code isn't so hard to read since we are applying only three functions: f(), then g(), then h() and each of the functions is short in its name. Further, each of these functions also only has one argument. However, you can imagine that this will get progressively harder to read as the number of functions applied in your sequence increases and the arguments in each function increase as well. This is where the pipe operator %>% comes in handy. %>% takes the output of one function and then "pipes" it to be the input of the next function. Furthermore, a helpful trick is to read %>% as "then" or "and then." For example, you can obtain the same output as the hypothetical sequence of functions as follows:

```
x %>%
  f() %>%
  g() %>%
  h()
```

You would read this sequence as:

1. Take x *then*
2. Use this output as the input to the next function f() *then*
3. Use this output as the input to the next function g() *then*
4. Use this output as the input to the next function h()

So while both approaches achieve the same goal, the latter is much more human-readable because you can clearly read the sequence of operations line-by-line. But what are the hypothetical x, f(), g(), and h()? Throughout this chapter on data wrangling:

1. The starting value x will be a data frame. For example, the flights data frame we explored in Section 1.4.
2. The sequence of functions, here f(), g(), and h(), will mostly be a sequence of any number of the six data wrangling verb-named functions we listed in the introduction to this chapter. For example, the filter(carrier == "AS") function and argument specified we previewed earlier.
3. The result will be the transformed/modified data frame that you want. In our example, we'll save the result in a new data frame by using the <- assignment operator with the name alaska_flights via alaska_flights <-.

```
alaska_flights <- flights %>%
  filter(carrier == "AS")
```

Much like when adding layers to a ggplot() using the + sign, you form a single *chain* of data wrangling operations by combining verb-named functions into a single sequence using the pipe operator %>%. Furthermore, much like how the + sign has to come at the end of lines when constructing plots, the pipe operator %>% has to come at the end of lines as well.

Keep in mind, there are many more advanced data wrangling functions than just the six listed in the introduction to this chapter; you'll see some examples of these in Section 3.8. However, just with these six verb-named functions you'll be able to perform a broad array of data wrangling tasks for the rest of this book.

3.2 `filter rows`

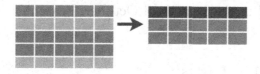

FIGURE 3.1: Diagram of filter() rows operation.

The `filter()` function here works much like the "Filter" option in Microsoft Excel; it allows you to specify criteria about the values of a variable in your dataset and then filters out only the rows that match that criteria.

We begin by focusing only on flights from New York City to Portland, Oregon. The `dest` destination code (or airport code) for Portland, Oregon is `"PDX"`. Run the following and look at the results in RStudio's spreadsheet viewer to ensure that only flights heading to Portland are chosen here:

```
portland_flights <- flights %>%
  filter(dest == "PDX")
View(portland_flights)
```

Note the order of the code. First, take the `flights` data frame `flights` *then* `filter()` the data frame so that only those where the `dest` equals `"PDX"` are included. We test for equality using the double equal sign == and not a single equal sign =. In other words `filter(dest = "PDX")` will yield an error. This is a convention across many programming languages. If you are new to coding, you'll probably forget to use the double equal sign == a few times before you get the hang of it.

You can use other operators beyond just the == operator that tests for equality:

- \> corresponds to "greater than"
- \< corresponds to "less than"
- >= corresponds to "greater than or equal to"
- <= corresponds to "less than or equal to"
- != corresponds to "not equal to." The ! is used in many programming languages to indicate "not."

Furthermore, you can combine multiple criteria using operators that make comparisons:

- | corresponds to "or"
- & corresponds to "and"

To see many of these in action, let's filter `flights` for all rows that departed from JFK *and* were heading to Burlington, Vermont (`"BTV"`) or Seattle, Washington (`"SEA"`) *and* departed in the months of October, November, or December. Run the following:

```
btv_sea_flights_fall <- flights %>%
  filter(origin == "JFK" & (dest == "BTV" | dest == "SEA") & month >= 10)
View(btv_sea_flights_fall)
```

Note that even though colloquially speaking one might say "all flights leaving Burlington, Vermont *and* Seattle, Washington," in terms of computer operations, we really mean "all flights leaving Burlington, Vermont *or* leaving Seattle, Washington." For a given row in the data, `dest` can be `"BTV"`, or `"SEA"`, or something else, but not both `"BTV"` and `"SEA"` at the same time. Furthermore, note the careful use of parentheses around `dest == "BTV" | dest == "SEA"`.

We can often skip the use of & and just separate our conditions with a comma. The previous code will return the identical output `btv_sea_flights_fall` as the following code:

```
btv_sea_flights_fall <- flights %>%
  filter(origin == "JFK", (dest == "BTV" | dest == "SEA"), month >= 10)
View(btv_sea_flights_fall)
```

Let's present another example that uses the ! "not" operator to pick rows that *don't* match a criteria. As mentioned earlier, the ! can be read as "not." Here we are filtering rows corresponding to flights that didn't go to Burlington, VT or Seattle, WA.

```
not_BTV_SEA <- flights %>%
  filter(!(dest == "BTV" | dest == "SEA"))
View(not_BTV_SEA)
```

Again, note the careful use of parentheses around the (`dest == "BTV" | dest == "SEA"`). If we didn't use parentheses as follows:

```
flights %>% filter(!dest == "BTV" | dest == "SEA")
```

We would be returning all flights not headed to `"BTV"` *or* those headed to `"SEA"`, which is an entirely different resulting data frame.

Now say we have a larger number of airports we want to filter for, say `"SEA"`, `"SFO"`, `"PDX"`, `"BTV"`, and `"BDL"`. We could continue to use the `|` (*or*) operator:

```
many_airports <- flights %>%
  filter(dest == "SEA" | dest == "SFO" | dest == "PDX" |
         dest == "BTV" | dest == "BDL")
```

but as we progressively include more airports, this will get unwieldy to write. A slightly shorter approach uses the `%in%` operator along with the `c()` function. Recall from Subsection 1.2.1 that the `c()` function "combines" or "concatenates" values into a single *vector* of values.

```
many_airports <- flights %>%
  filter(dest %in% c("SEA", "SFO", "PDX", "BTV", "BDL"))
View(many_airports)
```

What this code is doing is filtering `flights` for all flights where `dest` is in the vector of airports `c("BTV", "SEA", "PDX", "SFO", "BDL")`. Both outputs of `many_airports` are the same, but as you can see the latter takes much less energy to code. The `%in%` operator is useful for looking for matches commonly in one vector/variable compared to another.

As a final note, we recommend that `filter()` should often be among the first verbs you consider applying to your data. This cleans your dataset to only those rows you care about, or put differently, it narrows down the scope of your data frame to just the observations you care about.

Learning check

(LC3.1) What's another way of using the "not" operator `!` to filter only the rows that are not going to Burlington, VT nor Seattle, WA in the `flights` data frame? Test this out using the previous code.

3.3 `summarize` variables

The next common task when working with data frames is to compute *summary statistics*. Summary statistics are single numerical values that summarize a large number of values. Commonly known examples of summary statistics include the mean (also called the average) and the median (the middle value). Other examples of summary statistics that might not immediately come to mind include the *sum*, the smallest value also called the *minimum*, the largest value also called the *maximum*, and the *standard deviation*. See Appendix A.1 for a glossary of such summary statistics.

Let's calculate two summary statistics of the `temp` temperature variable in the `weather` data frame: the mean and standard deviation (recall from Section 1.4 that the `weather` data frame is included in the `nycflights13` package). To compute these summary statistics, we need the `mean()` and `sd()` *summary functions* in R. Summary functions in R take in many values and return a single value, as illustrated in Figure 3.2.

FIGURE 3.2: Diagram illustrating a summary function in R.

More precisely, we'll use the `mean()` and `sd()` summary functions within the `summarize()` function from the `dplyr` package. Note you can also use the British English spelling of `summarise()`. As shown in Figure 3.3, the `summarize()` function takes in a data frame and returns a data frame with only one row corresponding to the summary statistics.

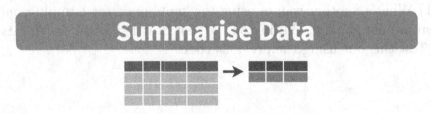

FIGURE 3.3: Diagram of summarize() rows.

We'll save the results in a new data frame called `summary_temp` that will have two columns/variables: the `mean` and the `std_dev`:

```
summary_temp <- weather %>%
  summarize(mean = mean(temp), std_dev = sd(temp))
summary_temp
```

```
# A tibble: 1 x 2
  mean std_dev
  <dbl>   <dbl>
1   NA      NA
```

Why are the values returned NA? As we saw in Subsection 2.3.1 when creating the scatterplot of departure and arrival delays for `alaska_flights`, NA is how R encodes *missing values* where NA indicates "not available" or "not applicable." If a value for a particular row and a particular column does not exist, NA is stored instead. Values can be missing for many reasons. Perhaps the data was collected but someone forgot to enter it? Perhaps the data was not collected at all because it was too difficult to do so? Perhaps there was an erroneous value that someone entered that has been corrected to read as missing? You'll often encounter issues with missing values when working with real data.

Going back to our `summary_temp` output, by default any time you try to calculate a summary statistic of a variable that has one or more NA missing values in R, NA is returned. To work around this fact, you can set the `na.rm` argument to TRUE, where `rm` is short for "remove"; this will ignore any NA missing values and only return the summary value for all non-missing values.

The code that follows computes the mean and standard deviation of all non-missing values of `temp`:

```
summary_temp <- weather %>%
  summarize(mean = mean(temp, na.rm = TRUE),
            std_dev = sd(temp, na.rm = TRUE))
summary_temp
```

```
# A tibble: 1 x 2
  mean std_dev
  <dbl>   <dbl>
1  55.3    17.8
```

Notice how the `na.rm = TRUE` are used as arguments to the `mean()` and `sd()` summary functions individually, and not to the `summarize()` function.

However, one needs to be cautious whenever ignoring missing values as we've just done. In the upcoming *Learning checks* questions, we'll consider the possible ramifications of blindly sweeping rows with missing values "under the rug." This is in fact why the `na.rm` argument to any summary statistic function in R is set to `FALSE` by default. In other words, R does not ignore rows with missing values by default. R is alerting you to the presence of missing data and you should be mindful of this missingness and any potential causes of this missingness throughout your analysis.

What are other summary functions we can use inside the `summarize()` verb to compute summary statistics? As seen in the diagram in Figure 3.2, you can use any function in R that takes many values and returns just one. Here are just a few:

- `mean()`: the average
- `sd()`: the standard deviation, which is a measure of spread
- `min()` and `max()`: the minimum and maximum values, respectively
- `IQR()`: interquartile range
- `sum()`: the total amount when adding multiple numbers
- `n()`: a count of the number of rows in each group. This particular summary function will make more sense when `group_by()` is covered in Section 3.4.

Learning check

(LC3.2) Say a doctor is studying the effect of smoking on lung cancer for a large number of patients who have records measured at five-year intervals. She notices that a large number of patients have missing data points because the patient has died, so she chooses to ignore these patients in her analysis. What is wrong with this doctor's approach?

(LC3.3) Modify the `summarize()` function to create `summary_temp` to also use the `n()` summary function: `summarize(count = n())`. What does the returned value correspond to?

(LC3.4) Why doesn't the following code work? Run the code line-by-line instead of all at once, and then look at the data. In other words, run `summary_temp <- weather %>% summarize(mean = mean(temp, na.rm = TRUE))` first.

```
summary_temp <- weather %>%
  summarize(mean = mean(temp, na.rm = TRUE)) %>%
  summarize(std_dev = sd(temp, na.rm = TRUE))
```

3.4 `group_by` rows

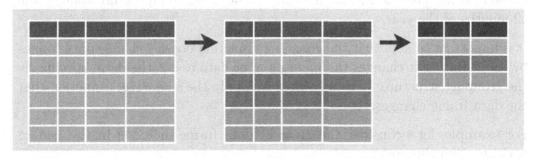

FIGURE 3.4: Diagram of group_by() and summarize().

Say instead of a single mean temperature for the whole year, you would like 12 mean temperatures, one for each of the 12 months separately. In other words, we would like to compute the mean temperature split by month. We can do this by "grouping" temperature observations by the values of another variable, in this case by the 12 values of the variable month. Run the following code:

```
summary_monthly_temp <- weather %>%
  group_by(month) %>%
  summarize(mean = mean(temp, na.rm = TRUE),
            std_dev = sd(temp, na.rm = TRUE))
summary_monthly_temp
```

```
# A tibble: 12 x 3
   month  mean std_dev
   <int> <dbl>   <dbl>
1      1  35.6    10.2
2      2  34.3    6.98
3      3  39.9    6.25
4      4  51.7    8.79
5      5  61.8    9.68
6      6  72.2    7.55
7      7  80.1    7.12
8      8  74.5    5.19
9      9  67.4    8.47
10    10  60.1    8.85
11    11  45.0    10.4
12    12  38.4    9.98
```

This code is identical to the previous code that created `summary_temp`, but with an extra `group_by(month)` added before the `summarize()`. Grouping the `weather` dataset by `month` and then applying the `summarize()` functions yields a data frame that displays the mean and standard deviation temperature split by the 12 months of the year.

It is important to note that the `group_by()` function doesn't change data frames by itself. Rather it changes the *meta-data*, or data about the data, specifically the grouping structure. It is only after we apply the `summarize()` function that the data frame changes.

For example, let's consider the `diamonds` data frame included in the `ggplot2` package. Run this code:

```
diamonds
```

```
# A tibble: 53,940 x 10
   carat cut        color clarity depth table price    x    y    z
   <dbl> <ord>      <ord> <ord>   <dbl> <dbl> <int> <dbl> <dbl> <dbl>
 1 0.23  Ideal      E     SI2      61.5    55   326  3.95  3.98  2.43
 2 0.21  Premium    E     SI1      59.8    61   326  3.89  3.84  2.31
 3 0.23  Good       E     VS1      56.9    65   327  4.05  4.07  2.31
 4 0.290 Premium    I     VS2      62.4    58   334  4.2   4.23  2.63
 5 0.31  Good       J     SI2      63.3    58   335  4.34  4.35  2.75
 6 0.24  Very Good  J     VVS2     62.8    57   336  3.94  3.96  2.48
 7 0.24  Very Good  I     VVS1     62.3    57   336  3.95  3.98  2.47
 8 0.26  Very Good  H     SI1      61.9    55   337  4.07  4.11  2.53
 9 0.22  Fair       E     VS2      65.1    61   337  3.87  3.78  2.49
10 0.23  Very Good  H     VS1      59.4    61   338  4     4.05  2.39
# ... with 53,930 more rows
```

Observe that the first line of the output reads `# A tibble: 53,940 x 10`. This is an example of meta-data, in this case the number of observations/rows and variables/columns in `diamonds`. The actual data itself are the subsequent table of values. Now let's pipe the `diamonds` data frame into `group_by(cut)`:

```
diamonds %>%
  group_by(cut)
```

```
# A tibble: 53,940 x 10
# Groups:   cut [5]
   carat cut        color clarity depth table price    x    y    z
   <dbl> <ord>      <ord> <ord>   <dbl> <dbl> <int> <dbl> <dbl> <dbl>
```

1	0.23	Ideal	E	SI2	61.5	55	326	3.95	3.98	2.43
2	0.21	Premium	E	SI1	59.8	61	326	3.89	3.84	2.31
3	0.23	Good	E	VS1	56.9	65	327	4.05	4.07	2.31
4	0.290	Premium	I	VS2	62.4	58	334	4.2	4.23	2.63
5	0.31	Good	J	SI2	63.3	58	335	4.34	4.35	2.75
6	0.24	Very Good	J	VVS2	62.8	57	336	3.94	3.96	2.48
7	0.24	Very Good	I	VVS1	62.3	57	336	3.95	3.98	2.47
8	0.26	Very Good	H	SI1	61.9	55	337	4.07	4.11	2.53
9	0.22	Fair	E	VS2	65.1	61	337	3.87	3.78	2.49
10	0.23	Very Good	H	VS1	59.4	61	338	4	4.05	2.39

```
# ... with 53,930 more rows
```

Observe that now there is additional meta-data: `# Groups: cut [5]` indicating that the grouping structure meta-data has been set based on the 5 possible levels of the categorical variable `cut`: `"Fair"`, `"Good"`, `"Very Good"`, `"Premium"`, and `"Ideal"`. On the other hand, observe that the data has not changed: it is still a table of $53,940 \times 10$ values.

Only by combining a `group_by()` with another data wrangling operation, in this case `summarize()`, will the data actually be transformed.

```
diamonds %>%
  group_by(cut) %>%
  summarize(avg_price = mean(price))
```

```
# A tibble: 5 x 2
  cut       avg_price
  <ord>       <dbl>
1 Fair        4359.
2 Good        3929.
3 Very Good   3982.
4 Premium     4584.
5 Ideal       3458.
```

If you would like to remove this grouping structure meta-data, we can pipe the resulting data frame into the `ungroup()` function:

```
diamonds %>%
  group_by(cut) %>%
  ungroup()
```

```
# A tibble: 53,940 x 10
   carat cut       color clarity depth table price   x   y   z
```

<dbl>	<ord>		<ord>	<ord>	<dbl>	<dbl>	<int>	<dbl>	<dbl>	<dbl>
1 0.23	Ideal	E		SI2	61.5	55	326	3.95	3.98	2.43
2 0.21	Premium	E		SI1	59.8	61	326	3.89	3.84	2.31
3 0.23	Good	E		VS1	56.9	65	327	4.05	4.07	2.31
4 0.290	Premium	I		VS2	62.4	58	334	4.2	4.23	2.63
5 0.31	Good	J		SI2	63.3	58	335	4.34	4.35	2.75
6 0.24	Very Good	J		VVS2	62.8	57	336	3.94	3.96	2.48
7 0.24	Very Good	I		VVS1	62.3	57	336	3.95	3.98	2.47
8 0.26	Very Good	H		SI1	61.9	55	337	4.07	4.11	2.53
9 0.22	Fair	E		VS2	65.1	61	337	3.87	3.78	2.49
10 0.23	Very Good	H		VS1	59.4	61	338	4	4.05	2.39

```
# ... with 53,930 more rows
```

Observe how the # Groups: cut [5] meta-data is no longer present.

Let's now revisit the n() counting summary function we briefly introduced previously. Recall that the n() function counts rows. This is opposed to the sum() summary function that returns the sum of a numerical variable. For example, suppose we'd like to count how many flights departed each of the three airports in New York City:

```
by_origin <- flights %>%
  group_by(origin) %>%
  summarize(count = n())
by_origin
```

```
# A tibble: 3 x 2
  origin   count
  <chr>    <int>
1 EWR     120835
2 JFK     111279
3 LGA     104662
```

We see that Newark ("EWR") had the most flights departing in 2013 followed by "JFK" and lastly by LaGuardia ("LGA"). Note there is a subtle but important difference between sum() and n(); while sum() returns the sum of a numerical variable, n() returns a count of the number of rows/observations.

3.4.1 Grouping by more than one variable

You are not limited to grouping by one variable. Say you want to know the number of flights leaving each of the three New York City airports *for each month*. We can also group by a second variable month using group_by(origin, month):

```
by_origin_monthly <- flights %>%
  group_by(origin, month) %>%
  summarize(count = n())
by_origin_monthly
```

```
# A tibble: 36 x 3
# Groups:   origin [3]
   origin month count
   <chr>  <int> <int>
 1 EWR        1  9893
 2 EWR        2  9107
 3 EWR        3 10420
 4 EWR        4 10531
 5 EWR        5 10592
 6 EWR        6 10175
 7 EWR        7 10475
 8 EWR        8 10359
 9 EWR        9  9550
10 EWR       10 10104
# ... with 26 more rows
```

Observe that there are 36 rows to by_origin_monthly because there are 12 months for 3 airports (EWR, JFK, and LGA).

Why do we group_by(origin, month) and not group_by(origin) and then group_by(month)? Let's investigate:

```
by_origin_monthly_incorrect <- flights %>%
  group_by(origin) %>%
  group_by(month) %>%
  summarize(count = n())
by_origin_monthly_incorrect
```

```
# A tibble: 12 x 2
   month count
   <int> <int>
 1     1 27004
 2     2 24951
 3     3 28834
 4     4 28330
 5     5 28796
 6     6 28243
```

```
 7       7 29425
 8       8 29327
 9       9 27574
10      10 28889
11      11 27268
12      12 28135
```

What happened here is that the second group_by(month) overwrote the grouping
structure meta-data of the earlier group_by(origin), so that in the end we are
only grouping by month. The lesson here is if you want to group_by() two or
more variables, you should include all the variables at the same time in the
same group_by() adding a comma between the variable names.

Learning check

(LC3.5) Recall from Chapter 2 when we looked at temperatures by months
in NYC. What does the standard deviation column in the summary_monthly_temp
data frame tell us about temperatures in NYC throughout the year?

(LC3.6) What code would be required to get the mean and standard deviation
temperature for each day in 2013 for NYC?

(LC3.7) Recreate by_monthly_origin, but instead of grouping via
group_by(origin, month), group variables in a different order group_by(month,
origin). What differs in the resulting dataset?

(LC3.8) How could we identify how many flights left each of the three airports
for each carrier?

(LC3.9) How does the filter() operation differ from a group_by() followed by
a summarize()?

3.5 mutate existing variables

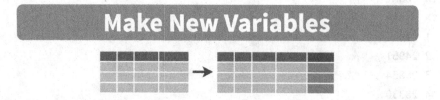

FIGURE 3.5: Diagram of mutate() columns.

Another common transformation of data is to create/compute new variables based on existing ones. For example, say you are more comfortable thinking of temperature in degrees Celsius (°C) instead of degrees Fahrenheit (°F). The formula to convert temperatures from °F to °C is

$$\text{temp in C} = \frac{\text{temp in F} - 32}{1.8}$$

We can apply this formula to the `temp` variable using the `mutate()` function from the `dplyr` package, which takes existing variables and mutates them to create new ones.

```
weather <- weather %>%
  mutate(temp_in_C = (temp - 32) / 1.8)
```

In this code, we `mutate()` the `weather` data frame by creating a new variable `temp_in_C = (temp - 32) / 1.8` and then *overwrite* the original `weather` data frame. Why did we overwrite the data frame `weather`, instead of assigning the result to a new data frame like `weather_new`? As a rough rule of thumb, as long as you are not losing original information that you might need later, it's acceptable practice to overwrite existing data frames with updated ones, as we did here. On the other hand, why did we not overwrite the variable `temp`, but instead created a new variable called `temp_in_C`? Because if we did this, we would have erased the original information contained in `temp` of temperatures in Fahrenheit that may still be valuable to us.

Let's now compute monthly average temperatures in both °F and °C using the `group_by()` and `summarize()` code we saw in Section 3.4:

```
summary_monthly_temp <- weather %>%
  group_by(month) %>%
  summarize(mean_temp_in_F = mean(temp, na.rm = TRUE),
            mean_temp_in_C = mean(temp_in_C, na.rm = TRUE))
summary_monthly_temp
```

```
# A tibble: 12 x 3
   month mean_temp_in_F mean_temp_in_C
   <int>          <dbl>          <dbl>
1      1           35.6           2.02
2      2           34.3           1.26
3      3           39.9           4.38
4      4           51.7          11.0
```

5	5	61.8	16.6
6	6	72.2	22.3
7	7	80.1	26.7
8	8	74.5	23.6
9	9	67.4	19.7
10	10	60.1	15.6
11	11	45.0	7.22
12	12	38.4	3.58

Let's consider another example. Passengers are often frustrated when their flight departs late, but aren't as annoyed if, in the end, pilots can make up some time during the flight. This is known in the airline industry as *gain*, and we will create this variable using the mutate() function:

```
flights <- flights %>%
  mutate(gain = dep_delay - arr_delay)
```

Let's take a look at only the dep_delay, arr_delay, and the resulting gain variables for the first 5 rows in our updated flights data frame in Table 3.1.

TABLE 3.1: First five rows of departure/arrival delay and gain variables

dep_delay	arr_delay	gain
2	11	-9
4	20	-16
2	33	-31
-1	-18	17
-6	-25	19

The flight in the first row departed 2 minutes late but arrived 11 minutes late, so its "gained time in the air" is a loss of 9 minutes, hence its gain is 2 - 11 = -9. On the other hand, the flight in the fourth row departed a minute early (dep_delay of -1) but arrived 18 minutes early (arr_delay of -18), so its "gained time in the air" is $-1 - (-18) = -1 + 18 = 17$ minutes, hence its gain is +17.

Let's look at some summary statistics of the gain variable by considering multiple summary functions at once in the same summarize() code:

```
gain_summary <- flights %>%
  summarize(
    min = min(gain, na.rm = TRUE),
    q1 = quantile(gain, 0.25, na.rm = TRUE),
```

```
    median = quantile(gain, 0.5, na.rm = TRUE),
    q3 = quantile(gain, 0.75, na.rm = TRUE),
    max = max(gain, na.rm = TRUE),
    mean = mean(gain, na.rm = TRUE),
    sd = sd(gain, na.rm = TRUE),
    missing = sum(is.na(gain))
  )
gain_summary
```

```
# A tibble: 1 x 8
    min    q1 median    q3   max  mean    sd missing
  <dbl> <dbl>  <dbl> <dbl> <dbl> <dbl> <dbl>   <int>
1  -196    -3      7    17   109  5.66  18.0    9430
```

We see for example that the average gain is +5 minutes, while the largest is
+109 minutes! However, this code would take some time to type out in practice.
We'll see later on in Subsection 5.1.1 that there is a much more succinct way
to compute a variety of common summary statistics: using the `skim()` function
from the `skimr` package.

Recall from Section 2.5 that since `gain` is a numerical variable, we can visualize
its distribution using a histogram.

```
ggplot(data = flights, mapping = aes(x = gain)) +
  geom_histogram(color = "white", bins = 20)
```

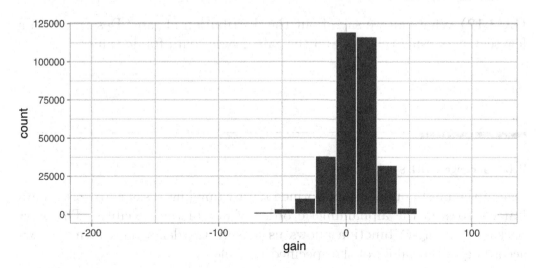

FIGURE 3.6: Histogram of gain variable.

The resulting histogram in Figure 3.6 provides a different perspective on the
gain variable than the summary statistics we computed earlier. For example,
note that most values of gain are right around 0.

To close out our discussion on the mutate() function to create new variables,
note that we can create multiple new variables at once in the same mutate()
code. Furthermore, within the same mutate() code we can refer to new variables
we just created. As an example, consider the mutate() code Hadley Wickham
and Garrett Grolemund show in Chapter 5 of *R for Data Science* (Grolemund
and Wickham, 2017):

```
flights <- flights %>%
  mutate(
    gain = dep_delay - arr_delay,
    hours = air_time / 60,
    gain_per_hour = gain / hours
  )
```

Learning check

(LC3.10) What do positive values of the gain variable in flights correspond
to? What about negative values? And what about a zero value?

(LC3.11) Could we create the dep_delay and arr_delay columns by simply
subtracting dep_time from sched_dep_time and similarly for arrivals? Try the
code out and explain any differences between the result and what actually
appears in flights.

(LC3.12) What can we say about the distribution of gain? Describe it in a
few sentences using the plot and the gain_summary data frame values.

3.6 arrange and sort rows

One of the most commonly performed data wrangling tasks is to sort a data
frame's rows in the alphanumeric order of one of the variables. The dplyr
package's arrange() function allows us to sort/reorder a data frame's rows
according to the values of the specified variable.

Suppose we are interested in determining the most frequent destination airports
for all domestic flights departing from New York City in 2013:

```
freq_dest <- flights %>%
  group_by(dest) %>%
  summarize(num_flights = n())
freq_dest
```

```
# A tibble: 105 x 2
   dest  num_flights
   <chr>       <int>
 1 ABQ           254
 2 ACK           265
 3 ALB           439
 4 ANC             8
 5 ATL         17215
 6 AUS          2439
 7 AVL           275
 8 BDL           443
 9 BGR           375
10 BHM           297
# ... with 95 more rows
```

Observe that by default the rows of the resulting `freq_dest` data frame are sorted in alphabetical order of `destination`. Say instead we would like to see the same data, but sorted from the most to the least number of flights (`num_flights`) instead:

```
freq_dest %>%
  arrange(num_flights)
```

```
# A tibble: 105 x 2
   dest  num_flights
   <chr>       <int>
 1 LEX             1
 2 LGA             1
 3 ANC             8
 4 SBN            10
 5 HDN            15
 6 MTJ            15
 7 EYW            17
 8 PSP            19
 9 JAC            25
10 BZN            36
```

```
# ... with 95 more rows
```

This is, however, the opposite of what we want. The rows are sorted with the least frequent destination airports displayed first. This is because arrange() always returns rows sorted in ascending order by default. To switch the ordering to be in "descending" order instead, we use the desc() function as so:

```
freq_dest %>%
  arrange(desc(num_flights))
```

```
# A tibble: 105 x 2
   dest   num_flights
   <chr>        <int>
 1 ORD          17283
 2 ATL          17215
 3 LAX          16174
 4 BOS          15508
 5 MCO          14082
 6 CLT          14064
 7 SFO          13331
 8 FLL          12055
 9 MIA          11728
10 DCA           9705
# ... with 95 more rows
```

3.7 join data frames

Another common data transformation task is "joining" or "merging" two different datasets. For example, in the flights data frame, the variable carrier lists the carrier code for the different flights. While the corresponding airline names for "UA" and "AA" might be somewhat easy to guess (United and American Airlines), what airlines have codes "VX", "HA", and "B6"? This information is provided in a separate data frame airlines.

```
View(airlines)
```

We see that in airports, carrier is the carrier code, while name is the full name of the airline company. Using this table, we can see that "VX", "HA", and "B6" correspond to Virgin America, Hawaiian Airlines, and JetBlue, respectively. However, wouldn't it be nice to have all this information in a single data frame instead of two separate data frames? We can do this by "joining" the `flights` and `airlines` data frames.

Note that the values in the variable `carrier` in the `flights` data frame match the values in the variable `carrier` in the `airlines` data frame. In this case, we can use the variable `carrier` as a *key variable* to match the rows of the two data frames. Key variables are almost always *identification variables* that uniquely identify the observational units as we saw in Subsection 1.4.4. This ensures that rows in both data frames are appropriately matched during the join. Hadley and Garrett (Grolemund and Wickham, 2017) created the diagram shown in Figure 3.7 to help us understand how the different data frames in the `nycflights13` package are linked by various key variables:

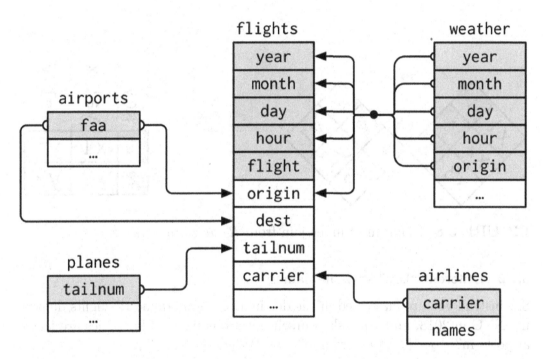

FIGURE 3.7: Data relationships in nycflights13 from *R for Data Science*.

3.7.1 Matching "key" variable names

In both the `flights` and `airlines` data frames, the key variable we want to join/merge/match the rows by has the same name: `carrier`. Let's use the `inner_join()` function to join the two data frames, where the rows will be matched by the variable `carrier`, and then compare the resulting data frames:

```
flights_joined <- flights %>%
  inner_join(airlines, by = "carrier")
```

```
View(flights)
View(flights_joined)
```

Observe that the `flights` and `flights_joined` data frames are identical except that `flights_joined` has an additional variable `name`. The values of `name` correspond to the airline companies' names as indicated in the `airlines` data frame.

A visual representation of the `inner_join()` is shown in Figure 3.8 (Grolemund and Wickham, 2017). There are other types of joins available (such as `left_join()`, `right_join()`, `outer_join()`, and `anti_join()`), but the `inner_join()` will solve nearly all of the problems you'll encounter in this book.

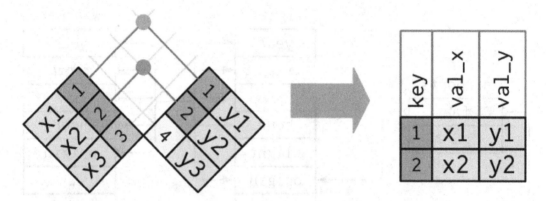

FIGURE 3.8: Diagram of inner join from *R for Data Science*.

3.7.2 Different "key" variable names

Say instead you are interested in the destinations of all domestic flights departing NYC in 2013, and you ask yourself questions like: "What cities are these airports in?", or "Is `"ORD"` Orlando?", or "Where is `"FLL"`?".

The `airports` data frame contains the airport codes for each airport:

```
View(airports)
```

However, if you look at both the `airports` and `flights` data frames, you'll find that the airport codes are in variables that have different names. In `airports` the airport code is in `faa`, whereas in `flights` the airport codes are in `origin` and `dest`. This fact is further highlighted in the visual representation of the relationships between these data frames in Figure 3.7.

In order to join these two data frames by airport code, our `inner_join()` operation will use the `by = c("dest" = "faa")` argument with modified code syntax allowing us to join two data frames where the key variable has a different name:

```
flights_with_airport_names <- flights %>%
  inner_join(airports, by = c("dest" = "faa"))
View(flights_with_airport_names)
```

Let's construct the chain of pipe operators %>% that computes the number of flights from NYC to each destination, but also includes information about each destination airport:

```
named_dests <- flights %>%
  group_by(dest) %>%
  summarize(num_flights = n()) %>%
  arrange(desc(num_flights)) %>%
  inner_join(airports, by = c("dest" = "faa")) %>%
  rename(airport_name = name)
named_dests
```

```
# A tibble: 101 x 9
   dest  num_flights airport_name      lat    lon   alt    tz dst   tzone
   <chr>       <int> <chr>           <dbl>  <dbl> <dbl> <dbl> <chr> <chr>
 1 ORD         17283 Chicago Ohare I~ 42.0  -87.9   668    -6 A     Ameri~
 2 ATL         17215 Hartsfield Jack~ 33.6  -84.4  1026    -5 A     Ameri~
 3 LAX         16174 Los Angeles Intl 33.9 -118.    126    -8 A     Ameri~
 4 BOS         15508 General Edward ~ 42.4  -71.0    19    -5 A     Ameri~
 5 MCO         14082 Orlando Intl     28.4  -81.3    96    -5 A     Ameri~
 6 CLT         14064 Charlotte Dougl~ 35.2  -80.9   748    -5 A     Ameri~
 7 SFO         13331 San Francisco I~ 37.6 -122.     13    -8 A     Ameri~
 8 FLL         12055 Fort Lauderdale~ 26.1  -80.2     9    -5 A     Ameri~
 9 MIA         11728 Miami Intl       25.8  -80.3     8    -5 A     Ameri~
10 DCA          9705 Ronald Reagan W~ 38.9  -77.0    15    -5 A     Ameri~
# ... with 91 more rows
```

In case you didn't know, "ORD" is the airport code of Chicago O'Hare airport and "FLL" is the main airport in Fort Lauderdale, Florida, which can be seen in the airport_name variable.

3.7.3 Multiple "key" variables

Say instead we want to join two data frames by *multiple key variables*. For example, in Figure 3.7, we see that in order to join the flights and weather data frames, we need more than one key variable: year, month, day, hour, and origin. This is because the combination of these 5 variables act to uniquely identify each observational unit in the weather data frame: hourly weather recordings at each of the 3 NYC airports.

We achieve this by specifying a *vector* of key variables to join by using the c() function. Recall from Subsection 1.2.1 that c() is short for "combine" or "concatenate."

```
flights_weather_joined <- flights %>%
  inner_join(weather, by = c("year", "month", "day", "hour", "origin"))
View(flights_weather_joined)
```

Learning check

(LC3.13) Looking at Figure 3.7, when joining flights and weather (or, in other words, matching the hourly weather values with each flight), why do we need to join by all of year, month, day, hour, and origin, and not just hour?

(LC3.14) What surprises you about the top 10 destinations from NYC in 2013?

3.7.4 Normal forms

The data frames included in the nycflights13 package are in a form that minimizes redundancy of data. For example, the flights data frame only saves the carrier code of the airline company; it does not include the actual name of the airline. For example, the first row of flights has carrier equal to UA, but it does not include the airline name of "United Air Lines Inc."

The names of the airline companies are included in the name variable of the airlines data frame. In order to have the airline company name included in flights, we could join these two data frames as follows:

```
joined_flights <- flights %>%
  inner_join(airlines, by = "carrier")
View(joined_flights)
```

We are capable of performing this join because each of the data frames have *keys* in common to relate one to another: the carrier variable in both the flights and airlines data frames. The *key* variable(s) that we base our joins on are often *identification variables* as we mentioned previously.

This is an important property of what's known as *normal forms* of data. The process of decomposing data frames into less redundant tables without losing information is called *normalization*. More information is available on Wikipedia[2].

Both `dplyr` and SQL[3] we mentioned in the introduction of this chapter use such *normal forms*. Given that they share such commonalities, once you learn either of these two tools, you can learn the other very easily.

Learning check

(LC3.15) What are some advantages of data in normal forms? What are some disadvantages?

3.8 Other verbs

Here are some other useful data wrangling verbs:

- `select()` only a subset of variables/columns.
- `rename()` variables/columns to have new names.
- Return only the `top_n()` values of a variable.

3.8.1 `select` variables

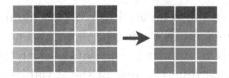

FIGURE 3.9: Diagram of select() columns.

We've seen that the `flights` data frame in the `nycflights13` package contains 19 different variables. You can identify the names of these 19 variables by running the `glimpse()` function from the `dplyr` package:

---

[2]https://en.wikipedia.org/wiki/Database_normalization
[3]https://en.wikipedia.org/wiki/SQL

```
glimpse(flights)
```

However, say you only need two of these 19 variables, say carrier and flight. You can select() these two variables:

```
flights %>%
  select(carrier, flight)
```

This function makes it easier to explore large datasets since it allows us to limit the scope to only those variables we care most about. For example, if we select() only a smaller number of variables as is shown in Figure 3.9, it will make viewing the dataset in RStudio's spreadsheet viewer more digestible.

Let's say instead you want to drop, or de-select, certain variables. For example, consider the variable year in the flights data frame. This variable isn't quite a "variable" because it is always 2013 and hence doesn't change. Say you want to remove this variable from the data frame. We can deselect year by using the - sign:

```
flights_no_year <- flights %>% select(-year)
```

Another way of selecting columns/variables is by specifying a range of columns:

```
flight_arr_times <- flights %>% select(month:day, arr_time:sched_arr_time)
flight_arr_times
```

This will select() all columns between month and day, as well as between arr_time and sched_arr_time, and drop the rest.

The select() function can also be used to reorder columns when used with the everything() helper function. For example, suppose we want the hour, minute, and time_hour variables to appear immediately after the year, month, and day variables, while not discarding the rest of the variables. In the following code, everything() will pick up all remaining variables:

```
flights_reorder <- flights %>%
  select(year, month, day, hour, minute, time_hour, everything())
glimpse(flights_reorder)
```

Lastly, the helper functions starts_with(), ends_with(), and contains() can be used to select variables/columns that match those conditions. As examples,

```
flights %>% select(starts_with("a"))
flights %>% select(ends_with("delay"))
flights %>% select(contains("time"))
```

3.8.2 `rename` variables

Another useful function is `rename()`, which as you may have guessed changes the name of variables. Suppose we want to only focus on `dep_time` and `arr_time` and change `dep_time` and `arr_time` to be `departure_time` and `arrival_time` instead in the `flights_time` data frame:

```
flights_time_new <- flights %>%
  select(dep_time, arr_time) %>%
  rename(departure_time = dep_time, arrival_time = arr_time)
glimpse(flights_time_new)
```

Note that in this case we used a single = sign within the `rename()`. For example, `departure_time = dep_time` renames the `dep_time` variable to have the new name `departure_time`. This is because we are not testing for equality like we would using ==. Instead we want to assign a new variable `departure_time` to have the same values as `dep_time` and then delete the variable `dep_time`. Note that new dplyr users often forget that the new variable name comes before the equal sign.

3.8.3 `top_n` values of a variable

We can also return the top n values of a variable using the `top_n()` function. For example, we can return a data frame of the top 10 destination airports using the example from Subsection 3.7.2. Observe that we set the number of values to return to n = 10 and wt = num_flights to indicate that we want the rows corresponding to the top 10 values of `num_flights`. See the help file for `top_n()` by running ?top_n for more information.

```
named_dests %>% top_n(n = 10, wt = num_flights)
```

Let's further `arrange()` these results in descending order of `num_flights`:

```
named_dests  %>%
  top_n(n = 10, wt = num_flights) %>%
  arrange(desc(num_flights))
```

(LC3.16) What are some ways to select all three of the `dest`, `air_time`, and `distance` variables from `flights`? Give the code showing how to do this in at least three different ways.

(LC3.17) How could one use `starts_with()`, `ends_with()`, and `contains()` to select columns from the `flights` data frame? Provide three different examples in total: one for `starts_with()`, one for `ends_with()`, and one for `contains()`.

(LC3.18) Why might we want to use the `select` function on a data frame?

(LC3.19) Create a new data frame that shows the top 5 airports with the largest arrival delays from NYC in 2013.

3.9   Conclusion

3.9.1   Summary table

Let's recap our data wrangling verbs in Table 3.2. Using these verbs and the pipe %>% operator from Section 3.1, you'll be able to write easily legible code to perform almost all the data wrangling and data transformation necessary for the rest of this book.

TABLE 3.2: Summary of data wrangling verbs

Verb	Data wrangling operation
filter()	Pick out a subset of rows
summarize()	Summarize many values to one using a summary statistic function like mean(), median(), etc.
group_by()	Add grouping structure to rows in data frame. Note this does not change values in data frame, rather only the meta-data
mutate()	Create new variables by mutating existing ones
arrange()	Arrange rows of a data variable in ascending (default) or descending order
inner_join()	Join/merge two data frames, matching rows by a key variable

Learning check

(LC3.20) Let's now put your newly acquired data wrangling skills to the test!

An airline industry measure of a passenger airline's capacity is the available seat miles[4], which is equal to the number of seats available multiplied by the number of miles or kilometers flown summed over all flights.

For example, let's consider the scenario in Figure 3.10. Since the airplane has 4 seats and it travels 200 miles, the available seat miles are $4 \times 200 = 800$.

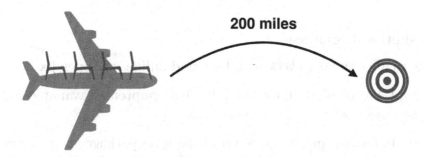

FIGURE 3.10: Example of available seat miles for one flight.

Extending this idea, let's say an airline had 2 flights using a plane with 10 seats that flew 500 miles and 3 flights using a plane with 20 seats that flew 1000 miles, the available seat miles would be $2 \times 10 \times 500 + 3 \times 20 \times 1000 = 70,000$ seat miles.

Using the datasets included in the `nycflights13` package, compute the available seat miles for each airline sorted in descending order. After completing all the necessary data wrangling steps, the resulting data frame should have 16 rows (one for each airline) and 2 columns (airline name and available seat miles). Here are some hints:

1. **Crucial**: Unless you are very confident in what you are doing, it is worthwhile not starting to code right away. Rather, first sketch out on paper all the necessary data wrangling steps not using exact code,

---

[4]https://en.wikipedia.org/wiki/Available_seat_miles

but rather high-level *pseudocode* that is informal yet detailed enough to articulate what you are doing. This way you won't confuse *what* you are trying to do (the algorithm) with *how* you are going to do it (writing `dplyr` code).

2. Take a close look at all the datasets using the `View()` function: `flights`, `weather`, `planes`, `airports`, and `airlines` to identify which variables are necessary to compute available seat miles.

3. Figure 3.7 showing how the various datasets can be joined will also be useful.

4. Consider the data wrangling verbs in Table 3.2 as your toolbox!

3.9.2   Additional resources

Solutions to all *Learning checks* can be found online in Appendix D[5].

An R script file of all R code used in this chapter is available at `https://www.moderndive.com/scripts/03-wrangling.R`.

If you want to further unlock the power of the `dplyr` package for data wrangling, we suggest that you check out RStudio's "Data Transformation with dplyr" cheatsheet. This cheatsheet summarizes much more than what we've discussed in this chapter, in particular more intermediate level and advanced data wrangling functions, while providing quick and easy-to-read visual descriptions. In fact, many of the diagrams illustrating data wrangling operations in this chapter, such as Figure 3.1 on `filter()`, originate from this cheatsheet.

In the current version of RStudio in late 2019, you can access this cheatsheet by going to the RStudio Menu Bar -> Help -> Cheatsheets -> "Data Transformation with dplyr."

On top of the data wrangling verbs and examples we presented in this section, if you'd like to see more examples of using the `dplyr` package for data wrangling, check out Chapter 5[6] of *R for Data Science* (Grolemund and Wickham, 2017).

3.9.3   What's to come?

So far in this book, we've explored, visualized, and wrangled data saved in data frames. These data frames were saved in a spreadsheet-like format: in a rectangular shape with a certain number of rows corresponding to observations

---

[5]https://moderndive.com/D-appendixD.html
[6]http://r4ds.had.co.nz/transform.html

and a certain number of columns corresponding to variables describing these observations.

We'll see in the upcoming Chapter 4 that there are actually two ways to represent data in spreadsheet-type rectangular format: (1) "wide" format and (2) "tall/narrow" format. The tall/narrow format is also known as *"tidy"* format in R user circles. While the distinction between "tidy" and non-"tidy" formatted data is subtle, it has immense implications for our data science work. This is because almost all the packages used in this book, including the `ggplot2` package for data visualization and the `dplyr` package for data wrangling, all assume that all data frames are in "tidy" format.

Furthermore, up until now we've only explored, visualized, and wrangled data saved within R packages. But what if you want to analyze data that you have saved in a Microsoft Excel, a Google Sheets, or a "Comma-Separated Values" (CSV) file? In Section 4.1, we'll show you how to import this data into R using the `readr` package.

4

Data Importing and "Tidy" Data

In Subsection 1.2.1, we introduced the concept of a data frame in R: a rectangular spreadsheet-like representation of data where the rows correspond to observations and the columns correspond to variables describing each observation. In Section 1.4, we started exploring our first data frame: the `flights` data frame included in the `nycflights13` package. In Chapter 2, we created visualizations based on the data included in `flights` and other data frames such as `weather`. In Chapter 3, we learned how to take existing data frames and transform/modify them to suit our ends.

In this final chapter of the "Data Science with `tidyverse`" portion of the book, we extend some of these ideas by discussing a type of data formatting called "tidy" data. You will see that having data stored in "tidy" format is about more than just what the everyday definition of the term "tidy" might suggest: having your data "neatly organized." Instead, we define the term "tidy" as it's used by data scientists who use R, outlining a set of rules by which data is saved.

Knowledge of this type of data formatting was not necessary for our treatment of data visualization in Chapter 2 and data wrangling in Chapter 3. This is because all the data used were already in "tidy" format. In this chapter, we'll now see that this format is essential to using the tools we covered up until now. Furthermore, it will also be useful for all subsequent chapters in this book when we cover regression and statistical inference. First, however, we'll show you how to import spreadsheet data in R.

Needed packages

Let's load all the packages needed for this chapter (this assumes you've already installed them). If needed, read Section 1.3 for information on how to install and load R packages.

```
library(dplyr)
library(ggplot2)
library(readr)
library(tidyr)
```

```
library(nycflights13)
library(fivethirtyeight)
```

4.1   Importing data

Up to this point, we've almost entirely used data stored inside of an R package. Say instead you have your own data saved on your computer or somewhere online. How can you analyze this data in R? Spreadsheet data is often saved in one of the following three formats:

First, a *Comma Separated Values* .csv file. You can think of a .csv file as a bare-bones spreadsheet where:

- Each line in the file corresponds to one row of data/one observation.
- Values for each line are separated with commas. In other words, the values of different variables are separated by commas in each row.
- The first line is often, but not always, a *header* row indicating the names of the columns/variables.

Second, an Excel .xlsx spreadsheet file. This format is based on Microsoft's proprietary Excel software. As opposed to bare-bones .csv files, .xlsx Excel files contain a lot of meta-data (data about data). Recall we saw a previous example of meta-data in Section 3.4 when adding "group structure" meta-data to a data frame by using the group_by() verb. Some examples of Excel spreadsheet meta-data include the use of bold and italic fonts, colored cells, different column widths, and formula macros.

Third, a Google Sheets[1] file, which is a "cloud" or online-based way to work with a spreadsheet. Google Sheets allows you to download your data in both comma separated values .csv and Excel .xlsx formats. One way to import Google Sheets data in R is to go to the Google Sheets menu bar -> File -> Download as -> Select "Microsoft Excel" or "Comma-separated values" and then load that data into R. A more advanced way to import Google Sheets data in R is by using the googlesheets[2] package, a method we leave to a more advanced data science book.

---

[1] https://www.google.com/sheets/about/

[2] https://cran.r-project.org/web/packages/googlesheets/vignettes/basic-usage.html

We'll cover two methods for importing .csv and .xlsx spreadsheet data in R: one using the console and the other using RStudio's graphical user interface, abbreviated as "GUI."

4.1.1 Using the console

First, let's import a Comma Separated Values .csv file that exists on the internet. The .csv file dem_score.csv contains ratings of the level of democracy in different countries spanning 1952 to 1992 and is accessible at https://moderndive.com/data/dem_score.csv. Let's use the read_csv() function from the readr (Wickham et al., 2018) package to read it off the web, import it into R, and save it in a data frame called dem_score.

```
library(readr)
dem_score <- read_csv("https://moderndive.com/data/dem_score.csv")
dem_score
```

```
# A tibble: 96 x 10
   country  `1952` `1957` `1962` `1967` `1972` `1977` `1982` `1987` `1992`
   <chr>    <dbl>  <dbl>  <dbl>  <dbl>  <dbl>  <dbl>  <dbl>  <dbl>  <dbl>
 1 Albania     -9     -9     -9     -9     -9     -9     -9     -9      5
 2 Argentina   -9     -1     -1     -9     -9     -9     -8      8      7
 3 Armenia     -9     -7     -7     -7     -7     -7     -7     -7      7
 4 Australia   10     10     10     10     10     10     10     10     10
 5 Austria     10     10     10     10     10     10     10     10     10
 6 Azerbaij~   -9     -7     -7     -7     -7     -7     -7     -7      1
 7 Belarus     -9     -7     -7     -7     -7     -7     -7     -7      7
 8 Belgium     10     10     10     10     10     10     10     10     10
 9 Bhutan     -10    -10    -10    -10    -10    -10    -10    -10    -10
10 Bolivia     -4     -3     -3     -4     -7     -7      8      9      9
# ... with 86 more rows
```

In this dem_score data frame, the minimum value of -10 corresponds to a highly autocratic nation, whereas a value of 10 corresponds to a highly democratic nation. Note also that backticks surround the different variable names. Variable names in R by default are not allowed to start with a number nor include spaces, but we can get around this fact by surrounding the column name with backticks. We'll revisit the dem_score data frame in a case study in the upcoming Section 4.3.

Note that the read_csv() function included in the readr package is different than the read.csv() function that comes installed with R. While the difference in the names might seem trivial (an _ instead of a .), the read_csv() function is, in our opinion, easier to use since it can more easily read data off the web and generally imports data at a much faster speed. Furthermore, the `read_csv()` function included in the `readr` saves data frames as `tibbles` by default.

4.1.2 Using RStudio's interface

Let's read in the exact same data, but this time from an Excel file saved on your computer. Furthermore, we'll do this using RStudio's graphical interface instead of running `read_csv()` in the console. First, download the Excel file `dem_score.xlsx` by going to https://moderndive.com/data/dem_score.xlsx, then

1. Go to the Files pane of RStudio.
2. Navigate to the directory (i.e., folder on your computer) where the downloaded `dem_score.xlsx` Excel file is saved. For example, this might be in your Downloads folder.
3. Click on `dem_score.xlsx`.
4. Click "Import Dataset..."

At this point, you should see a screen pop-up like in Figure 4.1. After clicking on the "Import" button on the bottom right of Figure 4.1, RStudio will save this spreadsheet's data in a data frame called `dem_score` and display its contents in the spreadsheet viewer.

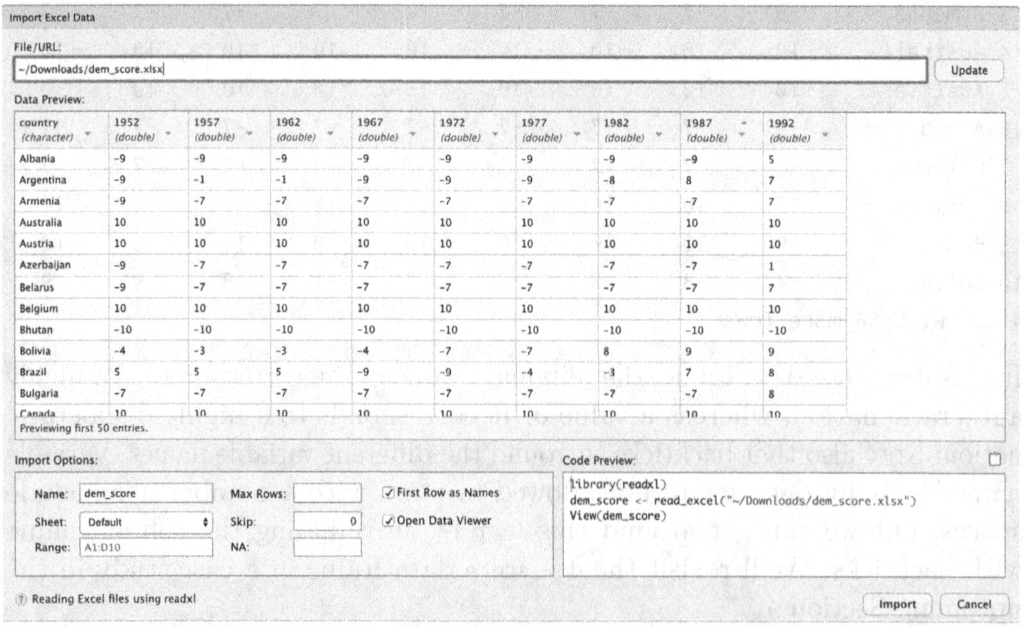

country (character)	1952 (double)	1957 (double)	1962 (double)	1967 (double)	1972 (double)	1977 (double)	1982 (double)	1987 (double)	1992 (double)
Albania	-9	-9	-9	-9	-9	-9	-9	-9	5
Argentina	-9	-1	-1	-9	-9	-9	-8	8	7
Armenia	-9	-7	-7	-7	-7	-7	-7	-7	7
Australia	10	10	10	10	10	10	10	10	10
Austria	10	10	10	10	10	10	10	10	10
Azerbaijan	-9	-7	-7	-7	-7	-7	-7	-7	1
Belarus	-9	-7	-7	-7	-7	-7	-7	-7	7
Belgium	10	10	10	10	10	10	10	10	10
Bhutan	-10	-10	-10	-10	-10	-10	-10	-10	-10
Bolivia	-4	-3	-3	-4	-7	-7	8	9	9
Brazil	5	5	5	-9	-9	-4	-3	7	8
Bulgaria	-7	-7	-7	-7	-7	-7	-7	-7	8
Canada	10	10	10	10	10	10	10	10	10

FIGURE 4.1: Importing an Excel file to R.

Furthermore, note the "Code Preview" block in the bottom right of Figure 4.1. You can copy and paste this code to reload your data again later programmatically, instead of repeating this manual point-and-click process.

4.2 "Tidy" data

Let's now switch gears and learn about the concept of "tidy" data format with a motivating example from the `fivethirtyeight` package. The `fivethirtyeight` package (Kim et al., 2019) provides access to the datasets used in many articles published by the data journalism website, FiveThirtyEight.com[3]. For a complete list of all 127 datasets included in the `fivethirtyeight` package, check out the package webpage by going to: `https://fivethirtyeight-r.netlify.com/articles/fivethirtyeight.html`.

Let's focus our attention on the `drinks` data frame and look at its first 5 rows:

```
# A tibble: 5 x 5
  country    beer_servings spirit_servings wine_servings total_litres_of_pu~
  <chr>              <int>           <int>         <int>                <dbl>
1 Afghanis~              0               0             0                    0
2 Albania               89             132            54                  4.9
3 Algeria               25               0            14                  0.7
4 Andorra              245             138           312                 12.4
5 Angola               217              57            45                  5.9
```

After reading the help file by running `?drinks`, you'll see that `drinks` is a data frame containing results from a survey of the average number of servings of beer, spirits, and wine consumed in 193 countries. This data was originally reported on FiveThirtyEight.com in Mona Chalabi's article: "Dear Mona Followup: Where Do People Drink The Most Beer, Wine And Spirits?"[4].

Let's apply some of the data wrangling verbs we learned in Chapter 3 on the `drinks` data frame:

1. `filter()` the `drinks` data frame to only consider 4 countries: the United States, China, Italy, and Saudi Arabia, *then*
2. `select()` all columns except `total_litres_of_pure_alcohol` by using the - sign, *then*

---

[3]`https://fivethirtyeight.com/`

[4]`https://fivethirtyeight.com/features/dear-mona-followup-where-do-people-drink-the-most-beer-wine-and-spirits/`

3. rename() the variables beer_servings, spirit_servings, and wine_servings
    to beer, spirit, and wine, respectively.

and save the resulting data frame in drinks_smaller:

```
drinks_smaller <- drinks %>%
  filter(country %in% c("USA", "China", "Italy", "Saudi Arabia")) %>%
  select(-total_litres_of_pure_alcohol) %>%
  rename(beer = beer_servings, spirit = spirit_servings, wine = wine_servings)
drinks_smaller
```

```
# A tibble: 4 x 4
  country      beer spirit  wine
  <chr>       <int>  <int> <int>
1 China          79    192     8
2 Italy          85     42   237
3 Saudi Arabia    0      5     0
4 USA           249    158    84
```

Let's now ask ourselves a question: "Using the drinks_smaller data frame, how
would we create the side-by-side barplot in Figure 4.2?". Recall we saw barplots
displaying two categorical variables in Subsection 2.8.3.

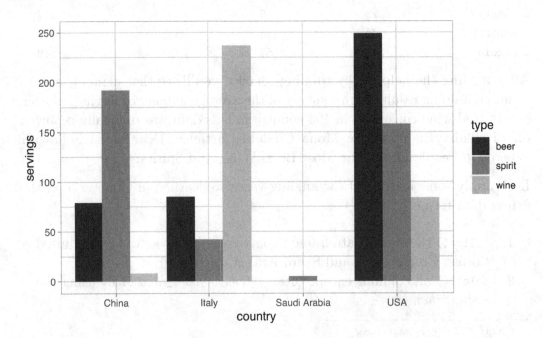

FIGURE 4.2: Comparing alcohol consumption in 4 countries.

Let's break down the grammar of graphics we introduced in Section 2.1:

1. The categorical variable country with four levels (China, Italy, Saudi Arabia, USA) would have to be mapped to the x-position of the bars.
2. The numerical variable servings would have to be mapped to the y-position of the bars (the height of the bars).
3. The categorical variable type with three levels (beer, spirit, wine) would have to be mapped to the fill color of the bars.

Observe, however, that drinks_smaller has three separate variables beer, spirit, and wine. In order to use the ggplot() function to recreate the barplot in Figure 4.2. However, we need a *single variable* type with three possible values: beer, spirit, and wine. We could then map this type variable to the fill aesthetic of our plot. In other words, to recreate the barplot in Figure 4.2, our data frame would have to look like this:

```
drinks_smaller_tidy
```

```
# A tibble: 12 x 3
   country      type    servings
   <chr>        <chr>      <int>
 1 China        beer          79
 2 Italy        beer          85
 3 Saudi Arabia beer           0
 4 USA          beer         249
 5 China        spirit       192
 6 Italy        spirit        42
 7 Saudi Arabia spirit         5
 8 USA          spirit       158
 9 China        wine           8
10 Italy        wine         237
11 Saudi Arabia wine           0
12 USA          wine          84
```

Observe that while drinks_smaller and drinks_smaller_tidy are both rectangular in shape and contain the same 12 numerical values (3 alcohol types by 4 countries), they are formatted differently. drinks_smaller is formatted in what's known as "wide"[5] format, whereas drinks_smaller_tidy is formatted in what's known as "long/narrow"[6] format.

---

[5]https://en.wikipedia.org/wiki/Wide_and_narrow_data
[6]https://en.wikipedia.org/wiki/Wide_and_narrow_data#Narrow

In the context of doing data science in R, long/narrow format is also known as "tidy" format. In order to use the `ggplot2` and `dplyr` packages for data visualization and data wrangling, your input data frames *must* be in "tidy" format. Thus, all non-"tidy" data must be converted to "tidy" format first. Before we convert non-"tidy" data frames like `drinks_smaller` to "tidy" data frames like `drinks_smaller_tidy`, let's define "tidy" data.

4.2.1 Definition of "tidy" data

You have surely heard the word "tidy" in your life:

- "Tidy up your room!"
- "Write your homework in a tidy way so it is easier to provide feedback."
- Marie Kondo's best-selling book, *The Life-Changing Magic of Tidying Up: The Japanese Art of Decluttering and Organizing*[7], and Netflix TV series *Tidying Up with Marie Kondo*[8].
- "I am not by any stretch of the imagination a tidy person, and the piles of unread books on the coffee table and by my bed have a plaintive, pleading quality to me - 'Read me, please!' " - Linda Grant

What does it mean for your data to be "tidy"? While "tidy" has a clear English meaning of "organized," the word "tidy" in data science using R means that your data follows a standardized format. We will follow Hadley Wickham's definition of *"tidy" data* (Wickham, 2014) shown also in Figure 4.3:

A *dataset* is a collection of values, usually either numbers (if quantitative) or strings AKA text data (if qualitative/categorical). Values are organised in two ways. Every value belongs to a variable and an observation. A variable contains all values that measure the same underlying attribute (like height, temperature, duration) across units. An observation contains all values measured on the same unit (like a person, or a day, or a city) across attributes.

"Tidy" data is a standard way of mapping the meaning of a dataset to its structure. A dataset is messy or tidy depending on how rows, columns and tables are matched up with observations, variables and types. In *tidy data*:

1. Each variable forms a column.
2. Each observation forms a row.
3. Each type of observational unit forms a table.

---

[7] https://www.powells.com/book/-9781607747307
[8] https://www.netflix.com/title/80209379

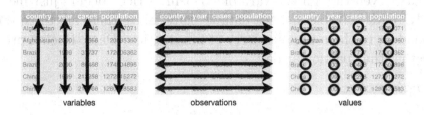

variables   observations   values

FIGURE 4.3: Tidy data graphic from *R for Data Science*.

For example, say you have the following table of stock prices in Table 4.1:

TABLE 4.1: Stock prices (non-tidy format)

Date	Boeing stock price	Amazon stock price	Google stock price
2009-01-01	$173.55	$174.90	$174.34
2009-01-02	$172.61	$171.42	$170.04

Although the data are neatly organized in a rectangular spreadsheet-type format, they do not follow the definition of data in "tidy" format. While there are three variables corresponding to three unique pieces of information (date, stock name, and stock price), there are not three columns. In "tidy" data format, each variable should be its own column, as shown in Table 4.2. Notice that both tables present the same information, but in different formats.

TABLE 4.2: Stock prices (tidy format)

Date	Stock Name	Stock Price
2009-01-01	Boeing	$173.55
2009-01-01	Amazon	$174.90
2009-01-01	Google	$174.34
2009-01-02	Boeing	$172.61
2009-01-02	Amazon	$171.42
2009-01-02	Google	$170.04

Now we have the requisite three columns Date, Stock Name, and Stock Price. On the other hand, consider the data in Table 4.3.

TABLE 4.3: Example of tidy data

Date	Boeing Price	Weather
2009-01-01	$173.55	Sunny
2009-01-02	$172.61	Overcast

In this case, even though the variable "Boeing Price" occurs just like in our non-"tidy" data in Table 4.1, the data *is* "tidy" since there are three variables corresponding to three unique pieces of information: Date, Boeing price, and the Weather that particular day.

Learning check

(**LC4.1**) What are common characteristics of "tidy" data frames?

(**LC4.2**) What makes "tidy" data frames useful for organizing data?

4.2.2   Converting to "tidy" data

In this book so far, you've only seen data frames that were already in "tidy" format. Furthermore, for the rest of this book, you'll mostly only see data frames that are already in "tidy" format as well. This is not always the case however with all datasets in the world. If your original data frame is in wide (non-"tidy") format and you would like to use the `ggplot2` or `dplyr` packages, you will first have to convert it to "tidy" format. To do so, we recommend using the `pivot_longer()` function in the `tidyr` package (Wickham and Henry, 2019).

Going back to our `drinks_smaller` data frame from earlier:

```
drinks_smaller
```

```
# A tibble: 4 x 4
  country       beer spirit  wine
  <chr>        <int>  <int> <int>
1 China           79    192     8
2 Italy           85     42   237
3 Saudi Arabia     0      5     0
4 USA            249    158    84
```

We convert it to "tidy" format by using the `pivot_longer()` function from the `tidyr` package as follows:

```
drinks_smaller_tidy <- drinks_smaller %>%
  pivot_longer(names_to = "type",
               values_to = "servings",
               cols = -country)
drinks_smaller_tidy
```

```
# A tibble: 12 x 3
   country       type    servings
   <chr>         <chr>      <int>
 1 China         beer          79
 2 China         spirit       192
 3 China         wine           8
 4 Italy         beer          85
 5 Italy         spirit        42
 6 Italy         wine         237
 7 Saudi Arabia  beer           0
 8 Saudi Arabia  spirit         5
 9 Saudi Arabia  wine           0
10 USA           beer         249
11 USA           spirit       158
12 USA           wine          84
```

We set the arguments to `pivot_longer()` as follows:

1. `names_to` here corresponds to the name of the variable in the new "tidy"/long data frame that will contain the *column names* of the original data. Observe how we set `names_to = "type"`. In the resulting `drinks_smaller_tidy`, the column `type` contains the three types of alcohol `beer`, `spirit`, and `wine`. Since `type` is a variable name that doesn't appear in `drinks_smaller`, we use quotation marks around it. You'll receive an error if you just use `names_to = type` here.

2. `values_to` here is the name of the variable in the new "tidy" data frame that will contain the *values* of the original data. Observe how we set `values_to = "servings"` since each of the numeric values in each of the `beer`, `wine`, and `spirit` columns of the `drinks_smaller` data corresponds to a value of `servings`. In the resulting `drinks_smaller_tidy`, the column `servings` contains the $4 \times 3 = 12$ numerical values. Note again that `servings` doesn't appear as a variable in `drinks_smaller` so it again needs quotation marks around it for the `values_to` argument.

3. The third argument `cols` is the columns in the `drinks_smaller` data frame you either want to or don't want to "tidy." Observe how we set this to `-country` indicating that we don't want to "tidy" the `country` variable in `drinks_smaller` and rather only `beer`, `spirit`, and `wine`. Since `country` is a column that appears in `drinks_smaller` we don't put quotation marks around it.

The third argument here of `cols` is a little nuanced, so let's consider code that's written slightly differently but that produces the same output:

```
drinks_smaller %>%
  pivot_longer(names_to = "type",
               values_to = "servings",
               cols = c(beer, spirit, wine))
```

Note that the third argument now specifies which columns we want to "tidy" with c(beer, spirit, wine), instead of the columns we don't want to "tidy" using -country. We use the c() function to create a vector of the columns in drinks_smaller that we'd like to "tidy." Note that since these three columns appear one after another in the drinks_smaller data frame, we could also do the following for the cols argument:

```
drinks_smaller %>%
  pivot_longer(names_to = "type",
               values_to = "servings",
               cols = beer:wine)
```

With our drinks_smaller_tidy "tidy" formatted data frame, we can now produce the barplot you saw in Figure 4.2 using geom_col(). This is done in Figure 4.4. Recall from Section 2.8 on barplots that we use geom_col() and not geom_bar(), since we would like to map the "pre-counted" servings variable to the y-aesthetic of the bars.

```
ggplot(drinks_smaller_tidy, aes(x = country, y = servings, fill = type)) +
  geom_col(position = "dodge")
```

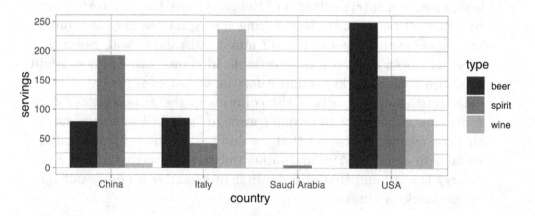

FIGURE 4.4: Comparing alcohol consumption in 4 countries using geom_col().

Converting "wide" format data to "tidy" format often confuses new R users. The only way to learn to get comfortable with the `pivot_longer()` function is with practice, practice, and more practice using different datasets. For example, run `?pivot_longer` and look at the examples in the bottom of the help file. We'll show another example of using `pivot_longer()` to convert a "wide" formatted data frame to "tidy" format in Section 4.3.

If however you want to convert a "tidy" data frame to "wide" format, you will need to use the `pivot_wider()` function instead. Run `?pivot_wider` and look at the examples in the bottom of the help file for examples.

You can also view examples of both `pivot_longer()` and `pivot_wider()` on the tidyverse.org[9] webpage. There's a nice example to check out the different functions available for data tidying and a case study using data from the World Health Organization on that webpage. Furthermore, each week the R4DS Online Learning Community posts a dataset in the weekly #TidyTuesday event[10] that might serve as a nice place for you to find other data to explore and transform.

Learning check

(LC4.3) Take a look at the `airline_safety` data frame included in the `fivethirtyeight` data package. Run the following:

```
airline_safety
```

After reading the help file by running `?airline_safety`, we see that `airline_safety` is a data frame containing information on different airline companies' safety records. This data was originally reported on the data journalism website, FiveThirtyEight.com, in Nate Silver's article, "Should Travelers Avoid Flying Airlines That Have Had Crashes in the Past?"[11]. Let's only consider the variables `airlines` and those relating to fatalities for simplicity:

```
airline_safety_smaller <- airline_safety %>%
  select(airline, starts_with("fatalities"))
airline_safety_smaller
```

```
# A tibble: 56 x 3
   airline                 fatalities_85_99 fatalities_00_14
```

---

[9]https://tidyr.tidyverse.org/dev/articles/pivot.html#pew
[10]https://github.com/rfordatascience/tidytuesday
[11]https://fivethirtyeight.com/features/should-travelers-avoid-flying-airlines-that-have-had-crashes-in-the-past/

	<chr>	<int>	<int>
1	Aer Lingus	0	0
2	Aeroflot	128	88
3	Aerolineas Argentinas	0	0
4	Aeromexico	64	0
5	Air Canada	0	0
6	Air France	79	337
7	Air India	329	158
8	Air New Zealand	0	7
9	Alaska Airlines	0	88
10	Alitalia	50	0

`# ... with 46 more rows`

This data frame is not in "tidy" format. How would you convert this data frame to be in "tidy" format, in particular so that it has a variable fatalities_years indicating the incident year and a variable count of the fatality counts?

4.2.3 nycflights13 package

Recall the nycflights13 package we introduced in Section 1.4 with data about all domestic flights departing from New York City in 2013. Let's revisit the flights data frame by running View(flights). We saw that flights has a rectangular shape, with each of its 336,776 rows corresponding to a flight and each of its 22 columns corresponding to different characteristics/measurements of each flight. This satisfied the first two criteria of the definition of "tidy" data from Subsection 4.2.1: that "Each variable forms a column" and "Each observation forms a row." But what about the third property of "tidy" data that "Each type of observational unit forms a table"?

Recall that we saw in Subsection 1.4.3 that the observational unit for the flights data frame is an individual flight. In other words, the rows of the flights data frame refer to characteristics/measurements of individual flights. Also included in the nycflights13 package are other data frames with their rows representing different observational units (Wickham, 2019a):

- airlines: translation between two letter IATA carrier codes and airline company names (16 in total). The observational unit is an airline company.
- planes: aircraft information about each of 3,322 planes used, i.e., the observational unit is an aircraft.
- weather: hourly meteorological data (about 8,705 observations) for each of the three NYC airports, i.e., the observational unit is an hourly measurement of weather at one of the three airports.

- `airports`: airport names and locations. The observational unit is an airport.

The organization of the information into these five data frames follows the third "tidy" data property: observations corresponding to the same observational unit should be saved in the same table, i.e., data frame. You could think of this property as the old English expression: "birds of a feather flock together."

4.3 Case study: Democracy in Guatemala

In this section, we'll show you another example of how to convert a data frame that isn't in "tidy" format ("wide" format) to a data frame that is in "tidy" format ("long/narrow" format). We'll do this using the `pivot_longer()` function from the `tidyr` package again.

Furthermore, we'll make use of functions from the `ggplot2` and `dplyr` packages to produce a *time-series plot* showing how the democracy scores have changed over the 40 years from 1952 to 1992 for Guatemala. Recall that we saw time-series plots in Section 2.4 on creating linegraphs using `geom_line()`.

Let's use the `dem_score` data frame we imported in Section 4.1, but focus on only data corresponding to Guatemala.

```
guat_dem <- dem_score %>%
  filter(country == "Guatemala")
guat_dem
```

```
# A tibble: 1 x 10
  country `1952` `1957` `1962` `1967` `1972` `1977` `1982` `1987` `1992`
  <chr>    <dbl>  <dbl>  <dbl>  <dbl>  <dbl>  <dbl>  <dbl>  <dbl>  <dbl>
1 Guatemala    2     -6     -5      3      1     -3     -7      3      3
```

Let's lay out the grammar of graphics we saw in Section 2.1.

First we know we need to set `data = guat_dem` and use a `geom_line()` layer, but what is the aesthetic mapping of variables? We'd like to see how the democracy score has changed over the years, so we need to map:

- `year` to the x-position aesthetic and
- `democracy_score` to the y-position aesthetic

Now we are stuck in a predicament, much like with our `drinks_smaller` example in Section 4.2. We see that we have a variable named `country`, but its only value is `"Guatemala"`. We have other variables denoted by different year values. Unfortunately, the `guat_dem` data frame is not "tidy" and hence is not in the appropriate format to apply the grammar of graphics, and thus we cannot use the `ggplot2` package just yet.

We need to take the values of the columns corresponding to years in `guat_dem` and convert them into a new "names" variable called `year`. Furthermore, we need to take the democracy score values in the inside of the data frame and turn them into a new "values" variable called `democracy_score`. Our resulting data frame will have three columns: `country`, `year`, and `democracy_score`. Recall that the `pivot_longer()` function in the `tidyr` package does this for us:

```
guat_dem_tidy <- guat_dem %>%
  pivot_longer(names_to = "year",
               values_to = "democracy_score",
               cols = -country,
               names_ptypes = list(year = integer()))
guat_dem_tidy
```

```
# A tibble: 9 x 3
  country    year democracy_score
  <chr>     <int>           <dbl>
1 Guatemala  1952               2
2 Guatemala  1957              -6
3 Guatemala  1962              -5
4 Guatemala  1967               3
5 Guatemala  1972               1
6 Guatemala  1977              -3
7 Guatemala  1982              -7
8 Guatemala  1987               3
9 Guatemala  1992               3
```

We set the arguments to `pivot_longer()` as follows:

1. `names_to` is the name of the variable in the new "tidy" data frame that will contain the *column names* of the original data. Observe how we set `names_to = "year"`. In the resulting `guat_dem_tidy`, the column `year` contains the years where Guatemala's democracy scores were measured.

2. `values_to` is the name of the variable in the new "tidy" data frame that will contain the *values* of the original data. Observe how we set `values_to = "democracy_score"`. In the resulting `guat_dem_tidy` the

column democracy_score contains the $1 \times 9 = 9$ democracy scores as numeric values.

3. The third argument is the columns you either want to or don't want to "tidy." Observe how we set this to cols = -country indicating that we don't want to "tidy" the country variable in guat_dem and rather only variables 1952 through 1992.

4. The last argument of names_ptypes tells R what type of variable year should be set to. Without specifying that it is an integer as we've done here, pivot_longer() will set it to be a character value by default.

We can now create the time-series plot in Figure 4.5 to visualize how democracy scores in Guatemala have changed from 1952 to 1992 using a geom_line(). Furthermore, we'll use the labs() function in the ggplot2 package to add informative labels to all the aes()thetic attributes of our plot, in this case the x and y positions.

```
ggplot(guat_dem_tidy, aes(x = year, y = democracy_score)) +
  geom_line() +
  labs(x = "Year", y = "Democracy Score")
```

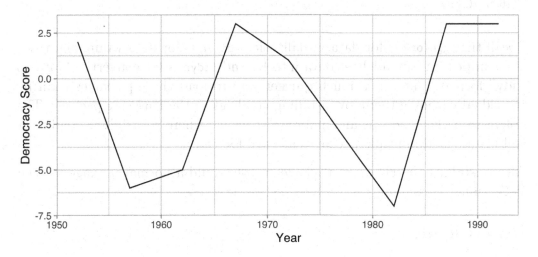

FIGURE 4.5: Democracy scores in Guatemala 1952-1992.

Note that if we forgot to include the names_ptypes argument specifying that year was not of character format, we would have gotten an error here since geom_line() wouldn't have known how to sort the character values in year in the right order.

Learning check

(LC4.4) Convert the `dem_score` data frame into a "tidy" data frame and assign the name of `dem_score_tidy` to the resulting long-formatted data frame.

(LC4.5) Read in the life expectancy data stored at `https://moderndive.com/data/le_mess.csv` and convert it to a "tidy" data frame.

4.4   `tidyverse` package

Notice at the beginning of the chapter we loaded the following four packages, which are among four of the most frequently used R packages for data science:

```
library(ggplot2)
library(dplyr)
library(readr)
library(tidyr)
```

Recall that `ggplot2` is for data visualization, `dplyr` is for data wrangling, `readr` is for importing spreadsheet data into R, and `tidyr` is for converting data to "tidy" format. There is a much quicker way to load these packages than by individually loading them: by installing and loading the `tidyverse` package. The `tidyverse` package acts as an "umbrella" package whereby installing/loading it will install/load multiple packages at once for you.

After installing the `tidyverse` package as you would a normal package as seen in Section 1.3, running:

```
library(tidyverse)
```

would be the same as running:

```
library(ggplot2)
library(dplyr)
library(readr)
library(tidyr)
library(purrr)
```

```
library(tibble)
library(stringr)
library(forcats)
```

The `purrr`, `tibble`, `stringr`, and `forcats` are left for a more advanced book; check out *R for Data Science*[12] to learn about these packages.

For the remainder of this book, we'll start every chapter by running `library(tidyverse)`, instead of loading the various component packages individually. The `tidyverse` "umbrella" package gets its name from the fact that all the functions in all its packages are designed to have common inputs and outputs: data frames are in "tidy" format. This standardization of input and output data frames makes transitions between different functions in the different packages as seamless as possible. For more information, check out the tidyverse.org[13] webpage for the package.

4.5 Conclusion

4.5.1 Additional resources

Solutions to all *Learning checks* can be found online in Appendix D[14].

An R script file of all R code used in this chapter is available at `https://www.moderndive.com/scripts/04-tidy.R`.

If you want to learn more about using the `readr` and `tidyr` package, we suggest that you check out RStudio's "Data Import Cheat Sheet." In the current version of RStudio in late 2019, you can access this cheatsheet by going to the RStudio Menu Bar -> Help -> Cheatsheets -> "Browse Cheatsheets" -> Scroll down the page to the "Data Import Cheat Sheet." The first page of this cheatsheet has information on using the `readr` package to import data, while the second page has information on using the `tidyr` package to "tidy" data.

4.5.2 What's to come?

Congratulations! You've completed the "Data Science with `tidyverse`" portion of this book. We'll now move to the "Data modeling with moderndive" portion of this book in Chapters 5 and 6, where you'll leverage your data visualization

---

[12]http://r4ds.had.co.nz/
[13]https://www.tidyverse.org/
[14]https://moderndive.com/D-appendixD.html

and wrangling skills to model relationships between different variables in data frames.

However, we're going to leave Chapter 10 on "Inference for Regression" until after we've covered statistical inference in Chapters 7, 8, and 9. Onwards and upwards into Data Modeling as shown in Figure 4.6!

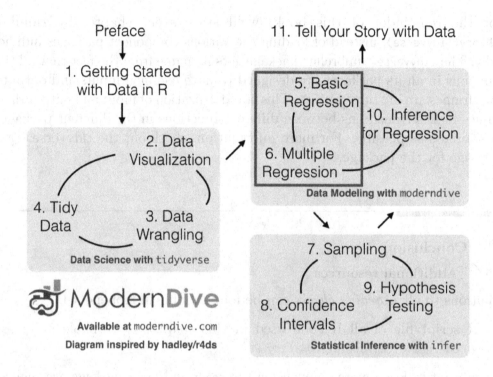

FIGURE 4.6: *ModernDive* flowchart - on to Part II!

Part II

Data Modeling with moderndive

5

Basic Regression

Now that we are equipped with data visualization skills from Chapter 2, data wrangling skills from Chapter 3, and an understanding of how to import data and the concept of a "tidy" data format from Chapter 4, let's now proceed with data modeling. The fundamental premise of data modeling is to make explicit the relationship between:

- an *outcome variable* y, also called a *dependent variable* or response variable, and
- an *explanatory/predictor variable* x, also called an *independent variable* or covariate.

Another way to state this is using mathematical terminology: we will model the outcome variable y "as a function" of the explanatory/predictor variable x. When we say "function" here, we aren't referring to functions in R like the ggplot() function, but rather as a mathematical function. But, why do we have two different labels, explanatory and predictor, for the variable x? That's because even though the two terms are often used interchangeably, roughly speaking data modeling serves one of two purposes:

1. **Modeling for explanation**: When you want to explicitly describe and quantify the relationship between the outcome variable y and a set of explanatory variables x, determine the significance of any relationships, have measures summarizing these relationships, and possibly identify any *causal* relationships between the variables.
2. **Modeling for prediction**: When you want to predict an outcome variable y based on the information contained in a set of predictor variables x. Unlike modeling for explanation, however, you don't care so much about understanding how all the variables relate and interact with one another, but rather only whether you can make good predictions about y using the information in x.

For example, say you are interested in an outcome variable y of whether patients develop lung cancer and information x on their risk factors, such as smoking habits, age, and socioeconomic status. If we are modeling for explanation, we would be interested in both describing and quantifying the effects of the different risk factors. One reason could be that you want to design an intervention to reduce lung cancer incidence in a population, such as targeting smokers of a specific age group with advertising for smoking cessation programs. If we are modeling for prediction, however, we wouldn't care so much about understanding how all the individual risk factors contribute to lung cancer, but rather only whether we can make good predictions of which people will contract lung cancer.

In this book, we'll focus on modeling for explanation and hence refer to x as *explanatory variables*. If you are interested in learning about modeling for prediction, we suggest you check out books and courses on the field of *machine learning* such as *An Introduction to Statistical Learning with Applications in R (ISLR)*[1] (James et al., 2017). Furthermore, while there exist many techniques for modeling, such as tree-based models and neural networks, in this book we'll focus on one particular technique: *linear regression*. Linear regression is one of the most commonly-used and easy-to-understand approaches to modeling.

Linear regression involves a *numerical* outcome variable y and explanatory variables x that are either *numerical* or *categorical*. Furthermore, the relationship between y and x is assumed to be linear, or in other words, a line. However, we'll see that what constitutes a "line" will vary depending on the nature of your explanatory variables x .

In Chapter 5 on basic regression, we'll only consider models with a single explanatory variable x. In Section 5.1, the explanatory variable will be numerical. This scenario is known as *simple linear regression*. In Section 5.2, the explanatory variable will be categorical.

In Chapter 6 on multiple regression, we'll extend the ideas behind basic regression and consider models with two explanatory variables x_1 and x_2. In Section 6.1, we'll have two numerical explanatory variables. In Section 6.2, we'll have one numerical and one categorical explanatory variable. In particular, we'll consider two such models: *interaction* and *parallel slopes* models.

In Chapter 10 on inference for regression, we'll revisit our regression models and analyze the results using the tools for *statistical inference* you'll develop in Chapters 7, 8, and 9 on sampling, bootstrapping and confidence intervals, and hypothesis testing and p-values, respectively.

Let's now begin with basic regression, which refers to linear regression models with a single explanatory variable x. We'll also discuss important statistical concepts like the *correlation coefficient*, that "correlation isn't necessarily causation," and what it means for a line to be "best-fitting."

---

[1] http://www-bcf.usc.edu/~gareth/ISL/

Needed packages

Let's now load all the packages needed for this chapter (this assumes you've already installed them). In this chapter, we introduce some new packages:

1. The `tidyverse` "umbrella" (Wickham, 2019b) package. Recall from our discussion in Section 4.4 that loading the `tidyverse` package by running `library(tidyverse)` loads the following commonly used data science packages all at once:
   - `ggplot2` for data visualization
   - `dplyr` for data wrangling
   - `tidyr` for converting data to "tidy" format
   - `readr` for importing spreadsheet data into R
   - As well as the more advanced `purrr`, `tibble`, `stringr`, and `forcats` packages
2. The `moderndive` package of datasets and functions for tidyverse-friendly introductory linear regression.
3. The `skimr` (Quinn et al., 2019) package, which provides a simple-to-use function to quickly compute a wide array of commonly used summary statistics.

If needed, read Section 1.3 for information on how to install and load R packages.

```
library(tidyverse)
library(moderndive)
library(skimr)
library(gapminder)
```

5.1 One numerical explanatory variable

Why do some professors and instructors at universities and colleges receive high teaching evaluations scores from students while others receive lower ones? Are there differences in teaching evaluations between instructors of different demographic groups? Could there be an impact due to student biases? These are all questions that are of interest to university/college administrators, as teaching evaluations are among the many criteria considered in determining which instructors and professors get promoted.

Researchers at the University of Texas in Austin, Texas (UT Austin) tried to answer the following research question: what factors explain differences in instructor teaching evaluation scores? To this end, they collected instructor and course information on 463 courses. A full description of the study can be found at openintro.org[2].

In this section, we'll keep things simple for now and try to explain differences in instructor teaching scores as a function of one numerical variable: the instructor's "beauty" score (we'll describe how this score was determined shortly). Could it be that instructors with higher "beauty" scores also have higher teaching evaluations? Could it be instead that instructors with higher "beauty" scores tend to have lower teaching evaluations? Or could it be that there is no relationship between "beauty" score and teaching evaluations? We'll answer these questions by modeling the relationship between teaching scores and "beauty" scores using *simple linear regression* where we have:

1. A numerical outcome variable y (the instructor's teaching score) and
2. A single numerical explanatory variable x (the instructor's "beauty" score).

5.1.1 Exploratory data analysis

The data on the 463 courses at UT Austin can be found in the `evals` data frame included in the `moderndive` package. However, to keep things simple, let's `select()` only the subset of the variables we'll consider in this chapter, and save this data in a new data frame called `evals_ch5`:

```
evals_ch5 <- evals %>%
  select(ID, score, bty_avg, age)
```

A crucial step before doing any kind of analysis or modeling is performing an *exploratory data analysis*, or EDA for short. EDA gives you a sense of the distributions of the individual variables in your data, whether any potential relationships exist between variables, whether there are outliers and/or missing values, and (most importantly) how to build your model. Here are three common steps in an EDA:

1. Most crucially, looking at the raw data values.
2. Computing summary statistics, such as means, medians, and interquartile ranges.
3. Creating data visualizations.

---

[2]https://www.openintro.org/stat/data/?data=evals

Let's perform the first common step in an exploratory data analysis: looking at the raw data values. Because this step seems so trivial, unfortunately many data analysts ignore it. However, getting an early sense of what your raw data looks like can often prevent many larger issues down the road.

You can do this by using RStudio's spreadsheet viewer or by using the `glimpse()` function as introduced in Subsection 1.4.3 on exploring data frames:

```
glimpse(evals_ch5)
```

```
Observations: 463
Variables: 4
$ ID      <int> 1, 2, 3, 4, 5, 6, 7, 8, 9, 10, 11, 12, 13, 14, 15, 16,...
$ score   <dbl> 4.7, 4.1, 3.9, 4.8, 4.6, 4.3, 2.8, 4.1, 3.4, 4.5, 3.8,...
$ bty_avg <dbl> 5.00, 5.00, 5.00, 5.00, 3.00, 3.00, 3.00, 3.33, 3.33, ...
$ age     <int> 36, 36, 36, 36, 59, 59, 59, 51, 51, 40, 40, 40, 40, 40...
```

Observe that `Observations: 463` indicates that there are 463 rows/observations in `evals_ch5`, where each row corresponds to one observed course at UT Austin. It is important to note that the *observational unit* is an individual course and not an individual instructor. Recall from Subsection 1.4.3 that the observational unit is the "type of thing" that is being measured by our variables. Since instructors teach more than one course in an academic year, the same instructor will appear more than once in the data. Hence there are fewer than 463 unique instructors being represented in `evals_ch5`. We'll revisit this idea in Section 10.3, when we talk about the "independence assumption" for inference for regression.

A full description of all the variables included in `evals` can be found at openintro.org[3] or by reading the associated help file (run `?evals` in the console). However, let's fully describe only the 4 variables we selected in `evals_ch5`:

1. `ID`: An identification variable used to distinguish between the 1 through 463 courses in the dataset.
2. `score`: A numerical variable of the course instructor's average teaching score, where the average is computed from the evaluation scores from all students in that course. Teaching scores of 1 are lowest and 5 are highest. This is the outcome variable y of interest.
3. `bty_avg`: A numerical variable of the course instructor's average "beauty" score, where the average is computed from a separate panel of six students. "Beauty" scores of 1 are lowest and 10 are highest. This is the explanatory variable x of interest.

---

[3]https://www.openintro.org/stat/data/?data=evals

4. `age`: A numerical variable of the course instructor's age. This will be another explanatory variable x that we'll use in the *Learning check* at the end of this subsection.

An alternative way to look at the raw data values is by choosing a random sample of the rows in `evals_ch5` by piping it into the `sample_n()` function from the `dplyr` package. Here we set the `size` argument to be 5, indicating that we want a random sample of 5 rows. We display the results in Table 5.1. Note that due to the random nature of the sampling, you will likely end up with a different subset of 5 rows.

```
evals_ch5 %>%
  sample_n(size = 5)
```

TABLE 5.1: A random sample of 5 out of the 463 courses at UT Austin

ID	score	bty_avg	age
129	3.7	3.00	62
109	4.7	4.33	46
28	4.8	5.50	62
434	2.8	2.00	62
330	4.0	2.33	64

Now that we've looked at the raw values in our `evals_ch5` data frame and got a preliminary sense of the data, let's move on to the next common step in an exploratory data analysis: computing summary statistics. Let's start by computing the mean and median of our numerical outcome variable `score` and our numerical explanatory variable "beauty" score denoted as `bty_avg`. We'll do this by using the `summarize()` function from `dplyr` along with the `mean()` and `median()` summary functions we saw in Section 3.3.

```
evals_ch5 %>%
  summarize(mean_bty_avg = mean(bty_avg), mean_score = mean(score),
            median_bty_avg = median(bty_avg), median_score = median(score))
```

```
# A tibble: 1 x 4
  mean_bty_avg mean_score median_bty_avg median_score
         <dbl>      <dbl>          <dbl>        <dbl>
1         4.42       4.17           4.33          4.3
```

However, what if we want other summary statistics as well, such as the standard deviation (a measure of spread), the minimum and maximum values, and various percentiles?

Typing out all these summary statistic functions in `summarize()` would be long and tedious. Instead, let's use the convenient `skim()` function from the `skimr` package. This function takes in a data frame, "skims" it, and returns commonly used summary statistics. Let's take our `evals_ch5` data frame, `select()` only the outcome and explanatory variables teaching `score` and `bty_avg`, and pipe them into the `skim()` function:

```
evals_ch5 %>% select(score, bty_avg) %>% skim()
```

```
Skim summary statistics
 n obs: 463
 n variables: 2

── Variable type:numeric
 variable missing complete    n mean   sd   p0  p25  p50 p75 p100
  bty_avg       0      463  463 4.42 1.53 1.67 3.17 4.33 5.5 8.17
    score       0      463  463 4.17 0.54 2.3  3.8  4.3  4.6 5
```

(For formatting purposes in this book, the inline histogram that is usually printed with `skim()` has been removed. This can be done by using `skim_with(numeric = list(hist = NULL))` prior to using the `skim()` function for version 1.0.6 of `skimr`.)

For the numerical variables teaching `score` and `bty_avg` it returns:

- `missing`: the number of missing values
- `complete`: the number of non-missing or complete values
- `n`: the total number of values
- `mean`: the average
- `sd`: the standard deviation
- `p0`: the 0th percentile: the value at which 0% of observations are smaller than it (the *minimum* value)
- `p25`: the 25th percentile: the value at which 25% of observations are smaller than it (the *1st quartile*)
- `p50`: the 50th percentile: the value at which 50% of observations are smaller than it (the *2nd* quartile and more commonly called the *median*)
- `p75`: the 75th percentile: the value at which 75% of observations are smaller than it (the *3rd quartile*)
- `p100`: the 100th percentile: the value at which 100% of observations are smaller than it (the *maximum* value)

Looking at this output, we can see how the values of both variables distribute. For example, the mean teaching score was 4.17 out of 5, whereas the mean "beauty" score was 4.42 out of 10. Furthermore, the middle 50% of teaching scores was between 3.80 and 4.6 (the first and third quartiles), whereas the middle 50% of "beauty" scores falls within 3.17 to 5.5 out of 10.

The skim() function only returns what are known as *univariate* summary statistics: functions that take a single variable and return some numerical summary of that variable. However, there also exist *bivariate* summary statistics: functions that take in two variables and return some summary of those two variables. In particular, when the two variables are numerical, we can compute the *correlation coefficient*. Generally speaking, *coefficients* are quantitative expressions of a specific phenomenon. A *correlation coefficient* is a quantitative expression of the *strength of the linear relationship between two numerical variables*. Its value ranges between -1 and 1 where:

- -1 indicates a perfect *negative relationship*: As one variable increases, the value of the other variable tends to go down, following a straight line.
- 0 indicates no relationship: The values of both variables go up/down independently of each other.
- +1 indicates a perfect *positive relationship*: As the value of one variable goes up, the value of the other variable tends to go up as well in a linear fashion.

Figure 5.1 gives examples of 9 different correlation coefficient values for hypothetical numerical variables x and y. For example, observe in the top right plot that for a correlation coefficient of -0.75 there is a negative linear relationship between x and y, but it is not as strong as the negative linear relationship between x and y when the correlation coefficient is -0.9 or -1.

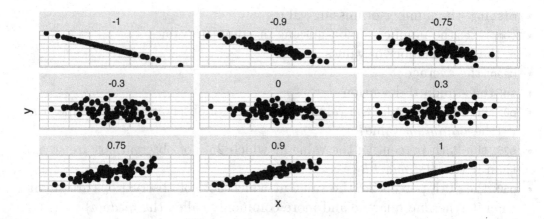

FIGURE 5.1: Nine different correlation coefficients.

The correlation coefficient can be computed using the `get_correlation()` function in the `moderndive` package. In this case, the inputs to the function are the two numerical variables for which we want to calculate the correlation coefficient.

We put the name of the outcome variable on the left-hand side of the ~ "tilde" sign, while putting the name of the explanatory variable on the right-hand side. This is known as R's *formula notation*. We will use this same "formula" syntax with regression later in this chapter.

```
evals_ch5 %>%
  get_correlation(formula = score ~ bty_avg)
```

```
# A tibble: 1 x 1
    cor
  <dbl>
1 0.187
```

An alternative way to compute correlation is to use the `cor()` summary function within a `summarize()`:

```
evals_ch5 %>%
  summarize(correlation = cor(score, bty_avg))
```

In our case, the correlation coefficient of 0.187 indicates that the relationship between teaching evaluation score and "beauty" average is "weakly positive." There is a certain amount of subjectivity in interpreting correlation coefficients, especially those that aren't close to the extreme values of -1, 0, and 1. To develop your intuition about correlation coefficients, play the "Guess the Correlation" 1980's style video game mentioned in Subsection 5.4.1.

Let's now perform the last of the steps in an exploratory data analysis: creating data visualizations. Since both the `score` and `bty_avg` variables are numerical, a scatterplot is an appropriate graph to visualize this data. Let's do this using `geom_point()` and display the result in Figure 5.2. Furthermore, let's highlight the six points in the top right of the visualization in a box.

```
ggplot(evals_ch5, aes(x = bty_avg, y = score)) +
  geom_point() +
  labs(x = "Beauty Score",
       y = "Teaching Score",
       title = "Scatterplot of relationship of teaching and beauty scores")
```

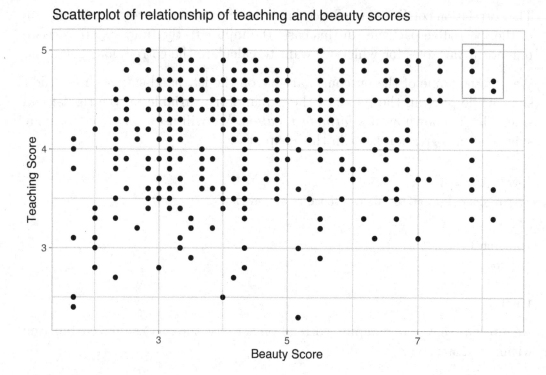

FIGURE 5.2: Instructor evaluation scores at UT Austin.

Observe that most "beauty" scores lie between 2 and 8, while most teaching scores lie between 3 and 5. Furthermore, while opinions may vary, it is our opinion that the relationship between teaching score and "beauty" score is "weakly positive." This is consistent with our earlier computed correlation coefficient of 0.187.

Furthermore, there appear to be six points in the top-right of this plot high-lighted in the box. However, this is not actually the case, as this plot suffers from *overplotting*. Recall from Subsection 2.3.2 that overplotting occurs when several points are stacked directly on top of each other, making it difficult to distinguish them. So while it may appear that there are only six points in the box, there are actually more. This fact is only apparent when using `geom_jitter()` in place of `geom_point()`. We display the resulting plot in Figure 5.3 along with the same small box as in Figure 5.2.

```
ggplot(evals_ch5, aes(x = bty_avg, y = score)) +
  geom_jitter() +
  labs(x = "Beauty Score", y = "Teaching Score",
       title = "Scatterplot of relationship of teaching and beauty scores")
```

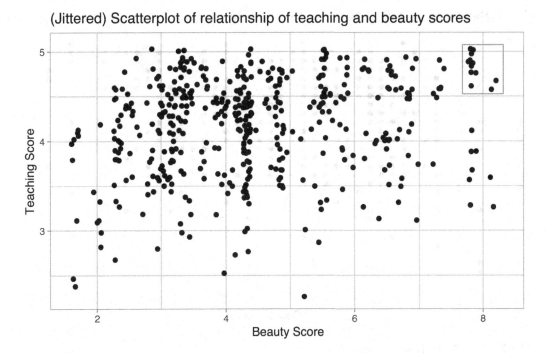

FIGURE 5.3: Instructor evaluation scores at UT Austin.

It is now apparent that there are 12 points in the area highlighted in the box and not six as originally suggested in Figure 5.2. Recall from Subsection 2.3.2 on overplotting that jittering adds a little random "nudge" to each of the points to break up these ties. Furthermore, recall that jittering is strictly a visualization tool; it does not alter the original values in the data frame evals_ch5. To keep things simple going forward, however, we'll only present regular scatterplots rather than their jittered counterparts.

Let's build on the unjittered scatterplot in Figure 5.2 by adding a "best-fitting" line: of all possible lines we can draw on this scatterplot, it is the line that "best" fits through the cloud of points. We do this by adding a new geom_smooth(method = "lm", se = FALSE) layer to the ggplot() code that created the scatterplot in Figure 5.2. The method = "lm" argument sets the line to be a "linear model." The se = FALSE argument suppresses *standard error* uncertainty bars. (We'll define the concept of *standard error* later in Subsection 7.3.2.)

```
ggplot(evals_ch5, aes(x = bty_avg, y = score)) +
  geom_point() +
  labs(x = "Beauty Score", y = "Teaching Score",
       title = "Relationship between teaching and beauty scores") +
  geom_smooth(method = "lm", se = FALSE)
```

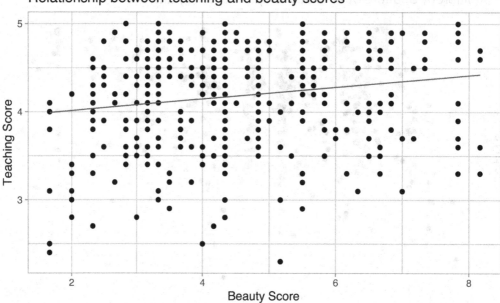

FIGURE 5.4: Regression line.

The line in the resulting Figure 5.4 is called a "regression line." The regression line is a visual summary of the relationship between two numerical variables, in our case the outcome variable score and the explanatory variable bty_avg. The positive slope of the blue line is consistent with our earlier observed correlation coefficient of 0.187 suggesting that there is a positive relationship between these two variables: as instructors have higher "beauty" scores, so also do they receive higher teaching evaluations. We'll see later, however, that while the correlation coefficient and the slope of a regression line always have the same sign (positive or negative), they typically do not have the same value.

Furthermore, a regression line is "best-fitting" in that it minimizes some mathematical criteria. We present these mathematical criteria in Subsection 5.3.2, but we suggest you read this subsection only after first reading the rest of this section on regression with one numerical explanatory variable.

Learning check

(LC5.1) Conduct a new exploratory data analysis with the same outcome variable y being score but with age as the new explanatory variable x. Remember, this involves three things:

(a) Looking at the raw data values.
(b) Computing summary statistics.

(c)   Creating data visualizations.

What can you say about the relationship between age and teaching scores based on this exploration?

.

5.1.2   Simple linear regression

You may recall from secondary/high school algebra that the equation of a line is $y = a + b \cdot x$. (Note that the $\cdot$ symbol is equivalent to the $\times$ "multiply by" mathematical symbol. We'll use the $\cdot$ symbol in the rest of this book as it is more succinct.) It is defined by two coefficients a and b. The intercept coefficient a is the value of y when $x = 0$. The slope coefficient b for x is the increase in y for every increase of one in x. This is also called the "rise over run."

However, when defining a regression line like the regression line in Figure 5.4, we use slightly different notation: the equation of the regression line is $\hat{y} = b_0 + b_1 \cdot x$. The intercept coefficient is b_0, so b_0 is the value of $\hat{y}$ when $x = 0$. The slope coefficient for x is b_1, i.e., the increase in $\hat{y}$ for every increase of one in x. Why do we put a "hat" on top of the y? It's a form of notation commonly used in regression to indicate that we have a "fitted value," or the value of y on the regression line for a given x value. We'll discuss this more in the upcoming Subsection 5.1.3.

We know that the regression line in Figure 5.4 has a positive slope b_1 corresponding to our explanatory x variable bty_avg. Why? Because as instructors tend to have higher bty_avg scores, so also do they tend to have higher teaching evaluation scores. However, what is the numerical value of the slope b_1? What about the intercept b_0? Let's not compute these two values by hand, but rather let's use a computer!

We can obtain the values of the intercept b_0 and the slope for btg_avg b_1 by outputting a *linear regression table*. This is done in two steps:

1.   We first "fit" the linear regression model using the lm() function and save it in score_model.
2.   We get the regression table by applying the get_regression_table() function from the moderndive package to score_model.

```
# Fit regression model:
score_model <- lm(score ~ bty_avg, data = evals_ch5)
```

```
# Get regression table:
get_regression_table(score_model)
```

TABLE 5.2: Linear regression table

term	estimate	std_error	statistic	p_value	lower_ci	upper_ci
intercept	3.880	0.076	50.96	0	3.731	4.030
bty_avg	0.067	0.016	4.09	0	0.035	0.099

Let's first focus on interpreting the regression table output in Table 5.2, and then we'll later revisit the code that produced it. In the `estimate` column of Table 5.2 are the intercept $b_0 = 3.88$ and the slope $b_1 = 0.067$ for `bty_avg`. Thus the equation of the regression line in Figure 5.4 follows:

$$\hat{y} = b_0 + b_1 \cdot x$$
$$\widehat{\text{score}} = b_0 + b_{\text{bty\_avg}} \cdot \text{bty\_avg}$$
$$= 3.880 + 0.067 \cdot \text{bty\_avg}$$

The intercept $b_0 = 3.88$ is the average teaching score $\hat{y} = \widehat{\text{score}}$ for those courses where the instructor had a "beauty" score `bty_avg` of 0. Or in graphical terms, it's where the line intersects the y axis when $x = 0$. Note, however, that while the intercept of the regression line has a mathematical interpretation, it has no *practical* interpretation here, since observing a `bty_avg` of 0 is impossible; it is the average of six panelists' "beauty" scores ranging from 1 to 10. Furthermore, looking at the scatterplot with the regression line in Figure 5.4, no instructors had a "beauty" score anywhere near 0.

Of greater interest is the slope $b_1 = b_{\text{bty\_avg}}$ for `bty_avg` of 0.067, as this summarizes the relationship between the teaching and "beauty" score variables. Note that the sign is positive, suggesting a positive relationship between these two variables, meaning teachers with higher "beauty" scores also tend to have higher teaching scores. Recall from earlier that the correlation coefficient is 0.187. They both have the same positive sign, but have a different value. Recall further that the correlation's interpretation is the "strength of linear association". The slope's interpretation is a little different:

For every increase of 1 unit in `bty_avg`, there is an *associated* increase of, *on average*, 0.067 units of `score`.

We only state that there is an *associated* increase and not necessarily a *causal* increase. For example, perhaps it's not that higher "beauty" scores directly cause higher teaching scores per se. Instead, the following could hold true: individuals from wealthier backgrounds tend to have stronger educational backgrounds and hence have higher teaching scores, while at the same time these wealthy individuals also tend to have higher "beauty" scores. In other words, just because two variables are strongly associated, it doesn't necessarily mean that one causes the other. This is summed up in the often quoted phrase, "correlation is not necessarily causation." We discuss this idea further in Subsection 5.3.1.

Furthermore, we say that this associated increase is *on average* 0.067 units of teaching score, because you might have two instructors whose bty_avg scores differ by 1 unit, but their difference in teaching scores won't necessarily be exactly 0.067. What the slope of 0.067 is saying is that across all possible courses, the *average* difference in teaching score between two instructors whose "beauty" scores differ by one is 0.067.

Now that we've learned how to compute the equation for the regression line in Figure 5.4 using the values in the estimate column of Table 5.2, and how to interpret the resulting intercept and slope, let's revisit the code that generated this table:

```
# Fit regression model:
score_model <- lm(score ~ bty_avg, data = evals_ch5)
# Get regression table:
get_regression_table(score_model)
```

First, we "fit" the linear regression model to the data using the lm() function and save this as score_model. When we say "fit", we mean "find the best fitting line to this data." lm() stands for "linear model" and is used as follows: lm(y ~ x, data = data_frame_name) where:

- y is the outcome variable, followed by a tilde ~. In our case, y is set to score.
- x is the explanatory variable. In our case, x is set to bty_avg.
- The combination of y ~ x is called a *model formula*. (Note the order of y and x.) In our case, the model formula is score ~ bty_avg. We saw such model formulas earlier when we computed the correlation coefficient using the get_correlation() function in Subsection 5.1.1.
- data_frame_name is the name of the data frame that contains the variables y and x. In our case, data_frame_name is the evals_ch5 data frame.

Second, we take the saved model in score_model and apply the get_regression_table() function from the moderndive package to it to obtain the regression table in Table 5.2. This function is an example of what's known in computer programming as a *wrapper function*. They take other pre-existing functions and "wrap" them into a single function that hides its inner workings. This concept is illustrated in Figure 5.5.

Wrapper Function

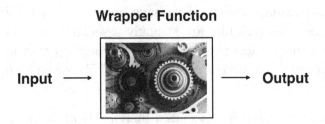

Input $\longrightarrow$ $\longrightarrow$ **Output**

FIGURE 5.5: The concept of a wrapper function.

So all you need to worry about is what the inputs look like and what the outputs look like; you leave all the other details "under the hood of the car." In our regression modeling example, the `get_regression_table()` function takes a saved `lm()` linear regression model as input and returns a data frame of the regression table as output. If you're interested in learning more about the `get_regression_table()` function's inner workings, check out Subsection 5.3.3.

Lastly, you might be wondering what the remaining five columns in Table 5.2 are: `std_error`, `statistic`, `p_value`, `lower_ci` and `upper_ci`. They are the *standard error, test statistic, p-value, lower 95% confidence interval bound*, and *upper 95% confidence interval bound*. They tell us about both the *statistical significance* and *practical significance* of our results. This is loosely the "meaningfulness" of our results from a statistical perspective. Let's put aside these ideas for now and revisit them in Chapter 10 on (statistical) inference for regression. We'll do this after we've had a chance to cover standard errors in Chapter 7, confidence intervals in Chapter 8, and hypothesis testing and *p*-values in Chapter 9.

Learning check

(LC5.2) Fit a new simple linear regression using `lm(score ~ age, data = evals_ch5)` where `age` is the new explanatory variable x. Get information about the "best-fitting" line from the regression table by applying the `get_regression_table()` function. How do the regression results match up with the results from your earlier exploratory data analysis?

5.1.3 Observed/fitted values and residuals

We just saw how to get the value of the intercept and the slope of a regression line from the estimate column of a regression table generated by the get_regression_table() function. Now instead say we want information on individual observations. For example, let's focus on the 21st of the 463 courses in the evals_ch5 data frame in Table 5.3:

TABLE 5.3: Data for the 21st course out of 463

ID	score	bty_avg	age
21	4.9	7.33	31

What is the value $\hat{y}$ on the regression line corresponding to this instructor's bty_avg "beauty" score of 7.333? In Figure 5.6 we mark three values corresponding to the instructor for this 21st course and give their statistical names:

- Circle: The *observed value* $y = 4.9$ is this course's instructor's actual teaching score.
- Square: The *fitted value* $\hat{y}$ is the value on the regression line for x = bty_avg = 7.333. This value is computed using the intercept and slope in the previous regression table:

$$\hat{y} = b_0 + b_1 \cdot x = 3.88 + 0.067 \cdot 7.333 = 4.369$$

- Arrow: The length of this arrow is the *residual* and is computed by subtracting the fitted value $\hat{y}$ from the observed value y. The residual can be thought of as a model's error or "lack of fit" for a particular observation. In the case of this course's instructor, it is $y - \hat{y}$ = 4.9 - 4.369 = 0.531.

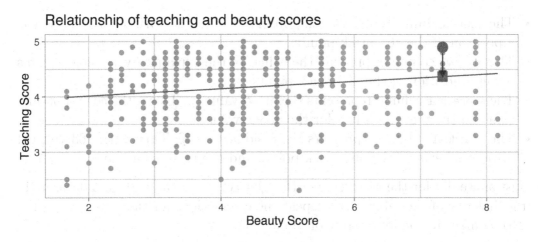

FIGURE 5.6: Example of observed value, fitted value, and residual.

Now say we want to compute both the fitted value $\hat{y} = b_0 + b_1 \cdot x$ and the residual $y - \hat{y}$ for *all* 463 courses in the study. Recall that each course corresponds to one of the 463 rows in the evals_ch5 data frame and also one of the 463 points in the regression plot in Figure 5.6.

We could repeat the previous calculations we performed by hand 463 times, but that would be tedious and time consuming. Instead, let's do this using a computer with the get_regression_points() function. Just like the get_regression_table() function, the get_regression_points() function is a "wrapper" function. However, this function returns a different output. Let's apply the get_regression_points() function to score_model, which is where we saved our lm() model in the previous section. In Table 5.4 we present the results of only the 21st through 24th courses for brevity's sake.

```
regression_points <- get_regression_points(score_model)
regression_points
```

TABLE 5.4: Regression points (for only the 21st through 24th courses)

ID	score	bty_avg	score_hat	residual
21	4.9	7.33	4.37	0.531
22	4.6	7.33	4.37	0.231
23	4.5	7.33	4.37	0.131
24	4.4	5.50	4.25	0.153

Let's inspect the individual columns and match them with the elements of Figure 5.6:

- The score column represents the observed outcome variable y. This is the y-position of the 463 black points.
- The bty_avg column represents the values of the explanatory variable x. This is the x-position of the 463 black points.
- The score_hat column represents the fitted values $\hat{y}$. This is the corresponding value on the regression line for the 463 x values.
- The residual column represents the residuals $y - \hat{y}$. This is the 463 vertical distances between the 463 black points and the regression line.

Just as we did for the instructor of the 21st course in the evals_ch5 dataset (in the first row of the table), let's repeat the calculations for the instructor of the 24th course (in the fourth row of Table 5.4):

- score = 4.4 is the observed teaching score y for this course's instructor.

- `bty_avg = 5.50` is the value of the explanatory variable `bty_avg` x for this course's instructor.
- `score_hat = 4.25 = 3.88 + 0.067 · 5.50` is the fitted value $\hat{y}$ on the regression line for this course's instructor.
- `residual = 0.153 = 4.4 - 4.25` is the value of the residual for this instructor. In other words, the model's fitted value was off by 0.153 teaching score units for this course's instructor.

At this point, you can skip ahead if you like to Subsection 5.3.2 to learn about the processes behind what makes "best-fitting" regression lines. As a primer, a "best-fitting" line refers to the line that minimizes the *sum of squared residuals* out of all possible lines we can draw through the points. In Section 5.2, we'll discuss another common scenario of having a categorical explanatory variable and a numerical outcome variable.

Learning check

(LC5.3) Generate a data frame of the residuals of the model where you used `age` as the explanatory x variable.

5.2   One categorical explanatory variable

It's an unfortunate truth that life expectancy is not the same across all countries in the world. International development agencies are interested in studying these differences in life expectancy in the hopes of identifying where governments should allocate resources to address this problem. In this section, we'll explore differences in life expectancy in two ways:

1. Differences between continents: Are there significant differences in average life expectancy between the five populated continents of the world: Africa, the Americas, Asia, Europe, and Oceania?
2. Differences within continents: How does life expectancy vary within the world's five continents? For example, is the spread of life expectancy among the countries of Africa larger than the spread of life expectancy among the countries of Asia?

To answer such questions, we'll use the `gapminder` data frame included in the `gapminder` package. This dataset has international development statistics such as life expectancy, GDP per capita, and population for 142 countries for 5-year intervals between 1952 and 2007. Recall we visualized some of this data in Figure 2.1 in Subsection 2.1.2 on the grammar of graphics.

We'll use this data for basic regression again, but now using an explanatory variable x that is categorical, as opposed to the numerical explanatory variable model we used in the previous Section 5.1.

1. A numerical outcome variable y (a country's life expectancy) and
2. A single categorical explanatory variable x (the continent that the country is a part of).

When the explanatory variable x is categorical, the concept of a "best-fitting" regression line is a little different than the one we saw previously in Section 5.1 where the explanatory variable x was numerical. We'll study these differences shortly in Subsection 5.2.2, but first we conduct an exploratory data analysis.

5.2.1   Exploratory data analysis

The data on the 142 countries can be found in the `gapminder` data frame included in the `gapminder` package. However, to keep things simple, let's `filter()` for only those observations/rows corresponding to the year 2007. Additionally, let's `select()` only the subset of the variables we'll consider in this chapter. We'll save this data in a new data frame called `gapminder2007`:

```
library(gapminder)
gapminder2007 <- gapminder %>%
  filter(year == 2007) %>%
  select(country, lifeExp, continent, gdpPercap)
```

Let's perform the first common step in an exploratory data analysis: looking at the raw data values. You can do this by using RStudio's spreadsheet viewer or by using the `glimpse()` command as introduced in Subsection 1.4.3 on exploring data frames:

```
glimpse(gapminder2007)
```

```
Observations: 142
Variables: 4
$ country   <fct> Afghanistan, Albania, Algeria, Angola, Argentina, Au...
$ lifeExp   <dbl> 43.8, 76.4, 72.3, 42.7, 75.3, 81.2, 79.8, 75.6, 64.1...
$ continent <fct> Asia, Europe, Africa, Africa, Americas, Oceania, Eur...
$ gdpPercap <dbl> 975, 5937, 6223, 4797, 12779, 34435, 36126, 29796, 1...
```

Observe that `Observations: 142` indicates that there are 142 rows/observations in `gapminder2007`, where each row corresponds to one country. In other words, the *observational unit* is an individual country. Furthermore, observe that the variable `continent` is of type `<fct>`, which stands for *factor*, which is R's way of encoding categorical variables.

A full description of all the variables included in `gapminder` can be found by reading the associated help file (run `?gapminder` in the console). However, let's fully describe only the 4 variables we selected in `gapminder2007`:

1. `country`: An identification variable of type character/text used to distinguish the 142 countries in the dataset.
2. `lifeExp`: A numerical variable of that country's life expectancy at birth. This is the outcome variable y of interest.
3. `continent`: A categorical variable with five levels. Here "levels" correspond to the possible categories: Africa, Asia, Americas, Europe, and Oceania. This is the explanatory variable x of interest.
4. `gdpPercap`: A numerical variable of that country's GDP per capita in US inflation-adjusted dollars that we'll use as another outcome variable y in the *Learning check* at the end of this subsection.

Let's look at a random sample of five out of the 142 countries in Table 5.5.

```
gapminder2007 %>% sample_n(size = 5)
```

TABLE 5.5: Random sample of 5 out of 142 countries

country	lifeExp	continent	gdpPercap
Togo	58.4	Africa	883
Sao Tome and Principe	65.5	Africa	1598
Congo, Dem. Rep.	46.5	Africa	278
Lesotho	42.6	Africa	1569
Bulgaria	73.0	Europe	10681

Note that random sampling will likely produce a different subset of 5 rows for you than what's shown. Now that we've looked at the raw values in our `gapminder2007` data frame and got a sense of the data, let's move on to computing summary statistics. Let's once again apply the `skim()` function from the `skimr` package. Recall from our previous EDA that this function takes in a data frame, "skims" it, and returns commonly used summary statistics. Let's take our `gapminder2007` data frame, `select()` only the outcome and explanatory variables `lifeExp` and `continent`, and pipe them into the `skim()` function:

```
gapminder2007 %>%
  select(lifeExp, continent) %>%
  skim()
```

```
Skim summary statistics
 n obs: 142
 n variables: 2
```

── Variable type:factor
```
 variable missing complete   n n_unique                 top_counts ordered
continent       0     142 142        5 Afr: 52, Asi: 33, Eur: 30, Ame: 25   FALSE
```

── Variable type:numeric
```
 variable missing complete   n  mean    sd    p0   p25   p50   p75 p100
  lifeExp       0        142 142 67.01 12.07 39.61 57.16 71.94 76.41 82.6
```

The skim() output now reports summaries for categorical variables (Variable type:factor) separately from the numerical variables (Variable type:numeric). For the categorical variable continent, it reports:

- missing, complete, and n, which are the number of missing, complete, and total number of values as before, respectively.
- n_unique: The number of unique levels to this variable, corresponding to Africa, Asia, Americas, Europe, and Oceania. This refers to how many countries are in the data for each continent.
- top_counts: In this case, the top four counts: Africa has 52 countries, Asia has 33, Europe has 30, and Americas has 25. Not displayed is Oceania with 2 countries.
- ordered: This tells us whether the categorical variable is "ordinal": whether there is an encoded hierarchy (like low, medium, high). In this case, continent is not ordered.

Turning our attention to the summary statistics of the numerical variable lifeExp, we observe that the global median life expectancy in 2007 was 71.94. Thus, half of the world's countries (71 countries) had a life expectancy less than 71.94. The mean life expectancy of 67.01 is lower, however. Why is the mean life expectancy lower than the median?

We can answer this question by performing the last of the three common steps in an exploratory data analysis: creating data visualizations. Let's visualize the distribution of our outcome variable $y = $ lifeExp in Figure 5.7.

```
ggplot(gapminder2007, aes(x = lifeExp)) +
  geom_histogram(binwidth = 5, color = "white") +
  labs(x = "Life expectancy", y = "Number of countries",
       title = "Histogram of distribution of worldwide life expectancies")
```

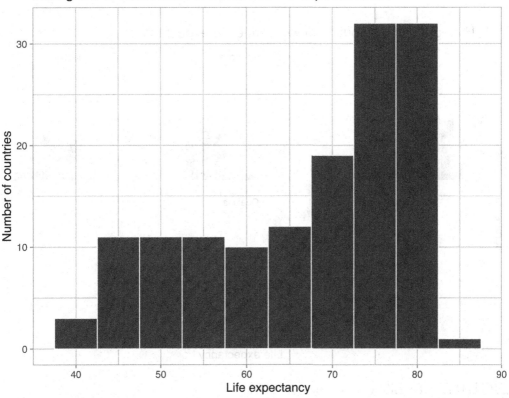

FIGURE 5.7: Histogram of life expectancy in 2007.

We see that this data is *left-skewed*, also known as *negatively* skewed: there are a few countries with low life expectancy that are bringing down the mean life expectancy. However, the median is less sensitive to the effects of such outliers; hence, the median is greater than the mean in this case.

Remember, however, that we want to compare life expectancies both between continents and within continents. In other words, our visualizations need to incorporate some notion of the variable continent. We can do this easily with a faceted histogram. Recall from Section 2.6 that facets allow us to split a visualization by the different values of another variable. We display the resulting visualization in Figure 5.8 by adding a facet_wrap(~ continent, nrow = 2) layer.

```
ggplot(gapminder2007, aes(x = lifeExp)) +
  geom_histogram(binwidth = 5, color = "white") +
  labs(x = "Life expectancy",
       y = "Number of countries",
       title = "Histogram of distribution of worldwide life expectancies") +
  facet_wrap(~ continent, nrow = 2)
```

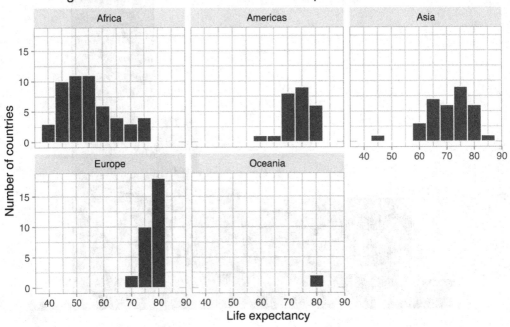

FIGURE 5.8: Life expectancy in 2007.

Observe that unfortunately the distribution of African life expectancies is much lower than the other continents, while in Europe life expectancies tend to be higher and furthermore do not vary as much. On the other hand, both Asia and Africa have the most variation in life expectancies. There is the least variation in Oceania, but keep in mind that there are only two countries in Oceania: Australia and New Zealand.

Recall that an alternative method to visualize the distribution of a numerical variable split by a categorical variable is by using a side-by-side boxplot. We map the categorical variable continent to the x-axis and the different life expectancies within each continent on the y-axis in Figure 5.9.

```
ggplot(gapminder2007, aes(x = continent, y = lifeExp)) +
   geom_boxplot() +
  labs(x = "Continent", y = "Life expectancy",
        title = "Life expectancy by continent")
```

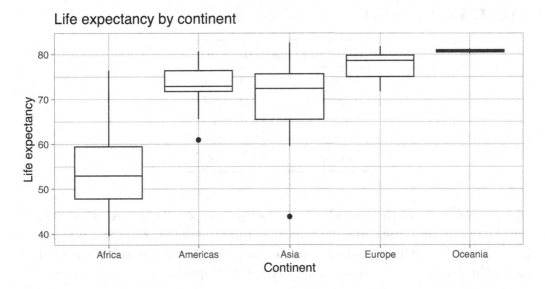

FIGURE 5.9: Life expectancy in 2007.

Some people prefer comparing the distributions of a numerical variable between different levels of a categorical variable using a boxplot instead of a faceted histogram. This is because we can make quick comparisons between the categorical variable's levels with imaginary horizontal lines. For example, observe in Figure 5.9 that we can quickly convince ourselves that Oceania has the highest median life expectancies by drawing an imaginary horizontal line at $y = 80$. Furthermore, as we observed in the faceted histogram in Figure 5.8, Africa and Asia have the largest variation in life expectancy as evidenced by their large interquartile ranges (the heights of the boxes).

It's important to remember, however, that the solid lines in the middle of the boxes correspond to the medians (the middle value) rather than the mean (the average). So, for example, if you look at Asia, the solid line denotes the median life expectancy of around 72 years. This tells us that half of all countries in Asia have a life expectancy below 72 years, whereas half have a life expectancy above 72 years.

Let's compute the median and mean life expectancy for each continent with a little more data wrangling and display the results in Table 5.6.

```
lifeExp_by_continent <- gapminder2007 %>%
  group_by(continent) %>%
  summarize(median = median(lifeExp),
            mean = mean(lifeExp))
```

TABLE 5.6: Life expectancy by continent

continent	median	mean
Africa	52.9	54.8
Americas	72.9	73.6
Asia	72.4	70.7
Europe	78.6	77.6
Oceania	80.7	80.7

Observe the order of the second column `median` life expectancy: Africa is lowest, the Americas and Asia are next with similar medians, then Europe, then Oceania. This ordering corresponds to the ordering of the solid black lines inside the boxes in our side-by-side boxplot in Figure 5.9.

Let's now turn our attention to the values in the third column `mean`. Using Africa's mean life expectancy of 54.8 as a *baseline for comparison*, let's start making comparisons to the mean life expectancies of the other four continents and put these values in Table 5.7, which we'll revisit later on in this section.

1. For the Americas, it is 73.6 - 54.8 = 18.8 years higher.
2. For Asia, it is 70.7 - 54.8 = 15.9 years higher.
3. For Europe, it is 77.6 - 54.8 = 22.8 years higher.
4. For Oceania, it is 80.7 - 54.8 = 25.9 years higher.

TABLE 5.7: Mean life expectancy by continent and relative differences from mean for Africa

continent	mean	Difference versus Africa
Africa	54.8	0.0
Americas	73.6	18.8
Asia	70.7	15.9
Europe	77.6	22.8
Oceania	80.7	25.9

Learning check

(LC5.4) Conduct a new exploratory data analysis with the same explanatory variable x being `continent` but with `gdpPercap` as the new outcome variable y. What can you say about the differences in GDP per capita between continents based on this exploration?

5.2.2 Linear regression

In Subsection 5.1.2 we introduced simple linear regression, which involves modeling the relationship between a numerical outcome variable y and a numerical explanatory variable x. In our life expectancy example, we now instead have a categorical explanatory variable `continent`. Our model will not yield a "best-fitting" regression line like in Figure 5.4, but rather *offsets* relative to a baseline for comparison.

As we did in Subsection 5.1.2 when studying the relationship between teaching scores and "beauty" scores, let's output the regression table for this model. Recall that this is done in two steps:

1. We first "fit" the linear regression model using the `lm(y ~ x, data)` function and save it in `lifeExp_model`.
2. We get the regression table by applying the `get_regression_table()` function from the `moderndive` package to `lifeExp_model`.

```
lifeExp_model <- lm(lifeExp ~ continent, data = gapminder2007)
get_regression_table(lifeExp_model)
```

TABLE 5.8: Linear regression table

term	estimate	std_error	statistic	p_value	lower_ci	upper_ci
intercept	54.8	1.02	53.45	0	52.8	56.8
continentAmericas	18.8	1.80	10.45	0	15.2	22.4
continentAsia	15.9	1.65	9.68	0	12.7	19.2
continentEurope	22.8	1.70	13.47	0	19.5	26.2
continentOceania	25.9	5.33	4.86	0	15.4	36.5

Let's once again focus on the values in the `term` and `estimate` columns of Table 5.8. Why are there now 5 rows? Let's break them down one-by-one:

1. `intercept` corresponds to the mean life expectancy of countries in Africa of 54.8 years.
2. `continentAmericas` corresponds to countries in the Americas and the value $+18.8$ is the same difference in mean life expectancy relative to Africa we displayed in Table 5.7. In other words, the mean life expectancy of countries in the Americas is $54.8 + 18.8 = 73.6$.
3. `continentAsia` corresponds to countries in Asia and the value $+15.9$ is the same difference in mean life expectancy relative to Africa we displayed in Table 5.7. In other words, the mean life expectancy of countries in Asia is $54.8 + 15.9 = 70.7$.
4. `continentEurope` corresponds to countries in Europe and the value $+22.8$ is the same difference in mean life expectancy relative to Africa we displayed in Table 5.7. In other words, the mean life expectancy of countries in Europe is $54.8 + 22.8 = 77.6$.
5. `continentOceania` corresponds to countries in Oceania and the value $+25.9$ is the same difference in mean life expectancy relative to Africa we displayed in Table 5.7. In other words, the mean life expectancy of countries in Oceania is $54.8 + 25.9 = 80.7$.

To summarize, the 5 values in the `estimate` column in Table 5.8 correspond to the "baseline for comparison" continent Africa (the intercept) as well as four "offsets" from this baseline for the remaining 4 continents: the Americas, Asia, Europe, and Oceania.

You might be asking at this point why was Africa chosen as the "baseline for comparison" group. This is the case for no other reason than it comes first alphabetically of the five continents; by default R arranges factors/categorical variables in alphanumeric order. You can change this baseline group to be another continent if you manipulate the variable `continent`'s factor "levels" using the `forcats` package. See Chapter 15^4 of *R for Data Science* (Grolemund and Wickham, 2017) for examples.

Let's now write the equation for our fitted values $\hat{y} = \widehat{\text{life exp}}$.

$$
\begin{aligned}
\hat{y} = \widehat{\text{life exp}} &= b_0 + b_{\text{Amer}} \cdot \mathbb{1}_{\text{Amer}}(x) + b_{\text{Asia}} \cdot \mathbb{1}_{\text{Asia}}(x) + \\
&\quad b_{\text{Euro}} \cdot \mathbb{1}_{\text{Euro}}(x) + b_{\text{Ocean}} \cdot \mathbb{1}_{\text{Ocean}}(x) \\
&= 54.8 + 18.8 \cdot \mathbb{1}_{\text{Amer}}(x) + 15.9 \cdot \mathbb{1}_{\text{Asia}}(x) + \\
&\quad 22.8 \cdot \mathbb{1}_{\text{Euro}}(x) + 25.9 \cdot \mathbb{1}_{\text{Ocean}}(x)
\end{aligned}
$$

Whoa! That looks daunting! Don't fret, however, as once you understand what all the elements mean, things simplify greatly. First, $\mathbb{1}_A(x)$ is what's known

---

[4] https://r4ds.had.co.nz/factors.html

in mathematics as an "indicator function." It returns only one of two possible values, 0 and 1, where

$$\mathbb{1}_A(x) = \begin{cases} 1 & \text{if } x \text{ is in } A \\ 0 & \text{if otherwise} \end{cases}$$

In a statistical modeling context, this is also known as a *dummy variable*. In our case, let's consider the first such indicator variable $\mathbb{1}_{\text{Amer}}(x)$. This indicator function returns 1 if a country is in the Americas, 0 otherwise:

$$\mathbb{1}_{\text{Amer}}(x) = \begin{cases} 1 & \text{if country } x \text{ is in the Americas} \\ 0 & \text{otherwise} \end{cases}$$

Second, b_0 corresponds to the intercept as before; in this case, it's the mean life expectancy of all countries in Africa. Third, the b_{Amer}, b_{Asia}, b_{Euro}, and b_{Ocean} represent the 4 "offsets relative to the baseline for comparison" in the regression table output in Table 5.8: `continentAmericas`, `continentAsia`, `continentEurope`, and `continentOceania`.

Let's put this all together and compute the fitted value $\hat{y} = \widehat{\text{life exp}}$ for a country in Africa. Since the country is in Africa, all four indicator functions $\mathbb{1}_{\text{Amer}}(x)$, $\mathbb{1}_{\text{Asia}}(x)$, $\mathbb{1}_{\text{Euro}}(x)$, and $\mathbb{1}_{\text{Ocean}}(x)$ will equal 0, and thus:

$$\begin{aligned} \widehat{\text{life exp}} &= b_0 + b_{\text{Amer}} \cdot \mathbb{1}_{\text{Amer}}(x) + b_{\text{Asia}} \cdot \mathbb{1}_{\text{Asia}}(x) + \\ & \quad b_{\text{Euro}} \cdot \mathbb{1}_{\text{Euro}}(x) + b_{\text{Ocean}} \cdot \mathbb{1}_{\text{Ocean}}(x) \\ &= 54.8 + 18.8 \cdot \mathbb{1}_{\text{Amer}}(x) + 15.9 \cdot \mathbb{1}_{\text{Asia}}(x) + \\ & \quad 22.8 \cdot \mathbb{1}_{\text{Euro}}(x) + 25.9 \cdot \mathbb{1}_{\text{Ocean}}(x) \\ &= 54.8 + 18.8 \cdot 0 + 15.9 \cdot 0 + 22.8 \cdot 0 + 25.9 \cdot 0 \\ &= 54.8 \end{aligned}$$

In other words, all that's left is the intercept b_0, corresponding to the average life expectancy of African countries of 54.8 years. Next, say we are considering a country in the Americas. In this case, only the indicator function $\mathbb{1}_{\text{Amer}}(x)$ for the Americas will equal 1, while all the others will equal 0, and thus:

$$\begin{aligned} \widehat{\text{life exp}} &= 54.8 + 18.8 \cdot \mathbb{1}_{\text{Amer}}(x) + 15.9 \cdot \mathbb{1}_{\text{Asia}}(x) + 22.8 \cdot \mathbb{1}_{\text{Euro}}(x) + \\ & \quad 25.9 \cdot \mathbb{1}_{\text{Ocean}}(x) \\ &= 54.8 + 18.8 \cdot 1 + 15.9 \cdot 0 + 22.8 \cdot 0 + 25.9 \cdot 0 \\ &= 54.8 + 18.8 \\ &= 73.6 \end{aligned}$$

which is the mean life expectancy for countries in the Americas of 73.6 years in Table 5.7. Note the "offset from the baseline for comparison" is +18.8 years.

Let's do one more. Say we are considering a country in Asia. In this case, only the indicator function $\mathbb{1}_{\text{Asia}}(x)$ for Asia will equal 1, while all the others will equal 0, and thus:

$$
\begin{aligned}
\widehat{\text{life exp}} &= 54.8 + 18.8 \cdot \mathbb{1}_{\text{Amer}}(x) + 15.9 \cdot \mathbb{1}_{\text{Asia}}(x) + 22.8 \cdot \mathbb{1}_{\text{Euro}}(x) + \\
& \quad 25.9 \cdot \mathbb{1}_{\text{Ocean}}(x) \\
&= 54.8 + 18.8 \cdot 0 + 15.9 \cdot 1 + 22.8 \cdot 0 + 25.9 \cdot 0 \\
&= 54.8 + 15.9 \\
&= 70.7
\end{aligned}
$$

which is the mean life expectancy for Asian countries of 70.7 years in Table 5.7. The "offset from the baseline for comparison" here is +15.9 years.

Let's generalize this idea a bit. If we fit a linear regression model using a categorical explanatory variable x that has k possible categories, the regression table will return an intercept and $k - 1$ "offsets." In our case, since there are $k = 5$ continents, the regression model returns an intercept corresponding to the baseline for comparison group of Africa and $k - 1 = 4$ offsets corresponding to the Americas, Asia, Europe, and Oceania.

Understanding a regression table output when you're using a categorical explanatory variable is a topic those new to regression often struggle with. The only real remedy for these struggles is practice, practice, practice. However, once you equip yourselves with an understanding of how to create regression models using categorical explanatory variables, you'll be able to incorporate many new variables into your models, given the large amount of the world's data that is categorical. If you feel like you're still struggling at this point, however, we suggest you closely compare Tables 5.7 and 5.8 and note how you can compute all the values from one table using the values in the other.

Learning check

(LC5.5) Fit a new linear regression using `lm(gdpPercap ~ continent, data = gapminder2007)` where `gdpPercap` is the new outcome variable y. Get information about the "best-fitting" line from the regression table by applying the `get_regression_table()` function. How do the regression results match up with the results from your previous exploratory data analysis?

5.2.3 Observed/fitted values and residuals

Recall in Subsection 5.1.3, we defined the following three concepts:

1. Observed values y, or the observed value of the outcome variable
2. Fitted values $\hat{y}$, or the value on the regression line for a given x value
3. Residuals $y - \hat{y}$, or the error between the observed value and the fitted value

We obtained these values and other values using the get_regression_points() function from the moderndive package. This time, however, let's add an argument setting ID = "country": this is telling the function to use the variable country in gapminder2007 as an *identification variable* in the output. This will help contextualize our analysis by matching values to countries.

```
regression_points <- get_regression_points(lifeExp_model, ID = "country")
regression_points
```

TABLE 5.9: Regression points (First 10 out of 142 countries)

country	lifeExp	continent	lifeExp_hat	residual
Afghanistan	43.8	Asia	70.7	-26.900
Albania	76.4	Europe	77.6	-1.226
Algeria	72.3	Africa	54.8	17.495
Angola	42.7	Africa	54.8	-12.075
Argentina	75.3	Americas	73.6	1.712
Australia	81.2	Oceania	80.7	0.515
Austria	79.8	Europe	77.6	2.180
Bahrain	75.6	Asia	70.7	4.907
Bangladesh	64.1	Asia	70.7	-6.666
Belgium	79.4	Europe	77.6	1.792

Observe in Table 5.9 that lifeExp_hat contains the fitted values $\hat{y} = \widehat{\text{lifeExp}}$. If you look closely, there are only 5 possible values for lifeExp_hat. These correspond to the five mean life expectancies for the 5 continents that we displayed in Table 5.7 and computed using the values in the estimate column of the regression table in Table 5.8.

The residual column is simply $y - \hat{y}$ = lifeExp - lifeExp_hat. These values can be interpreted as the deviation of a country's life expectancy from its continent's average life expectancy. For example, look at the first row of Table 5.9 corresponding to Afghanistan. The residual of $y - \hat{y} = 43.8 - 70.7 = -26.9$ is telling us that Afghanistan's life expectancy is a whopping 26.9 years lower than the mean life expectancy of all Asian countries. This can in part be explained by the many years of war that country has suffered.

Learning check

(LC5.6) Using either the sorting functionality of RStudio's spreadsheet viewer or using the data wrangling tools you learned in Chapter 3, identify the five countries with the five smallest (most negative) residuals? What do these negative residuals say about their life expectancy relative to their continents' life expectancy?

(LC5.7) Repeat this process, but identify the five countries with the five largest (most positive) residuals. What do these positive residuals say about their life expectancy relative to their continents' life expectancy?

5.3 Related topics

5.3.1 Correlation is not necessarily causation

Throughout this chapter we've been cautious when interpreting regression slope coefficients. We always discussed the "associated" effect of an explanatory variable x on an outcome variable y. For example, our statement from Subsection 5.1.2 that "for every increase of 1 unit in bty_avg, there is an *associated* increase of on average 0.067 units of score." We include the term "associated" to be extra careful not to suggest we are making a *causal* statement. So while "beauty" score of bty_avg is positively correlated with teaching score, we can't necessarily make any statements about "beauty" scores' direct causal effect on teaching score without more information on how this study was conducted. Here is another example: a not-so-great medical doctor goes through medical records and finds that patients who slept with their shoes on tended to wake up more with headaches. So this doctor declares, "Sleeping with shoes on causes headaches!"

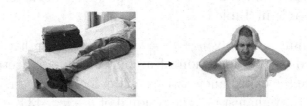

FIGURE 5.10: Does sleeping with shoes on cause headaches?

However, there is a good chance that if someone is sleeping with their shoes on, it's potentially because they are intoxicated from alcohol. Furthermore, higher levels of drinking leads to more hangovers, and hence more headaches. The amount of alcohol consumption here is what's known as a *confounding/lurking* variable. It "lurks" behind the scenes, confounding the causal relationship (if any) of "sleeping with shoes on" with "waking up with a headache." We can summarize this in Figure 5.11 with a *causal graph* where:

- Y is a *response* variable; here it is "waking up with a headache."
- X is a *treatment* variable whose causal effect we are interested in; here it is "sleeping with shoes on."

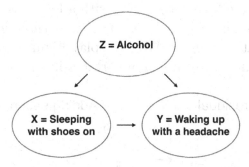

FIGURE 5.11: Causal graph.

To study the relationship between Y and X, we could use a regression model where the outcome variable is set to Y and the explanatory variable is set to be X, as you've been doing throughout this chapter. However, Figure 5.11 also includes a third variable with arrows pointing at both X and Y:

- Z is a *confounding* variable that affects both X and Y, thereby "confounding" their relationship. Here the confounding variable is alcohol.

Alcohol will cause people to be both more likely to sleep with their shoes on as well as be more likely to wake up with a headache. Thus any regression model of the relationship between X and Y should also use Z as an explanatory variable. In other words, our doctor needs to take into account who had been drinking the night before. In the next chapter, we'll start covering multiple regression models that allow us to incorporate more than one variable in our regression models.

Establishing causation is a tricky problem and frequently takes either carefully designed experiments or methods to control for the effects of confounding variables. Both these approaches attempt, as best they can, either to take all possible confounding variables into account or negate their impact. This allows researchers to focus only on the relationship of interest: the relationship between the outcome variable Y and the treatment variable X.

As you read news stories, be careful not to fall into the trap of thinking that correlation necessarily implies causation. Check out the Spurious Correlations[5] website for some rather comical examples of variables that are correlated, but are definitely not causally related.

5.3.2 Best-fitting line

Regression lines are also known as "best-fitting" lines. But what do we mean by "best"? Let's unpack the criteria that is used in regression to determine "best." Recall Figure 5.6, where for an instructor with a beauty score of $x = 7.333$ we mark the *observed value* y with a circle, the *fitted value* $\hat{y}$ with a square, and the *residual* $y - \hat{y}$ with an arrow. We re-display Figure 5.6 in the top-left plot of Figure 5.12 in addition to three more arbitrarily chosen course instructors:

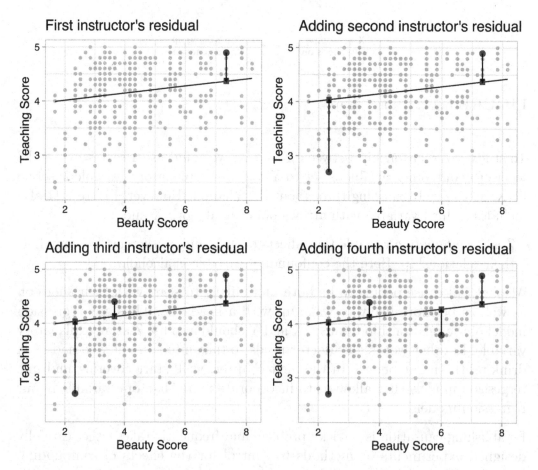

FIGURE 5.12: Example of observed value, fitted value, and residual.

---

[5]http://www.tylervigen.com/spurious-correlations

The three other plots refer to:

1. A course whose instructor had a "beauty" score $x = 2.333$ and teaching score $y = 2.7$. The residual in this case is $2.7 - 4.036 = -1.336$, which we mark with a new arrow in the top-right plot.
2. A course whose instructor had a "beauty" score $x = 3.667$ and teaching score $y = 4.4$. The residual in this case is $4.4 - 4.125 = 0.2753$, which we mark with a new arrow in the bottom-left plot.
3. A course whose instructor had a "beauty" score $x = 6$ and teaching score $y = 3.8$. The residual in this case is $3.8 - 4.28 = -0.4802$, which we mark with a new arrow in the bottom-right plot.

Now say we repeated this process of computing residuals for all 463 courses' instructors, then we squared all the residuals, and then we summed them. We call this quantity the *sum of squared residuals*; it is a measure of the *lack of fit* of a model. Larger values of the sum of squared residuals indicate a bigger lack of fit. This corresponds to a worse fitting model.

If the regression line fits all the points perfectly, then the sum of squared residuals is 0. This is because if the regression line fits all the points perfectly, then the fitted value $\hat{y}$ equals the observed value y in all cases, and hence the residual $y - \hat{y} = 0$ in all cases, and the sum of even a large number of 0's is still 0.

Furthermore, of all possible lines we can draw through the cloud of 463 points, the regression line minimizes this value. In other words, the regression and its corresponding fitted values $\hat{y}$ minimizes the sum of the squared residuals:

$$\sum_{i=1}^{n} (y_i - \hat{y}_i)^2$$

Let's use our data wrangling tools from Chapter 3 to compute the sum of squared residuals exactly:

```
# Fit regression model:
score_model <- lm(score ~ bty_avg,
                  data = evals_ch5)


# Get regression points:
regression_points <- get_regression_points(score_model)
regression_points
```

```
# A tibble: 463 x 5
```

```
      ID score bty_avg score_hat residual
    <int> <dbl>   <dbl>     <dbl>    <dbl>
1     1   4.7       5        4.21    0.486
2     2   4.1       5        4.21   -0.114
3     3   3.9       5        4.21   -0.314
4     4   4.8       5        4.21    0.586
5     5   4.6       3        4.08    0.52
6     6   4.3       3        4.08    0.22
7     7   2.8       3        4.08   -1.28
8     8   4.1    3.33        4.10   -0.002
9     9   3.4    3.33        4.10   -0.702
10   10   4.5    3.17        4.09    0.409
# ... with 453 more rows
```

```
# Compute sum of squared residuals
regression_points %>%
  mutate(squared_residuals = residual^2) %>%
  summarize(sum_of_squared_residuals = sum(squared_residuals))
```

```
# A tibble: 1 x 1
  sum_of_squared_residuals
                     <dbl>
1                     132.
```

Any other straight line drawn in the figure would yield a sum of squared residuals greater than 132. This is a mathematically guaranteed fact that you can prove using calculus and linear algebra. That's why alternative names for the linear regression line are the *best-fitting line* and the *least-squares line*. Why do we square the residuals (i.e., the arrow lengths)? So that both positive and negative deviations of the same amount are treated equally.

Learning check

(**LC5.8**) Note in Figure 5.13 there are 3 points marked with dots and:

- The "best" fitting solid regression line
- An arbitrarily chosen dotted line
- Another arbitrarily chosen dashed line

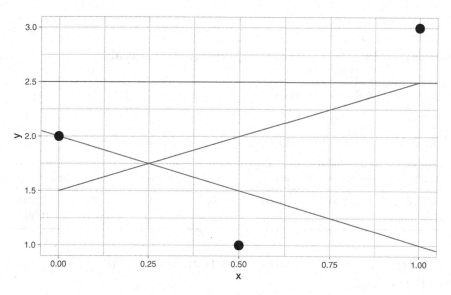

FIGURE 5.13: Regression line and two others.

Compute the sum of squared residuals by hand for each line and show that of these three lines, the regression line has the smallest value.

5.3.3 `get_regression_x()` functions

Recall in this chapter we introduced two functions from the `moderndive` package:

1. `get_regression_table()` that returns a regression table in Subsection 5.1.2 and
2. `get_regression_points()` that returns point-by-point information from a regression model in Subsection 5.1.3.

What is going on behind the scenes with the `get_regression_table()` and `get_regression_points()` functions? We mentioned in Subsection 5.1.2 that these were examples of *wrapper functions*. Such functions take other pre-existing functions and "wrap" them into single functions that hide the user from their inner workings. This way all the user needs to worry about is what the inputs look like and what the outputs look like. In this subsection, we'll "get under the hood" of these functions and see how the "engine" of these wrapper functions works.

Recall our two-step process to generate a regression table from Subsection 5.1.2:

```
# Fit regression model:
score_model <- lm(formula = score ~ bty_avg, data = evals_ch5)
# Get regression table:
get_regression_table(score_model)
```

TABLE 5.10: Regression table

term	estimate	std_error	statistic	p_value	lower_ci	upper_ci
intercept	3.880	0.076	50.96	0	3.731	4.030
bty_avg	0.067	0.016	4.09	0	0.035	0.099

The `get_regression_table()` wrapper function takes two pre-existing functions in other R packages:

- `tidy()` from the `broom` package[6] (Robinson and Hayes, 2019) and
- `clean_names()` from the `janitor` package[7] (Firke, 2019)

and "wraps" them into a single function that takes in a saved `lm()` linear model model, here `score_model`, and returns a regression table saved as a "tidy" data frame. Here is how we used the `tidy()` and `clean_names()` functions to produce Table 5.11:

```
library(broom)
library(janitor)
score_model %>%
  tidy(conf.int = TRUE) %>%
  mutate_if(is.numeric, round, digits = 3) %>%
  clean_names() %>%
  rename(lower_ci = conf_low, upper_ci = conf_high)
```

TABLE 5.11: Regression table using tidy() from broom package

term	estimate	std_error	statistic	p_value	lower_ci	upper_ci
(Intercept)	3.880	0.076	50.96	0	3.731	4.030
bty_avg	0.067	0.016	4.09	0	0.035	0.099

Yikes! That's a lot of code! So, in order to simplify your lives, we made the editorial decision to "wrap" all the code into `get_regression_table()`, freeing

---

[6]https://broom.tidyverse.org/

[7]https://github.com/sfirke/janitor

you from the need to understand the inner workings of the function. Note that the `mutate_if()` function is from the `dplyr` package and applies the `round()` function to three significant digits precision only to those variables that are numerical.

Similarly, the `get_regression_points()` function is another wrapper function, but this time returning information about the individual points involved in a regression model like the fitted values, observed values, and the residuals. `get_regression_points()` uses the `augment()` function in the `broom` package[8] instead of the `tidy()` function as with `get_regression_table()` to produce the data shown in Table 5.12:

```
library(broom)
library(janitor)
score_model %>%
  augment() %>%
  mutate_if(is.numeric, round, digits = 3) %>%
  clean_names() %>%
  select(-c("se_fit", "hat", "sigma", "cooksd", "std_resid"))
```

TABLE 5.12: Regression points using augment() from broom package

score	bty_avg	fitted	resid
4.7	5.00	4.21	0.486
4.1	5.00	4.21	-0.114
3.9	5.00	4.21	-0.314
4.8	5.00	4.21	0.586
4.6	3.00	4.08	0.520
4.3	3.00	4.08	0.220
2.8	3.00	4.08	-1.280
4.1	3.33	4.10	-0.002
3.4	3.33	4.10	-0.702
4.5	3.17	4.09	0.409

In this case, it outputs only the variables of interest to students learning regression: the outcome variable y (`score`), all explanatory/predictor variables (`bty_avg`), all resulting `fitted` values $\hat{y}$ used by applying the equation of the regression line to `bty_avg`, and the `residual` $y - \hat{y}$.

If you're even more curious about how these and other wrapper functions work, take a look at the source code for these functions on GitHub[9].

---

[8]https://broom.tidyverse.org/

[9]https://github.com/moderndive/moderndive/blob/master/R/regression_functions.R

5.4   Conclusion

5.4.1   Additional resources

Solutions to all *Learning checks* can be found online in Appendix D[10].

An R script file of all R code used in this chapter is available at `https://www.moderndive.com/scripts/05-regression.R`.

As we suggested in Subsection 5.1.1, interpreting coefficients that are not close to the extreme values of -1, 0, and 1 can be somewhat subjective. To help develop your sense of correlation coefficients, we suggest you play the 80s-style video game called, "Guess the Correlation", at `http://guessthecorrelation.com/`.

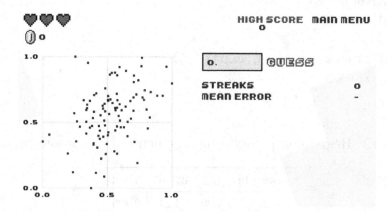

FIGURE 5.14: Preview of "Guess the Correlation" game.

5.4.2   What's to come?

In this chapter, you've studied the term *basic regression*, where you fit models that only have one explanatory variable. In Chapter 6, we'll study *multiple regression*, where our regression models can now have more than one explanatory variable! In particular, we'll consider two scenarios: regression models with one numerical and one categorical explanatory variable and regression models with two numerical explanatory variables. This will allow you to construct more sophisticated and more powerful models, all in the hopes of better explaining your outcome variable y.

---

[10]`https://moderndive.com/D-appendixD.html`

6

Multiple Regression

In Chapter 5 we introduced ideas related to modeling for explanation, in particular that the goal of modeling is to make explicit the relationship between some outcome variable y and some explanatory variable x. While there are many approaches to modeling, we focused on one particular technique: *linear regression*, one of the most commonly used and easy-to-understand approaches to modeling. Furthermore to keep things simple, we only considered models with one explanatory x variable that was either numerical in Section 5.1 or categorical in Section 5.2.

In this chapter on multiple regression, we'll start considering models that include more than one explanatory variable x. You can imagine when trying to model a particular outcome variable, like teaching evaluation scores as in Section 5.1 or life expectancy as in Section 5.2, that it would be useful to include more than just one explanatory variable's worth of information.

Since our regression models will now consider more than one explanatory variable, the interpretation of the associated effect of any one explanatory variable must be made in conjunction with the other explanatory variables included in your model. Let's begin!

Needed packages

Let's load all the packages needed for this chapter (this assumes you've already installed them). Recall from our discussion in Section 4.4 that loading the tidyverse package by running library(tidyverse) loads the following commonly used data science packages all at once:

- ggplot2 for data visualization
- dplyr for data wrangling
- tidyr for converting data to "tidy" format
- readr for importing spreadsheet data into R
- As well as the more advanced purrr, tibble, stringr, and forcats packages

If needed, read Section 1.3 for information on how to install and load R packages.

```
library(tidyverse)
library(moderndive)
library(skimr)
library(ISLR)
```

6.1   One numerical and one categorical explanatory variable

Let's revisit the instructor evaluation data from UT Austin we introduced in Section 5.1. We studied the relationship between teaching evaluation scores as given by students and "beauty" scores. The variable teaching score was the numerical outcome variable y, and the variable "beauty" score (bty_avg) was the numerical explanatory x variable.

In this section, we are going to consider a different model. Our outcome variable will still be teaching score, but we'll now include two different explanatory variables: age and (binary) gender. Could it be that instructors who are older receive better teaching evaluations from students? Or could it instead be that younger instructors receive better evaluations? Are there differences in evaluations given by students for instructors of different genders? We'll answer these questions by modeling the relationship between these variables using *multiple regression*, where we have:

1. A numerical outcome variable y, the instructor's teaching score, and
2. Two explanatory variables:
    1. A numerical explanatory variable x_1, the instructor's age.
    2. A categorical explanatory variable x_2, the instructor's (binary) gender.

It is important to note that at the time of this study due to then commonly held beliefs about gender, this variable was often recorded as a binary variable. While the results of a model that oversimplifies gender this way may be imperfect, we still found the results to be pertinent and relevant today.

6.1.1   Exploratory data analysis

Recall that data on the 463 courses at UT Austin can be found in the evals data frame included in the moderndive package. However, to keep things simple, let's select() only the subset of the variables we'll consider in this chapter, and save this data in a new data frame called `evals_ch6`. Note that these are different than the variables chosen in Chapter 5.

```
evals_ch6 <- evals %>%
  select(ID, score, age, gender)
```

Recall the three common steps in an exploratory data analysis we saw in Subsection 5.1.1:

1. Looking at the raw data values.
2. Computing summary statistics.
3. Creating data visualizations.

Let's first look at the raw data values by either looking at `evals_ch6` using RStudio's spreadsheet viewer or by using the `glimpse()` function from the `dplyr` package:

```
glimpse(evals_ch6)
```

```
Observations: 463
Variables: 4
$ ID     <int> 1, 2, 3, 4, 5, 6, 7, 8, 9, 10, 11, 12, 13, 14, 15, 16, ...
$ score  <dbl> 4.7, 4.1, 3.9, 4.8, 4.6, 4.3, 2.8, 4.1, 3.4, 4.5, 3.8, ...
$ age    <int> 36, 36, 36, 36, 59, 59, 59, 51, 51, 40, 40, 40, 40, 40,...
$ gender <fct> female, female, female, female, male, male, male, male,...
```

Let's also display a random sample of 5 rows of the 463 rows corresponding to different courses in Table 6.1. Remember due to the random nature of the sampling, you will likely end up with a different subset of 5 rows.

```
evals_ch6 %>% sample_n(size = 5)
```

TABLE 6.1: A random sample of 5 out of the 463 courses at UT Austin

ID	score	age	gender
129	3.7	62	male
109	4.7	46	female
28	4.8	62	male
434	2.8	62	male
330	4.0	64	male

Now that we've looked at the raw values in our `evals_ch6` data frame and got a sense of the data, let's compute summary statistics. As we did in our exploratory data analyses in Sections 5.1.1 and 5.2.1 from the previous chapter, let's use the `skim()` function from the `skimr` package, being sure to only `select()` the variables of interest in our model:

```
evals_ch6 %>% select(score, age, gender) %>% skim()
```

```
Skim summary statistics
 n obs: 463
 n variables: 3

── Variable type:factor
 variable missing complete   n n_unique                  top_counts ordered
   gender       0      463 463        2 mal: 268, fem: 195, NA: 0   FALSE

── Variable type:integer
 variable missing complete   n  mean  sd p0 p25 p50 p75 p100
      age       0      463 463 48.37 9.8 29  42  48  57   73

── Variable type:numeric
 variable missing complete   n mean   sd  p0 p25 p50 p75 p100
    score       0      463 463 4.17 0.54 2.3 3.8 4.3 4.6    5
```

Observe that we have no missing data, that there are 268 courses taught by male instructors and 195 courses taught by female instructors, and that the average instructor age is 48.37. Recall that each row represents a particular course and that the same instructor often teaches more than one course. Therefore, the average age of the unique instructors may differ.

Furthermore, let's compute the correlation coefficient between our two numerical variables: `score` and `age`. Recall from Subsection 5.1.1 that correlation coefficients only exist between numerical variables. We observe that they are "weakly negatively" correlated.

```
evals_ch6 %>%
  get_correlation(formula = score ~ age)
```

```
# A tibble: 1 x 1
     cor
   <dbl>
1 -0.107
```

Let's now perform the last of the three common steps in an exploratory data analysis: creating data visualizations. Given that the outcome variable score and explanatory variable age are both numerical, we'll use a scatterplot to display their relationship. How can we incorporate the categorical variable gender, however? By mapping the variable gender to the color aesthetic, thereby creating a *colored* scatterplot. The following code is similar to the code that created the scatterplot of teaching score over "beauty" score in Figure 5.2, but with color = gender added to the aes()thetic mapping.

```
ggplot(evals_ch6, aes(x = age, y = score, color = gender)) +
  geom_point() +
  labs(x = "Age", y = "Teaching Score", color = "Gender") +
  geom_smooth(method = "lm", se = FALSE)
```

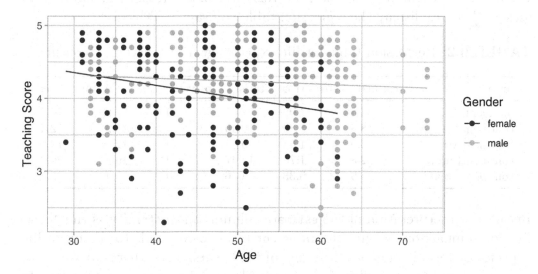

FIGURE 6.1: Colored scatterplot of relationship of teaching and beauty scores.

In the resulting Figure 6.1, observe that ggplot() assigns a default color scheme to the points and to the lines associated with the two levels of gender: female and male. Furthermore, the geom_smooth(method = "lm", se = FALSE) layer automatically fits a different regression line for each group.

We notice some interesting trends. First, there are almost no women faculty over the age of 60 as evidenced by lack of darker-colored dots above $x = 60$. Second, while both regression lines are negatively sloped with age (i.e., older instructors tend to have lower scores), the slope for age for the female instructors is *more* negative. In other words, female instructors are paying a harsher penalty for advanced age than the male instructors.

6.1.2 Interaction model

Let's now quantify the relationship of our outcome variable y and the two explanatory variables using one type of multiple regression model known as an *interaction model*. We'll explain where the term "interaction" comes from at the end of this section.

In particular, we'll write out the equation of the two regression lines in Figure 6.1 using the values from a regression table. Before we do this, however, let's go over a brief refresher of regression when you have a categorical explanatory variable x.

Recall in Subsection 5.2.2 we fit a regression model for countries' life expectancies as a function of which continent the country was in. In other words, we had a numerical outcome variable $y =$ lifeExp and a categorical explanatory variable $x =$ continent which had 5 levels: Africa, Americas, Asia, Europe, and Oceania. Let's re-display the regression table you saw in Table 5.8:

TABLE 6.2: Regression table for life expectancy as a function of continent

term	estimate	std_error	statistic	p_value	lower_ci	upper_ci
intercept	54.8	1.02	53.45	0	52.8	56.8
continentAmericas	18.8	1.80	10.45	0	15.2	22.4
continentAsia	15.9	1.65	9.68	0	12.7	19.2
continentEurope	22.8	1.70	13.47	0	19.5	26.2
continentOceania	25.9	5.33	4.86	0	15.4	36.5

Recall our interpretation of the estimate column. Since Africa was the "baseline for comparison" group, the intercept term corresponds to the mean life expectancy for all countries in Africa of 54.8 years. The other four values of estimate correspond to "offsets" relative to the baseline group. So, for example, the "offset" corresponding to the Americas is +18.8 as compared to the baseline for comparison group Africa. In other words, the average life expectancy for countries in the Americas is 18.8 years *higher*. Thus the mean life expectancy for all countries in the Americas is $54.8 + 18.8 = 73.6$. The same interpretation holds for Asia, Europe, and Oceania.

Going back to our multiple regression model for teaching score using age and gender in Figure 6.1, we generate the regression table using the same two-step approach from Chapter 5: we first "fit" the model using the lm() "linear model" function and then we apply the get_regression_table() function. This time, however, our model formula won't be of the form y ~ x, but rather of the form y ~ x1 * x2. In other words, our two explanatory variables x1 and x2 are separated by a * sign:

```
# Fit regression model:
score_model_interaction <- lm(score ~ age * gender, data = evals_ch6)

# Get regression table:
get_regression_table(score_model_interaction)
```

TABLE 6.3: Regression table for interaction model

term	estimate	std_error	statistic	p_value	lower_ci	upper_ci
intercept	4.883	0.205	23.80	0.000	4.480	5.286
age	-0.018	0.004	-3.92	0.000	-0.026	-0.009
gendermale	-0.446	0.265	-1.68	0.094	-0.968	0.076
age:gendermale	0.014	0.006	2.45	0.015	0.003	0.024

Looking at the regression table output in Table 6.3, there are four rows of values in the estimate column. While it is not immediately apparent, using these four values we can write out the equations of both lines in Figure 6.1. First, since the word female comes alphabetically before male, female instructors are the "baseline for comparison" group. Thus, intercept is the intercept *for only the female instructors.*

This holds similarly for age. It is the slope for age *for only the female instructors.* Thus, the darker-colored regression line in Figure 6.1 has an intercept of 4.883 and slope for age of -0.018. Remember that for this data, while the intercept has a mathematical interpretation, it has no *practical* interpretation since instructors can't have zero age.

What about the intercept and slope for age of the male instructors in the lighter-colored line in Figure 6.1? This is where our notion of "offsets" comes into play once again.

The value for gendermale of -0.446 is not the intercept for the male instructors, but rather the *offset* in intercept for male instructors relative to female instructors. The intercept for the male instructors is intercept + gendermale = 4.883 + (-0.446) = 4.883 - 0.446 = 4.437.

Similarly, age:gendermale = 0.014 is not the slope for age for the male instructors, but rather the *offset* in slope for the male instructors. Therefore, the slope for age for the male instructors is age + age:gendermale = $-0.018 + 0.014 = -0.004$. Thus, the lighter-colored regression line in Figure 6.1 has intercept 4.437 and slope for age of -0.004. Let's summarize these values in Table 6.4 and focus on the two slopes for age:

TABLE 6.4: Comparison of intercepts and slopes for interaction model

Gender	Intercept	Slope for age
Female instructors	4.883	-0.018
Male instructors	4.437	-0.004

Since the slope for age for the female instructors was -0.018, it means that on average, a female instructor who is a year older would have a teaching score that is 0.018 units **lower**. For the male instructors, however, the corresponding associated decrease was on average only 0.004 units. While both slopes for age were negative, the slope for age for the female instructors is *more negative*. This is consistent with our observation from Figure 6.1, that this model is suggesting that age impacts teaching scores for female instructors more than for male instructors.

Let's now write the equation for our regression lines, which we can use to compute our fitted values $\hat{y} = \widehat{score}$.

$$\hat{y} = \widehat{score} = b_0 + b_{age} \cdot age + b_{male} \cdot \mathbb{1}_{is\ male}(x) + b_{age,male} \cdot age \cdot \mathbb{1}_{is\ male}$$
$$= 4.883 - 0.018 \cdot age - 0.446 \cdot \mathbb{1}_{is\ male}(x) + 0.014 \cdot age \cdot \mathbb{1}_{is\ male}$$

Whoa! That's even more daunting than the equation you saw for the life expectancy as a function of continent in Subsection 5.2.2! However, if you recall what an "indicator function" does, the equation simplifies greatly. In the previous equation, we have one indicator function of interest:

$$\mathbb{1}_{is\ male}(x) = \begin{cases} 1 & \text{if instructor } x \text{ is male} \\ 0 & \text{otherwise} \end{cases}$$

Second, let's match coefficients in the previous equation with values in the estimate column in our regression table in Table 6.3:

1. b_0 is the intercept = 4.883 for the female instructors
2. b_{age} is the slope for age = -0.018 for the female instructors
3. b_{male} is the offset in intercept = -0.446 for the male instructors
4. $b_{age,male}$ is the offset in slope for age = 0.014 for the male instructors

Let's put this all together and compute the fitted value $\hat{y} = \widehat{score}$ for female instructors. Since for female instructors $\mathbb{1}_{is\ male}(x) = 0$, the previous equation becomes

$$\hat{y} = \widehat{\text{score}} = 4.883 - 0.018 \cdot \text{age} - 0.446 \cdot 0 + 0.014 \cdot \text{age} \cdot 0$$
$$= 4.883 - 0.018 \cdot \text{age} - 0 + 0$$
$$= 4.883 - 0.018 \cdot \text{age}$$

which is the equation of the darker-colored regression line in Figure 6.1 corresponding to the female instructors in Table 6.4. Correspondingly, since for male instructors $\mathbb{1}_{\text{is male}}(x) = 1$, the previous equation becomes

$$\hat{y} = \widehat{\text{score}} = 4.883 - 0.018 \cdot \text{age} - 0.446 + 0.014 \cdot \text{age}$$
$$= (4.883 - 0.446) + (-0.018 + 0.014) * \text{age}$$
$$= 4.437 - 0.004 \cdot \text{age}$$

which is the equation of the lighter-colored regression line in Figure 6.1 corresponding to the male instructors in Table 6.4.

Phew! That was a lot of arithmetic! Don't fret, however, this is as hard as modeling will get in this book. If you're still a little unsure about using indicator functions and using categorical explanatory variables in a regression model, we *highly* suggest you re-read Subsection 5.2.2. This involves only a single categorical explanatory variable and thus is much simpler.

Before we end this section, we explain why we refer to this type of model as an "interaction model." The $b_{\text{age,male}}$ term in the equation for the fitted value $\hat{y} = \widehat{\text{score}}$ is what's known in statistical modeling as an "interaction effect." The interaction term corresponds to the `age:gendermale` = 0.014 in the final row of the regression table in Table 6.3.

We say there is an interaction effect if the associated effect of one variable *depends on the value of another variable.* That is to say, the two variables are "interacting" with each other. Here, the associated effect of the variable age *depends* on the value of the other variable gender. The difference in slopes for age of +0.014 of male instructors relative to female instructors shows this.

Another way of thinking about interaction effects on teaching scores is as follows. For a given instructor at UT Austin, there might be an associated effect of their age *by itself*, there might be an associated effect of their gender *by itself*, but when age and gender are considered *together* there might be an *additional effect* above and beyond the two individual effects.

6.1.3 Parallel slopes model

When creating regression models with one numerical and one categorical explanatory variable, we are not just limited to interaction models as we just saw. Another type of model we can use is known as a *parallel slopes* model. Unlike interaction models where the regression lines can have different intercepts and different slopes, parallel slopes models still allow for different intercepts but *force* all lines to have the same slope. The resulting regression lines are thus parallel. Let's visualize the best-fitting parallel slopes model to evals_ch6.

Unfortunately, the geom_smooth() function in the ggplot2 package does not have a convenient way to plot parallel slopes models. Evgeni Chasnovski thus created a special purpose function called geom_parallel_slopes() that is included in the moderndive package. You won't find geom_parallel_slopes() in the ggplot2 package, but rather the moderndive package. Thus, if you want to be able to use it, you will need to load both the ggplot2 and moderndive packages. Using this function, let's now plot the parallel slopes model for teaching score. Notice how the code is identical to the code that produced the visualization of the interaction model in Figure 6.1, but now the geom_smooth(method = "lm", se = FALSE) layer is replaced with geom_parallel_slopes(se = FALSE).

```
ggplot(evals_ch6, aes(x = age, y = score, color = gender)) +
  geom_point() +
  labs(x = "Age", y = "Teaching Score", color = "Gender") +
  geom_parallel_slopes(se = FALSE)
```

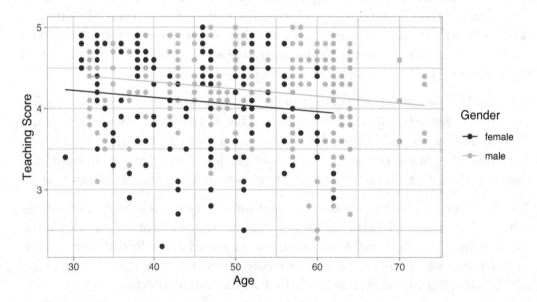

FIGURE 6.2: Parallel slopes model of score with age and gender.

Observe in Figure 6.2 that we now have parallel lines corresponding to the female and male instructors, respectively: here they have the same negative slope. This is telling us that instructors who are older will tend to receive lower teaching scores than instructors who are younger. Furthermore, since the lines are parallel, the associated penalty for being older is assumed to be the same for both female and male instructors.

However, observe also in Figure 6.2 that these two lines have different intercepts as evidenced by the fact that the lighter-colored line corresponding to the male instructors is higher than the darker-colored line corresponding to the female instructors. This is telling us that irrespective of age, female instructors tended to receive lower teaching scores than male instructors.

In order to obtain the precise numerical values of the two intercepts and the single common slope, we once again "fit" the model using the lm() "linear model" function and then apply the get_regression_table() function. However, unlike the interaction model which had a model formula of the form y ~ x1 * x2, our model formula is now of the form y ~ x1 + x2. In other words, our two explanatory variables x1 and x2 are separated by a + sign:

```
# Fit regression model:
score_model_parallel_slopes <- lm(score ~ age + gender, data = evals_ch6)
# Get regression table:
get_regression_table(score_model_parallel_slopes)
```

TABLE 6.5: Regression table for parallel slopes model

term	estimate	std_error	statistic	p_value	lower_ci	upper_ci
intercept	4.484	0.125	35.79	0.000	4.238	4.730
age	-0.009	0.003	-3.28	0.001	-0.014	-0.003
gendermale	0.191	0.052	3.63	0.000	0.087	0.294

Similarly to the regression table for the interaction model from Table 6.3, we have an intercept term corresponding to the intercept for the "baseline for comparison" female instructor group and a gendermale term corresponding to the *offset* in intercept for the male instructors relative to female instructors. In other words, in Figure 6.2 the darker-colored regression line corresponding to the female instructors has an intercept of 4.484 while the lighter-colored regression line corresponding to the male instructors has an intercept of 4.484 + 0.191 = 4.675. Once again, since there aren't any instructors of age 0, the intercepts only have a mathematical interpretation but no practical one.

Unlike in Table 6.3, however, we now only have a single slope for age of -0.009. This is because the model dictates that both the female and male instructors have a common slope for age. This is telling us that an instructor who is a year older than another instructor received a teaching score that is on average 0.009 units *lower*. This penalty for being of advanced age applies equally to both female and male instructors.

Let's summarize these values in Table 6.6, noting the different intercepts but common slopes:

TABLE 6.6: Comparison of intercepts and slope for parallel slopes model

Gender	Intercept	Slope for age
Female instructors	4.484	-0.009
Male instructors	4.675	-0.009

Let's now write the equation for our regression lines, which we can use to compute our fitted values $\hat{y} = \widehat{\text{score}}$.

$$\hat{y} = \widehat{\text{score}} = b_0 + b_{\text{age}} \cdot \text{age} + b_{\text{male}} \cdot \mathbb{1}_{\text{is male}}(x)$$
$$= 4.484 - 0.009 \cdot \text{age} + 0.191 \cdot \mathbb{1}_{\text{is male}}(x)$$

Let's put this all together and compute the fitted value $\hat{y} = \widehat{\text{score}}$ for female instructors. Since for female instructors the indicator function $\mathbb{1}_{\text{is male}}(x) = 0$, the previous equation becomes

$$\hat{y} = \widehat{\text{score}} = 4.484 - 0.009 \cdot \text{age} + 0.191 \cdot 0$$
$$= 4.484 - 0.009 \cdot \text{age}$$

which is the equation of the darker-colored regression line in Figure 6.2 corresponding to the female instructors. Correspondingly, since for male instructors the indicator function $\mathbb{1}_{\text{is male}}(x) = 1$, the previous equation becomes

$$\hat{y} = \widehat{\text{score}} = 4.484 - 0.009 \cdot \text{age} + 0.191 \cdot 1$$
$$= (4.484 + 0.191) - 0.009 \cdot \text{age}$$
$$= 4.675 - 0.009 \cdot \text{age}$$

which is the equation of the lighter-colored regression line in Figure 6.2 corresponding to the male instructors.

Great! We've considered both an interaction model and a parallel slopes model for our data. Let's compare the visualizations for both models side-by-side in Figure 6.3.

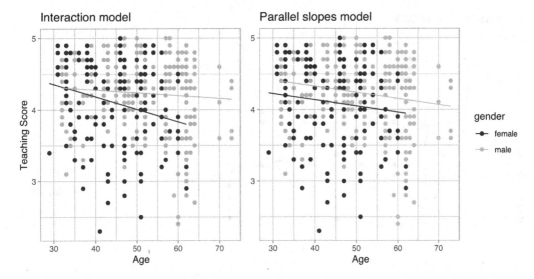

FIGURE 6.3: Comparison of interaction and parallel slopes models.

At this point, you might be asking yourself: "Why would we ever use a parallel slopes model?". Looking at the left-hand plot in Figure 6.3, the two lines definitely do not appear to be parallel, so why would we *force* them to be parallel? For this data, we agree! It can easily be argued that the interaction model on the left is more appropriate. However, in the upcoming Subsection 6.3.1 on model selection, we'll present an example where it can be argued that the case for a parallel slopes model might be stronger.

6.1.4   Observed/fitted values and residuals

For brevity's sake, in this section we'll only compute the observed values, fitted values, and residuals for the interaction model which we saved in `score_model_interaction`. You'll have an opportunity to study the corresponding values for the parallel slopes model in the upcoming *Learning check*.

Say, you have an instructor who identifies as female and is 36 years old. What fitted value $\hat{y} = \widehat{score}$ would our model yield? Say, you have another instructor who identifies as male and is 59 years old. What would their fitted value $\hat{y}$ be?

We answer this question visually first for the female instructor by finding the intersection of the darker-colored regression line and the vertical line at $x = $ age $= 36$. We mark this value with a large darker-colored dot in Figure 6.4. Similarly, we can identify the fitted value $\hat{y} = \widehat{score}$ for the male instructor by finding the intersection of the lighter-colored regression line and the vertical line at $x = $ age $= 59$. We mark this value with a large lighter-colored dot in Figure 6.4.

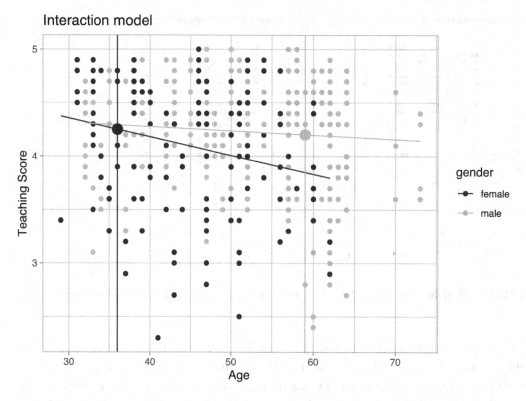

FIGURE 6.4: Fitted values for two new professors.

What are these two values of $\hat{y} = \widehat{score}$ precisely? We can use the equations of the two regression lines we computed in Subsection 6.1.2, which in turn were based on values from the regression table in Table 6.3:

- For all female instructors: $\hat{y} = \widehat{score} = 4.883 - 0.018 \cdot age$
- For all male instructors: $\hat{y} = \widehat{score} = 4.437 - 0.004 \cdot age$

So our fitted values would be: $4.883 - 0.018 \cdot 36 = 4.25$ and $4.437 - 0.004 \cdot 59 = 4.20$, respectively.

Now what if we want the fitted values not just for these two instructors, but for the instructors of all 463 courses included in the evals_ch6 data frame? Doing this by hand would be long and tedious! This is where the get_regression_points() function from the moderndive package can help: it will quickly automate the above calculations for all 463 courses. We present a preview of just the first 10 rows out of 463 in Table 6.7.

```
regression_points <- get_regression_points(score_model_interaction)
regression_points
```

TABLE 6.7: Regression points (First 10 out of 463 courses)

ID	score	age	gender	score_hat	residual
1	4.7	36	female	4.25	0.448
2	4.1	36	female	4.25	-0.152
3	3.9	36	female	4.25	-0.352
4	4.8	36	female	4.25	0.548
5	4.6	59	male	4.20	0.399
6	4.3	59	male	4.20	0.099
7	2.8	59	male	4.20	-1.401
8	4.1	51	male	4.23	-0.133
9	3.4	51	male	4.23	-0.833
10	4.5	40	female	4.18	0.318

It turns out that the female instructor of age 36 taught the first four courses, while the male instructor taught the next 3. The resulting $\hat{y} = \widehat{score}$ fitted values are in the score_hat column. Furthermore, the get_regression_points() function also returns the residuals $y - \hat{y}$. Notice, for example, the first and fourth courses the female instructor of age 36 taught had positive residuals, indicating that the actual teaching scores they received from students were greater than their fitted score of 4.25. On the other hand, the second and third courses this instructor taught had negative residuals, indicating that the actual teaching scores they received from students were less than 4.25.

Learning check

(LC6.1) Compute the observed values, fitted values, and residuals not for the interaction model as we just did, but rather for the parallel slopes model we saved in score_model_interaction.

6.2 Two numerical explanatory variables

Let's now switch gears and consider multiple regression models where instead of one numerical and one categorical explanatory variable, we now have two numerical explanatory variables. The dataset we'll use is from *An Introduction to Statistical Learning with Applications in R (ISLR)*[1], an intermediate-level textbook on statistical and machine learning (James et al., 2017). Its accompanying ISLR R package contains the datasets to which the authors apply various machine learning methods.

---

[1]http://www-bcf.usc.edu/~gareth/ISL/

One frequently used dataset in this book is the Credit dataset, where the outcome variable of interest is the credit card debt of 400 individuals. Other variables like income, credit limit, credit rating, and age are included as well. Note that the Credit data is not based on real individuals' financial information, but rather is a simulated dataset used for educational purposes.

In this section, we'll fit a regression model where we have

1. A numerical outcome variable y, the cardholder's credit card debt
2. Two explanatory variables:
    1. One numerical explanatory variable x_1, the cardholder's credit limit
    2. Another numerical explanatory variable x_2, the cardholder's income (in thousands of dollars).

6.2.1 Exploratory data analysis

Let's load the Credit dataset. To keep things simple let's select() the subset of the variables we'll consider in this chapter, and save this data in the new data frame credit_ch6. Notice our slightly different use of the select() verb here than we introduced in Subsection 3.8.1. For example, we'll select the Balance variable from Credit but then save it with a new variable name debt. We do this because here the term "debt" is easier to interpret than "balance."

```
library(ISLR)
credit_ch6 <- Credit %>% as_tibble() %>%
  select(ID, debt = Balance, credit_limit = Limit,
         income = Income, credit_rating = Rating, age = Age)
```

You can observe the effect of our use of select() in the first common step of an exploratory data analysis: looking at the raw values either in RStudio's spreadsheet viewer or by using glimpse().

```
glimpse(credit_ch6)
```

```
Observations: 400
Variables: 6
$ ID            <int> 1, 2, 3, 4, 5, 6, 7, 8, 9, 10, 11, 12, 13, 14, 1...
$ debt          <int> 333, 903, 580, 964, 331, 1151, 203, 872, 279, 13...
$ credit_limit  <int> 3606, 6645, 7075, 9504, 4897, 8047, 3388, 7114, ...
$ income        <dbl> 14.9, 106.0, 104.6, 148.9, 55.9, 80.2, 21.0, 71....
$ credit_rating <int> 283, 483, 514, 681, 357, 569, 259, 512, 266, 491...
```

```
$ age              <int> 34, 82, 71, 36, 68, 77, 37, 87, 66, 41, 30, 64, ...
```

Furthermore, let's look at a random sample of five out of the 400 credit card holders in Table 6.8. Once again, note that due to the random nature of the sampling, you will likely end up with a different subset of five rows.

```
credit_ch6 %>% sample_n(size = 5)
```

TABLE 6.8: Random sample of 5 credit card holders

ID	debt	credit_limit	income	credit_rating	age
272	436	4866	45.0	347	30
239	52	2910	26.5	236	58
87	815	6340	55.4	448	33
108	0	3189	39.1	263	72
149	0	2420	15.2	192	69

Now that we've looked at the raw values in our credit_ch6 data frame and got a sense of the data, let's move on to the next common step in an exploratory data analysis: computing summary statistics. Let's use the skim() function from the skimr package, being sure to only select() the columns of interest for our model:

```
credit_ch6 %>% select(debt, credit_limit, income) %>% skim()
```

```
Skim summary statistics
 n obs: 400
 n variables: 3

── Variable type:integer
  variable missing complete    n    mean      sd   p0    p25     p50     p75  p100
credit_limit    0      400  400  4735.6  2308.2  855   3088  4622.5 5872.75 13913
        debt    0      400  400  520.01  459.76    0  68.75   459.5     863   1999

── Variable type:numeric
  variable missing complete    n   mean     sd    p0    p25    p50    p75   p100
    income       0      400  400  45.22  35.24 10.35  21.01  33.12  57.47 186.63
```

Observe the summary statistics for the outcome variable debt: the mean and median credit card debt are $520.01 and $459.50, respectively, and that 25% of card holders had debts of $68.75 or less. Let's now look at one of the explanatory variables credit_limit: the mean and median credit card limit are

$4735.6 and $4622.50, respectively, while 75% of card holders had incomes of $57,470 or less.

Since our outcome variable debt and the explanatory variables credit_limit and income are numerical, we can compute the correlation coefficient between the different possible pairs of these variables. First, we can run the get_correlation() command as seen in Subsection 5.1.1 twice, once for each explanatory variable:

```
credit_ch6 %>% get_correlation(debt ~ credit_limit)
credit_ch6 %>% get_correlation(debt ~ income)
```

Or we can simultaneously compute them by returning a *correlation matrix* which we display in Table 6.9. We can see the correlation coefficient for any pair of variables by looking them up in the appropriate row/column combination.

```
credit_ch6 %>%
  select(debt, credit_limit, income) %>%
  cor()
```

TABLE 6.9: Correlation coefficients between credit card debt, credit limit, and income

	debt	credit_limit	income
debt	1.000	0.862	0.464
credit_limit	0.862	1.000	0.792
income	0.464	0.792	1.000

For example, the correlation coefficient of:

1. debt with itself is 1 as we would expect based on the definition of the correlation coefficient.
2. debt with credit_limit is 0.862. This indicates a strong positive linear relationship, which makes sense as only individuals with large credit limits can accrue large credit card debts.
3. debt with income is 0.464. This is suggestive of another positive linear relationship, although not as strong as the relationship between debt and credit_limit.
4. As an added bonus, we can read off the correlation coefficient between the two explanatory variables of credit_limit and income as 0.792.

We say there is a high degree of *collinearity* between the `credit_limit` and `income` explanatory variables. Collinearity (or multicollinearity) is a phenomenon where one explanatory variable in a multiple regression model is highly correlated with another.

So in our case since `credit_limit` and `income` are highly correlated, if we knew someone's `credit_limit`, we could make pretty good guesses about their `income` as well. Thus, these two variables provide somewhat redundant information. However, we'll leave discussion on how to work with collinear explanatory variables to a more intermediate-level book on regression modeling.

Let's visualize the relationship of the outcome variable with each of the two explanatory variables in two separate plots in Figure 6.5.

```
ggplot(credit_ch6, aes(x = credit_limit, y = debt)) +
  geom_point() +
  labs(x = "Credit limit (in $)", y = "Credit card debt (in $)",
      title = "Debt and credit limit") +
  geom_smooth(method = "lm", se = FALSE)

ggplot(credit_ch6, aes(x = income, y = debt)) +
  geom_point() +
  labs(x = "Income (in $1000)", y = "Credit card debt (in $)",
      title = "Debt and income") +
  geom_smooth(method = "lm", se = FALSE)
```

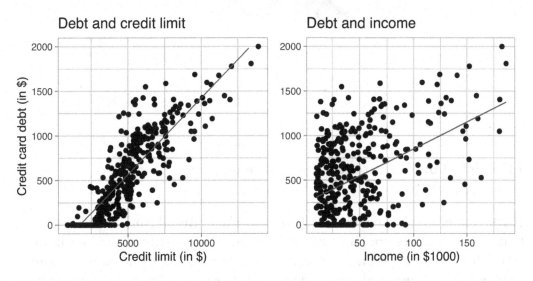

FIGURE 6.5: Relationship between credit card debt and credit limit/income.

Observe there is a positive relationship between credit limit and credit card debt: as credit limit increases so also does credit card debt. This is consistent with the strongly positive correlation coefficient of 0.862 we computed earlier. In the case of income, the positive relationship doesn't appear as strong, given the weakly positive correlation coefficient of 0.464.

However, the two plots in Figure 6.5 only focus on the relationship of the outcome variable with each of the two explanatory variables *separately*. To visualize the *joint* relationship of all three variables simultaneously, we need a 3-dimensional (3D) scatterplot as seen in Figure 6.6. Each of the 400 observations in the credit_ch6 data frame are marked with a point where

1. The numerical outcome variable y debt is on the vertical axis.
2. The two numerical explanatory variables, x_1 income and x_2 credit_limit, are on the two axes that form the bottom plane.

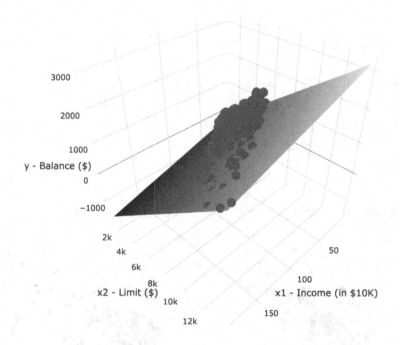

FIGURE 6.6: 3D scatterplot and regression plane.

Furthermore, we also include the *regression plane*. Recall from Subsection 5.3.2 that regression lines are "best-fitting" in that of all possible lines we can draw through a cloud of points, the regression line minimizes the *sum of squared residuals*. This concept also extends to models with two numerical explanatory variables. The difference is instead of a "best-fitting" line, we now have a "best-fitting" plane that similarly minimizes the sum of squared residuals. Head to this website[2] to open an interactive version of this plot in your browser.

> ### *Learning check*
>
> **(LC6.2)** Conduct a new exploratory data analysis with the same outcome variable y debt but with credit_rating and age as the new explanatory variables x_1 and x_2. What can you say about the relationship between a credit card holder's debt and their credit rating and age?

6.2.2 Regression plane

Let's now fit a regression model and get the regression table corresponding to the regression plane in Figure 6.6. To keep things brief in this subsection, we won't consider an interaction model for the two numerical explanatory variables income and credit_limit like we did in Subsection 6.1.2 using the model formula score ~ age * gender. Rather we'll only consider a model fit with a formula of the form y ~ x1 + x2. Confusingly, however, since we now have a regression plane instead of multiple lines, the label "parallel slopes" doesn't apply when you have two numerical explanatory variables. Just as we have done multiple times throughout Chapters 5 and this chapter, the regression table for this model using our two-step process is in Table 6.10.

```
# Fit regression model:
debt_model <- lm(debt ~ credit_limit + income, data = credit_ch6)
# Get regression table:
get_regression_table(debt_model)
```

TABLE 6.10: Multiple regression table

term	estimate	std_error	statistic	p_value	lower_ci	upper_ci
intercept	-385.179	19.465	-19.8	0	-423.446	-346.912
credit_limit	0.264	0.006	45.0	0	0.253	0.276
income	-7.663	0.385	-19.9	0	-8.420	-6.906

1. We first "fit" the linear regression model using the lm(y ~ x1 + x2, data) function and save it in debt_model.
2. We get the regression table by applying the get_regression_table() function from the moderndive package to debt_model.

---

[2]https://moderndive.com/regression-plane

Let's interpret the three values in the `estimate` column. First, the `intercept` value is -\$385.179. This intercept represents the credit card debt for an individual who has `credit_limit` of \$0 and `income` of \$0. In our data, however, the intercept has no practical interpretation since no individuals had `credit_limit` or `income` values of \$0. Rather, the intercept is used to situate the regression plane in 3D space.

Second, the `credit_limit` value is \$0.264. Taking into account all the other explanatory variables in our model, for every increase of one dollar in `credit_limit`, there is an associated increase of on average \$0.26 in credit card debt. Just as we did in Subsection 5.1.2, we are cautious *not* to imply causality as we saw in Subsection 5.3.1 that "correlation is not necessarily causation." We do this merely stating there was an *associated* increase.

Furthermore, we preface our interpretation with the statement, "taking into account all the other explanatory variables in our model." Here, by all other explanatory variables we mean `income`. We do this to emphasize that we are now jointly interpreting the associated effect of multiple explanatory variables in the same model at the same time.

Third, `income` = -\$7.66. Taking into account all other explanatory variables in our model, for every increase of one unit of `income` (\$1000 in actual income), there is an associated decrease of, on average, \$7.66 in credit card debt.

Putting these results together, the equation of the regression plane that gives us fitted values $\hat{y} = \widehat{\text{debt}}$ is:

$$\hat{y} = b_0 + b_1 \cdot x_1 + b_2 \cdot x_2$$
$$\widehat{\text{debt}} = b_0 + b_{\text{limit}} \cdot \text{limit} + b_{\text{income}} \cdot \text{income}$$
$$= -385.179 + 0.263 \cdot \text{limit} - 7.663 \cdot \text{income}$$

Recall however in the right-hand plot of Figure 6.5 that when plotting the relationship between `debt` and `income` in isolation, there appeared to be a *positive* relationship. In the last discussed multiple regression, however, when *jointly* modeling the relationship between `debt`, `credit_limit`, and `income`, there appears to be a *negative* relationship of `debt` and `income` as evidenced by the negative slope for `income` of -\$7.663. What explains these contradictory results? A phenomenon known as *Simpson's Paradox*, whereby overall trends that exist in aggregate either disappear or reverse when the data are broken down into groups. In Subsection 6.3.3 we elaborate on this idea by looking at the relationship between `credit_limit` and credit card `debt`, but split along different `income` brackets.

Learning check

(LC6.3) Fit a new simple linear regression using `lm(debt ~ credit_rating + age, data = credit_ch6)` where `credit_rating` and `age` are the new numerical explanatory variables x_1 and x_2. Get information about the "best-fitting" regression plane from the regression table by applying the `get_regression_table()` function. How do the regression results match up with the results from your previous exploratory data analysis?

6.2.3 Observed/fitted values and residuals

Let's also compute all fitted values and residuals for our regression model using the `get_regression_points()` function and present only the first 10 rows of output in Table 6.11. Remember that the coordinates of each of the points in our 3D scatterplot in Figure 6.6 can be found in the `income`, `credit_limit`, and `debt` columns. The fitted values on the regression plane are found in the `debt_hat` column and are computed using our equation for the regression plane in the previous section:

$$\hat{y} = \widehat{\text{debt}} = -385.179 + 0.263 \cdot \text{limit} - 7.663 \cdot \text{income}$$

```
get_regression_points(debt_model)
```

TABLE 6.11: Regression points (First 10 credit card holders out of 400)

ID	debt	credit_limit	income	debt_hat	residual
1	333	3606	14.9	454	-120.8
2	903	6645	106.0	559	344.3
3	580	7075	104.6	683	-103.4
4	964	9504	148.9	986	-21.7
5	331	4897	55.9	481	-150.0
6	1151	8047	80.2	1127	23.6
7	203	3388	21.0	349	-146.4
8	872	7114	71.4	948	-76.0
9	279	3300	15.1	371	-92.2
10	1350	6819	71.1	873	477.3

6.3 Related topics

6.3.1 Model selection

When should we use an interaction model versus a parallel slopes model? Recall in Sections 6.1.2 and 6.1.3 we fit both interaction and parallel slopes models for the outcome variable y (teaching score) using a numerical explanatory variable x_1 (age) and a categorical explanatory variable x_2 (gender recorded as a binary variable). We compared these models in Figure 6.3, which we display again now.

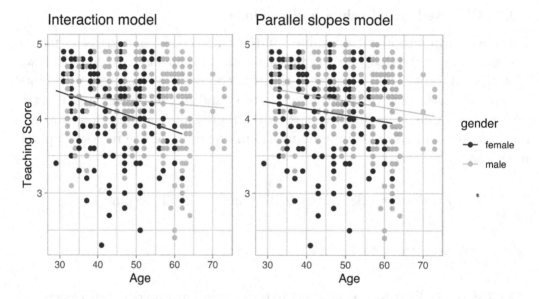

FIGURE 6.7: Previously seen comparison of interaction and parallel slopes models.

A lot of you might have asked yourselves: "Why would I force the lines to have parallel slopes (as seen in the right-hand plot) when they clearly have different slopes (as seen in the left-hand plot)?".

The answer lies in a philosophical principle known as "Occam's Razor." It states that, "all other things being equal, simpler solutions are more likely to be correct than complex ones." When viewed in a modeling framework, Occam's Razor can be restated as, "all other things being equal, simpler models are to be preferred over complex ones." In other words, we should only favor the more complex model if the additional complexity is *warranted*.

Let's revisit the equations for the regression line for both the interaction and parallel slopes model:

$$\text{Interaction} : \hat{y} = \widehat{\text{score}} = b_0 + b_{\text{age}} \cdot \text{age} + b_{\text{male}} \cdot \mathbb{1}_{\text{is male}}(x) +$$
$$b_{\text{age,male}} \cdot \text{age} \cdot \mathbb{1}_{\text{is male}}$$
$$\text{Parallel slopes} : \hat{y} = \widehat{\text{score}} = b_0 + b_{\text{age}} \cdot \text{age} + b_{\text{male}} \cdot \mathbb{1}_{\text{is male}}(x)$$

The interaction model is "more complex" in that there is an additional $b_{\text{age,male}} \cdot$ age $\cdot \mathbb{1}_{\text{is male}}$ interaction term in the equation not present for the parallel slopes model. Or viewed alternatively, the regression table for the interaction model in Table 6.3 has *four* rows, whereas the regression table for the parallel slopes model in Table 6.5 has *three* rows. The question becomes: "Is this additional complexity warranted?". In this case, it can be argued that this additional complexity is warranted, as evidenced by the clear x-shaped pattern of the two regression lines in the left-hand plot of Figure 6.7.

However, let's consider an example where the additional complexity might *not* be warranted. Let's consider the MA_schools data included in the moderndive package which contains 2017 data on Massachusetts public high schools provided by the Massachusetts Department of Education. For more details, read the help file for this data by running ?MA_schools in the console.

Let's model the numerical outcome variable y, average SAT math score for a given high school, as a function of two explanatory variables:

1. A numerical explanatory variable x_1, the percentage of that high school's student body that are economically disadvantaged and
2. A categorical explanatory variable x_2, the school size as measured by enrollment: small (13-341 students), medium (342-541 students), and large (542-4264 students).

Let's create visualizations of both the interaction and parallel slopes model once again and display the output in Figure 6.8. Recall from Subsection 6.1.3 that the geom_parallel_slopes() function is a special purpose function included in the moderndive package, since the geom_smooth() method in the ggplot2 package does not have a convenient way to plot parallel slopes models.

```
# Interaction model
ggplot(MA_schools,
       aes(x = perc_disadvan, y = average_sat_math, color = size)) +
  geom_point(alpha = 0.25) +
  geom_smooth(method = "lm", se = FALSE) +
  labs(x = "Percent economically disadvantaged", y = "Math SAT Score",
       color = "School size", title = "Interaction model")
```

```
# Parallel slopes model
ggplot(MA_schools,
       aes(x = perc_disadvan, y = average_sat_math, color = size)) +
  geom_point(alpha = 0.25) +
  geom_parallel_slopes(se = FALSE) +
  labs(x = "Percent economically disadvantaged", y = "Math SAT Score",
       color = "School size", title = "Parallel slopes model")
```

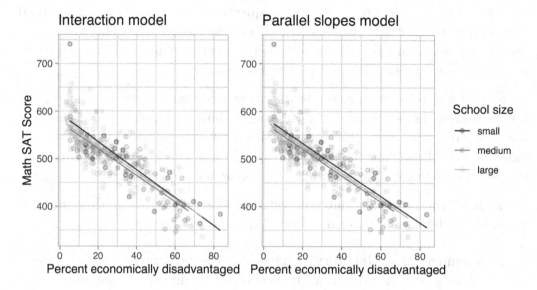

FIGURE 6.8: Comparison of interaction and parallel slopes models for Massachusetts schools.

Look closely at the left-hand plot of Figure 6.8 corresponding to an interaction model. While the slopes are indeed different, they do not differ *by much* and are nearly identical. Now compare the left-hand plot with the right-hand plot corresponding to a parallel slopes model. The two models don't appear all that different. So in this case, it can be argued that the additional complexity of the interaction model is *not warranted*. Thus following Occam's Razor, we should prefer the "simpler" parallel slopes model. Let's explicitly define what "simpler" means in this case. Let's compare the regression tables for the interaction and parallel slopes models in Tables 6.12 and 6.13.

```
model_2_interaction <- lm(average_sat_math ~ perc_disadvan * size,
                          data = MA_schools)
get_regression_table(model_2_interaction)
```

TABLE 6.12: Interaction model regression table

term	estimate	std_error	statistic	p_value	lower_ci	upper_ci
intercept	594.327	13.288	44.726	0.000	568.186	620.469
perc_disadvan	-2.932	0.294	-9.961	0.000	-3.511	-2.353
sizemedium	-17.764	15.827	-1.122	0.263	-48.899	13.371
sizelarge	-13.293	13.813	-0.962	0.337	-40.466	13.880
perc_disadvan:sizemedium	0.146	0.371	0.393	0.694	-0.585	0.877
perc_disadvan:sizelarge	0.189	0.323	0.586	0.559	-0.446	0.824

```
model_2_parallel_slopes <- lm(average_sat_math ~ perc_disadvan + size,
                              data = MA_schools)
get_regression_table(model_2_parallel_slopes)
```

TABLE 6.13: Parallel slopes regression table

term	estimate	std_error	statistic	p_value	lower_ci	upper_ci
intercept	588.19	7.607	77.325	0.000	573.23	603.15
perc_disadvan	-2.78	0.106	-26.120	0.000	-2.99	-2.57
sizemedium	-11.91	7.535	-1.581	0.115	-26.74	2.91
sizelarge	-6.36	6.923	-0.919	0.359	-19.98	7.26

Observe how the regression table for the interaction model has 2 more rows (6 versus 4). This reflects the additional "complexity" of the interaction model over the parallel slopes model.

Furthermore, note in Table 6.12 how the *offsets for the slopes* perc_disadvan:sizemedium being 0.146 and perc_disadvan:sizelarge being 0.189 are small relative to the *slope for the baseline group* of small schools of -2.932. In other words, all three slopes are similarly negative: -2.932 for small schools, -2.786 $(= -2.932+0.146)$ for medium schools, and -2.743 $(= -2.932+0.189)$ for large schools. These results are suggesting that irrespective of school size, the relationship between average math SAT scores and the percent of the student body that is economically disadvantaged is similar and, alas, quite negative.

What you have just performed is a rudimentary *model selection*: choosing which model fits data best among a set of candidate models. While the model selection approach we just took was visual in nature and hence somewhat qualitative, more statistically rigorous methods for model selection exist in the fields of multiple regression and statistical/machine learning.

6.3.2 Correlation coefficient

Recall from Table 6.9 that the correlation coefficient between income in thousands of dollars and credit card debt was 0.464. What if instead we looked at the correlation coefficient between income and credit card debt, but where income was in dollars and not thousands of dollars? This can be done by multiplying income by 1000.

```
credit_ch6 %>% select(debt, income) %>%
  mutate(income = income * 1000) %>%
  cor()
```

TABLE 6.14: Correlation between income (in dollars) and credit card debt

	debt	income
debt	1.000	0.464
income	0.464	1.000

We see it is the same! We say that the correlation coefficient is *invariant to linear transformations*. The correlation between x and y will be the same as the correlation between $a \cdot x + b$ and y for any numerical values a and b.

6.3.3 Simpson's Paradox

Recall in Section 6.2, we saw the two seemingly contradictory results when studying the relationship between credit card debt and income. On the one hand, the right hand plot of Figure 6.5 suggested that the relationship between credit card debt and income was *positive*. We re-display this in Figure 6.9.

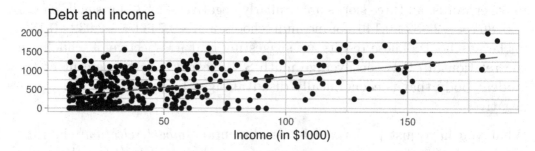

FIGURE 6.9: Relationship between credit card debt and income.

On the other hand, the multiple regression results in Table 6.10 suggested that the relationship between debt and income was *negative*. We re-display this information in Table 6.15.

TABLE 6.15: Multiple regression results

term	estimate	std_error	statistic	p_value	lower_ci	upper_ci
intercept	-385.179	19.465	-19.8	0	-423.446	-346.912
credit_limit	0.264	0.006	45.0	0	0.253	0.276
income	-7.663	0.385	-19.9	0	-8.420	-6.906

Observe how the slope for income is −7.663 and, most importantly for now, it is negative. This contradicts our observation in Figure 6.9 that the relationship is positive. How can this be? Recall the interpretation of the slope for income in the context of a multiple regression model: *taking into account all the other explanatory variables in our model,* for every increase of one unit in income (i.e., $1000), there is an associated decrease of on average $7.663 in debt.

In other words, while in *isolation,* the relationship between debt and income may be positive, when taking into account credit_limit as well, this relationship becomes negative. These seemingly paradoxical results are due to a phenomenon aptly named *Simpson's Paradox*[3]. Simpson's Paradox occurs when trends that exist for the data in aggregate either disappear or reverse when the data are broken down into groups.

Let's show how Simpson's Paradox manifests itself in the credit_ch6 data. Let's first visualize the distribution of the numerical explanatory variable credit_limit with a histogram in Figure 6.10.

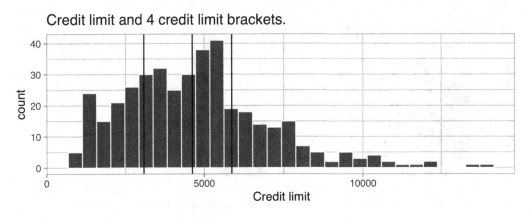

FIGURE 6.10: Histogram of credit limits and brackets.

The vertical dashed lines are the *quartiles* that cut up the variable credit_limit into four equally sized groups. Let's think of these quartiles as converting our numerical variable credit_limit into a categorical variable "credit_limit bracket" with four levels. This means that

---

[3]https://en.wikipedia.org/wiki/Simpson%27s_paradox

1. 25% of credit limits were between $0 and $3088. Let's assign these 100 people to the "low" credit_limit bracket.
2. 25% of credit limits were between $3088 and $4622. Let's assign these 100 people to the "medium-low" credit_limit bracket.
3. 25% of credit limits were between $4622 and $5873. Let's assign these 100 people to the "medium-high" credit_limit bracket.
4. 25% of credit limits were over $5873. Let's assign these 100 people to the "high" credit_limit bracket.

Now in Figure 6.11 let's re-display two versions of the scatterplot of debt and income from Figure 6.9, but with a slight twist:

1. The left-hand plot shows the regular scatterplot and the single regression line, just as you saw in Figure 6.9.
2. The right-hand plot shows the *colored scatterplot*, where the color aesthetic is mapped to "credit_limit bracket." Furthermore, there are now four separate regression lines.

In other words, the location of the 400 points are the same in both scatterplots, but the right-hand plot shows an additional variable of information: credit_limit bracket.

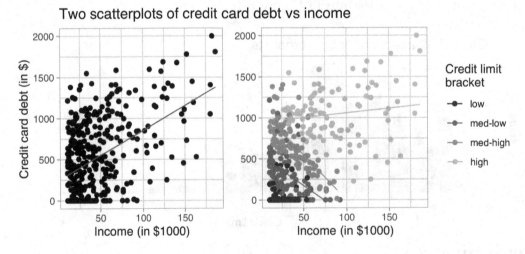

FIGURE 6.11: Relationship between credit card debt and income by credit limit bracket.

The left-hand plot of Figure 6.11 focuses on the relationship between debt and income in *aggregate*. It is suggesting that overall there exists a positive relationship between debt and income. However, the right-hand plot of Figure 6.11 focuses on the relationship between debt and income *broken down by* credit_limit *bracket*. In other words, we focus on four *separate* relationships between debt and income: one for the "low" credit_limit bracket, one for the "medium-low" credit_limit bracket, and so on.

Observe in the right-hand plot that the relationship between debt and income is clearly negative for the "medium-low" and "medium-high" credit_limit brackets, while the relationship is somewhat flat for the "low" credit_limit bracket. The only credit_limit bracket where the relationship remains positive is for the "high" credit_limit bracket. However, this relationship is less positive than in the relationship in aggregate, since the slope is shallower than the slope of the regression line in the left-hand plot.

In this example of Simpson's Paradox, the credit_limit is a *confounding variable* of the relationship between credit card debt and income as we defined in Subsection 5.3.1. Thus, credit_limit needs to be accounted for in any appropriate model for the relationship between debt and income.

6.4 Conclusion

6.4.1 Additional resources

Solutions to all *Learning checks* can be found online in Appendix D^4.

An R script file of all R code used in this chapter is available at https://www.moderndive.com/scripts/06-multiple-regression.R.

6.4.2 What's to come?

Congratulations! We've completed the "Data Modeling with moderndive" portion of this book. We're ready to proceed to Part III of this book: "Statistical Inference with infer." Statistical inference is the science of inferring about some unknown quantity using sampling.

For example, among the most well-known examples of sampling involves *polls*. Because asking an entire population about their opinions would be a long and arduous task, pollsters often take a smaller sample that is hopefully representative of the population. Based on the results of this sample, pollsters hope to make claims about the entire population.

Once we've covered Chapters 7 on sampling, 8 on confidence intervals, and 9 on hypothesis testing, we'll revisit the regression models we studied in Chapters 5 and 6 in Chapter 10 on inference for regression. So far, we've only studied

---

[4]https://moderndive.com/D-appendixD.html

the `estimate` column of all our regression tables. The next four chapters focus on what the remaining columns mean: the standard error (`std_error`), the test statistic, the `p_value`, and the lower and upper bounds of confidence intervals (`lower_ci` and `upper_ci`).

Furthermore in Chapter 10, we'll revisit the concept of residuals $y - \hat{y}$ and discuss their importance when interpreting the results of a regression model. We'll perform what is known as a *residual analysis* of the `residual` variable of all `get_regression_points()` outputs. Residual analyses allow you to verify what are known as the *conditions for inference for regression*. On to Chapter 7 on sampling in Part III as shown in Figure 6.12!

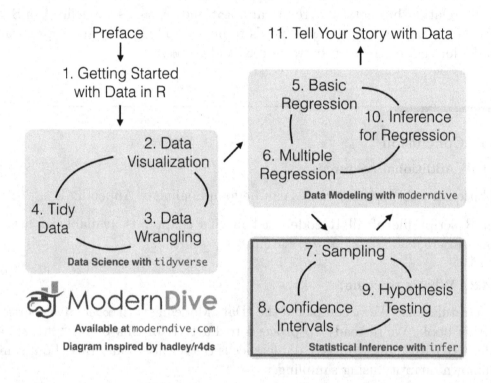

FIGURE 6.12: *ModernDive* flowchart - on to Part III!

Part III

Statistical Inference with infer

7

Sampling

In this chapter, we kick off the third portion of this book on statistical inference by learning about *sampling*. The concepts behind sampling form the basis of confidence intervals and hypothesis testing, which we'll cover in Chapters 8 and 9. We will see that the tools that you learned in the data science portion of this book, in particular data visualization and data wrangling, will also play an important role in the development of your understanding. As mentioned before, the concepts throughout this text all build into a culmination allowing you to "tell your story with data."

Needed packages

Let's load all the packages needed for this chapter (this assumes you've already installed them). Recall from our discussion in Section 4.4 that loading the `tidyverse` package by running `library(tidyverse)` loads the following commonly used data science packages all at once:

- `ggplot2` for data visualization
- `dplyr` for data wrangling
- `tidyr` for converting data to "tidy" format
- `readr` for importing spreadsheet data into R
- As well as the more advanced `purrr`, `tibble`, `stringr`, and `forcats` packages

If needed, read Section 1.3 for information on how to install and load R packages.

```
library(tidyverse)
library(moderndive)
```

7.1 Sampling bowl activity

Let's start with a hands-on activity.

7.1.1 What proportion of this bowl's balls are red?

Take a look at the bowl in Figure 7.1. It has a certain number of red and a certain number of white balls all of equal size. (Note that in this printed version of the book "red" corresponds to the darker-colored balls, and "white" corresponds to the lighter-colored balls. We kept the reference to "red" and "white" throughout this book since those are the actual colors of the balls as seen in the background of the image on our book's cover[1].) Furthermore, it appears the bowl has been mixed beforehand, as there does not seem to be any coherent pattern to the spatial distribution of the red and white balls.

Let's now ask ourselves, what proportion of this bowl's balls are red?

FIGURE 7.1: A bowl with red and white balls.

One way to answer this question would be to perform an exhaustive count: remove each ball individually, count the number of red balls and the number of white balls, and divide the number of red balls by the total number of balls. However, this would be a long and tedious process.

7.1.2 Using the shovel once

Instead of performing an exhaustive count, let's insert a shovel into the bowl as seen in Figure 7.2. Using the shovel, let's remove $5 \cdot 10 = 50$ balls, as seen in Figure 7.3.

---

[1]https://moderndive.com/images/logos/book_cover.png

FIGURE 7.2: Inserting a shovel into the bowl.

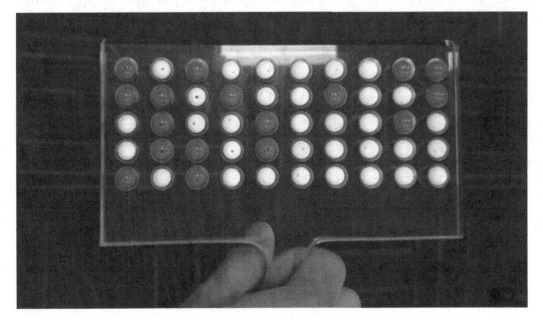

FIGURE 7.3: Removing 50 balls from the bowl.

Observe that 17 of the balls are red and thus $0.34 = 34\%$ of the shovel's balls are red. We can view the proportion of balls that are red in this shovel as a guess of the proportion of balls that are red in the entire bowl. While not as exact as doing an exhaustive count of all the balls in the bowl, our guess of 34% took much less time and energy to make.

However, say, we started this activity over from the beginning. In other words, we replace the 50 balls back into the bowl and start over. Would we remove exactly 17 red balls again? In other words, would our guess at the proportion of the bowl's balls that are red be exactly 34% again? Maybe?

What if we repeated this activity several times following the process shown in Figure 7.4? Would we obtain exactly 17 red balls each time? In other words, would our guess at the proportion of the bowl's balls that are red be exactly 34% every time? Surely not. Let's repeat this exercise several times with the help of 33 groups of friends to understand how the value differs with repetition.

7.1.3   Using the shovel 33 times

Each of our 33 groups of friends will do the following:

- Use the shovel to remove 50 balls each.
- Count the number of red balls and thus compute the proportion of the 50 balls that are red.
- Return the balls into the bowl.
- Mix the contents of the bowl a little to not let a previous group's results influence the next group's.

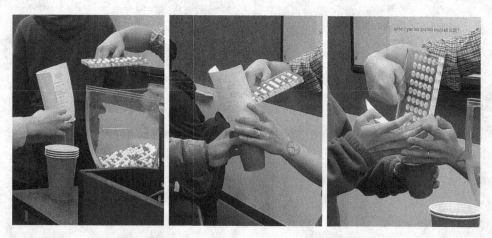

FIGURE 7.4: Repeating sampling activity 33 times.

Each of our 33 groups of friends make note of their proportion of red balls from their sample collected. Each group then marks their proportion of their 50 balls that were red in the appropriate bin in a hand-drawn histogram as seen in Figure 7.5.

FIGURE 7.5: Constructing a histogram of proportions.

Recall from Section 2.5 that histograms allow us to visualize the *distribution* of a numerical variable. In particular, where the center of the values falls and how the values vary. A partially completed histogram of the first 10 out of 33 groups of friends' results can be seen in Figure 7.6.

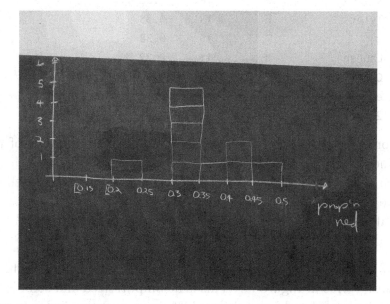

FIGURE 7.6: Hand-drawn histogram of first 10 out of 33 proportions.

Observe the following in the histogram in Figure 7.6:

- At the low end, one group removed 50 balls from the bowl with proportion red between 0.20 and 0.25.
- At the high end, another group removed 50 balls from the bowl with proportion between 0.45 and 0.5 red.
- However, the most frequently occurring proportions were between 0.30 and 0.35 red, right in the middle of the distribution.
- The shape of this distribution is somewhat bell-shaped.

Let's construct this same hand-drawn histogram in R using your data visualization skills that you honed in Chapter 2. We saved our 33 groups of friends' results in the `tactile_prop_red` data frame included in the `moderndive` package. Run the following to display the first 10 of 33 rows:

```
tactile_prop_red
```

```
# A tibble: 33 x 4
   group            replicate red_balls prop_red
   <chr>                <int>     <int>    <dbl>
 1 Ilyas, Yohan             1        21     0.42
 2 Morgan, Terrance         2        17     0.34
 3 Martin, Thomas           3        21     0.42
 4 Clark, Frank             4        21     0.42
 5 Riddhi, Karina           5        18     0.36
 6 Andrew, Tyler            6        19     0.38
 7 Julia                    7        19     0.38
 8 Rachel, Lauren           8        11     0.22
 9 Daniel, Caroline         9        15     0.3
10 Josh, Maeve             10        17     0.34
# ... with 23 more rows
```

Observe for each `group` that we have their names, the number of `red_balls` they obtained, and the corresponding proportion out of 50 balls that were red named `prop_red`. We also have a `replicate` variable enumerating each of the 33 groups. We chose this name because each row can be viewed as one instance of a replicated (in other words repeated) activity: using the shovel to remove 50 balls and computing the proportion of those balls that are red.

Let's visualize the distribution of these 33 proportions using `geom_histogram()` with `binwidth = 0.05` in Figure 7.7. This is a computerized and complete version of the partially completed hand-drawn histogram you saw in Figure 7.6. Note that setting `boundary = 0.4` indicates that we want a binning scheme such that

one of the bins' boundary is at 0.4. This helps us to more closely align this
histogram with the hand-drawn histogram in Figure 7.6.

```
ggplot(tactile_prop_red, aes(x = prop_red)) +
  geom_histogram(binwidth = 0.05, boundary = 0.4, color = "white") +
  labs(x = "Proportion of 50 balls that were red",
       title = "Distribution of 33 proportions red")
```

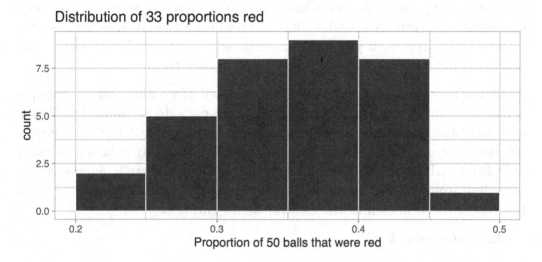

FIGURE 7.7: Distribution of 33 proportions based on 33 samples of size 50.

7.1.4 What did we just do?

What we just demonstrated in this activity is the statistical concept of *sampling*.
We would like to know the proportion of the bowl's balls that are red. Because
the bowl has a large number of balls, performing an exhaustive count of the
red and white balls would be time-consuming. We thus extracted a *sample*
of 50 balls using the shovel to make an *estimate*. Using this sample of 50
balls, we estimated the proportion of the *bowl's* balls that are red to be
34%.

Moreover, because we mixed the balls before each use of the shovel, the samples
were randomly drawn. Because each sample was drawn at random, the samples
were different from each other. Because the samples were different from each
other, we obtained the different proportions red observed in Figure 7.7. This is
known as the concept of *sampling variation*.

The purpose of this sampling activity was to develop an understanding of two
key concepts relating to sampling:

1. Understanding the effect of sampling variation.
2. Understanding the effect of sample size on sampling variation.

In Section 7.2, we'll mimic the hands-on sampling activity we just performed on a computer. This will allow us not only to repeat the sampling exercise much more than 33 times, but it will also allow us to use shovels with different numbers of slots than just 50.

Afterwards, we'll present you with definitions, terminology, and notation related to sampling in Section 7.3. As in many disciplines, such necessary background knowledge may seem inaccessible and even confusing at first. However, as with many difficult topics, if you truly understand the underlying concepts and practice, practice, practice, you'll be able to master them.

To tie the contents of this chapter to the real world, we'll present an example of one of the most recognizable uses of sampling: polls. In Section 7.4 we'll look at a particular case study: a 2013 poll on then U.S. President Barack Obama's popularity among young Americans, conducted by Kennedy School's Institute of Politics at Harvard University. To close this chapter, we'll generalize the "sampling from a bowl" exercise to other sampling scenarios and present a theoretical result known as the *Central Limit Theorem*.

Learning check

(**LC7.1**) Why was it important to mix the bowl before we sampled the balls?

(**LC7.2**) Why is it that our 33 groups of friends did not all have the same numbers of balls that were red out of 50, and hence different proportions red?

7.2 Virtual sampling

In the previous Section 7.1, we performed a *tactile* sampling activity by hand. In other words, we used a physical bowl of balls and a physical shovel. We performed this sampling activity by hand first so that we could develop a firm understanding of the root ideas behind sampling. In this section, we'll mimic this tactile sampling activity with a *virtual* sampling activity using a computer. In other words, we'll use a virtual analog to the bowl of balls and a virtual analog to the shovel.

7.2.1   Using the virtual shovel once

Let's start by performing the virtual analog of the tactile sampling exercise we performed in Section 7.1. We first need a virtual analog of the bowl seen in Figure 7.1. To this end, we included a data frame named `bowl` in the `moderndive` package. The rows of `bowl` correspond exactly with the contents of the actual bowl.

```
bowl
```

```
# A tibble: 2,400 x 2
   ball_ID color
     <int> <chr>
1        1 white
2        2 white
3        3 white
4        4 red
5        5 white
6        6 white
7        7 red
8        8 white
9        9 red
10      10 white
# ... with 2,390 more rows
```

Observe that `bowl` has 2400 rows, telling us that the bowl contains 2400 equally sized balls. The first variable `ball_ID` is used as an *identification variable* as discussed in Subsection 1.4.4; none of the balls in the actual bowl are marked with numbers. The second variable `color` indicates whether a particular virtual ball is red or white. View the contents of the bowl in RStudio's data viewer and scroll through the contents to convince yourself that `bowl` is indeed a virtual analog of the actual bowl in Figure 7.1.

Now that we have a virtual analog of our bowl, we now need a virtual analog to the shovel seen in Figure 7.2 to generate virtual samples of 50 balls. We're going to use the `rep_sample_n()` function included in the `moderndive` package. This function allows us to take repeated, or replicated, `samples` of size n.

```
virtual_shovel <- bowl %>%
  rep_sample_n(size = 50)
virtual_shovel
```

```
# A tibble: 50 x 3
```

```
# Groups:   replicate [1]
   replicate ball_ID color
       <int>   <int> <chr>
1         1    1970 white
2         1     842 red
3         1    2287 white
4         1     599 white
5         1     108 white
6         1     846 red
7         1     390 red
8         1     344 white
9         1     910 white
10        1    1485 white
# ... with 40 more rows
```

Observe that `virtual_shovel` has 50 rows corresponding to our virtual sample of size 50. The `ball_ID` variable identifies which of the 2400 balls from `bowl` are included in our sample of 50 balls while `color` denotes its color. However, what does the `replicate` variable indicate? In `virtual_shovel`'s case, `replicate` is equal to 1 for all 50 rows. This is telling us that these 50 rows correspond to the first repeated/replicated use of the shovel, in our case our first sample. We'll see shortly that when we "virtually" take 33 samples, `replicate` will take values between 1 and 33.

Let's compute the proportion of balls in our virtual sample that are red using the `dplyr` data wrangling verbs you learned in Chapter 3. First, for each of our 50 sampled balls, let's identify if it is red or not using a test for equality with `==`. Let's create a new Boolean variable `is_red` using the `mutate()` function from Section 3.5:

```
virtual_shovel %>%
  mutate(is_red = (color == "red"))
```

```
# A tibble: 50 x 4
# Groups:   replicate [1]
   replicate ball_ID color is_red
       <int>   <int> <chr> <lgl>
1         1    1970 white FALSE
2         1     842 red    TRUE
3         1    2287 white FALSE
4         1     599 white FALSE
5         1     108 white FALSE
6         1     846 red    TRUE
```

```
7          1      390 red    TRUE
8          1      344 white FALSE
9          1      910 white FALSE
10         1     1485 white FALSE
# ... with 40 more rows
```

Observe that for every row where `color == "red"`, the Boolean (logical) value `TRUE` is returned and for every row where `color` is not equal to `"red"`, the Boolean `FALSE` is returned.

Second, let's compute the number of balls out of 50 that are red using the `summarize()` function. Recall from Section 3.3 that `summarize()` takes a data frame with many rows and returns a data frame with a single row containing summary statistics, like the `mean()` or `median()`. In this case, we use the `sum()`:

```
virtual_shovel %>%
  mutate(is_red = (color == "red")) %>%
  summarize(num_red = sum(is_red))
```

```
# A tibble: 1 x 2
  replicate num_red
      <int>   <int>
1         1      12
```

Why does this work? Because R treats `TRUE` like the number 1 and `FALSE` like the number 0. So summing the number of `TRUE`s and `FALSE`s is equivalent to summing 1's and 0's. In the end, this operation counts the number of balls where `color` is red. In our case, 12 of the 50 balls were red. However, you might have gotten a different number red because of the randomness of the virtual sampling.

Third and lastly, let's compute the proportion of the 50 sampled balls that are red by dividing `num_red` by 50:

```
virtual_shovel %>%
  mutate(is_red = color == "red") %>%
  summarize(num_red = sum(is_red)) %>%
  mutate(prop_red = num_red / 50)
```

```
# A tibble: 1 x 3
  replicate num_red prop_red
      <int>   <int>    <dbl>
1         1      12     0.24
```

In other words, 24% of this virtual sample's balls were red. Let's make this code a little more compact and succinct by combining the first `mutate()` and the `summarize()` as follows:

```
virtual_shovel %>%
  summarize(num_red = sum(color == "red")) %>%
  mutate(prop_red = num_red / 50)
```

```
# A tibble: 1 x 3
  replicate num_red prop_red
      <int>   <int>    <dbl>
1         1      12     0.24
```

Great! 24% of `virtual_shovel`'s 50 balls were red! So based on this particular sample of 50 balls, our guess at the proportion of the `bowl`'s balls that are red is 24%. But remember from our earlier tactile sampling activity that if we repeat this sampling, we will not necessarily obtain the same value of 24% again. There will likely be some variation. In fact, our 33 groups of friends computed 33 such proportions whose distribution we visualized in Figure 7.6. We saw that these estimates *varied*. Let's now perform the virtual analog of having 33 groups of students use the sampling shovel!

7.2.2   Using the virtual shovel 33 times

Recall that in our tactile sampling exercise in Section 7.1, we had 33 groups of students each use the shovel, yielding 33 samples of size 50 balls. We then used these 33 samples to compute 33 proportions. In other words, we repeated/replicated using the shovel 33 times. We can perform this repeated/replicated sampling virtually by once again using our virtual shovel function `rep_sample_n()`, but by adding the `reps = 33` argument. This is telling R that we want to repeat the sampling 33 times.

We'll save these results in a data frame called `virtual_samples`. While we provide a preview of the first 10 rows of `virtual_samples` in what follows, we highly suggest you scroll through its contents using RStudio's spreadsheet viewer by running `View(virtual_samples)`.

```
virtual_samples <- bowl %>%
  rep_sample_n(size = 50, reps = 33)
virtual_samples
```

```
# A tibble: 1,650 x 3
# Groups:   replicate [33]
```

```
   replicate ball_ID color
      <int>   <int> <chr>
1         1     875 white
2         1    1851 red
3         1    1548 red
4         1    1975 white
5         1     835 white
6         1      16 white
7         1     327 white
8         1    1803 red
9         1     740 red
10        1     179 red
# ... with 1,640 more rows
```

Observe in the spreadsheet viewer that the first 50 rows of replicate are equal to 1 while the next 50 rows of replicate are equal to 2. This is telling us that the first 50 rows correspond to the first sample of 50 balls while the next 50 rows correspond to the second sample of 50 balls. This pattern continues for all reps = 33 replicates and thus virtual_samples has $33 \cdot 50 = 1650$ rows.

Let's now take virtual_samples and compute the resulting 33 proportions red. We'll use the same dplyr verbs as before, but this time with an additional group_by() of the replicate variable. Recall from Section 3.4 that by assigning the grouping variable "meta-data" before we summarize(), we'll obtain 33 different proportions red. We display a preview of the first 10 out of 33 rows:

```
virtual_prop_red <- virtual_samples %>%
  group_by(replicate) %>%
  summarize(red = sum(color == "red")) %>%
  mutate(prop_red = red / 50)
virtual_prop_red
```

```
# A tibble: 33 x 3
   replicate   red prop_red
      <int> <int>    <dbl>
1         1    23     0.46
2         2    19     0.38
3         3    18     0.36
4         4    19     0.38
5         5    15     0.3
6         6    21     0.42
7         7    21     0.42
8         8    16     0.32
```

```
 9          9   24    0.48
10         10   14    0.28
# ... with 23 more rows
```

As with our 33 groups of friends' tactile samples, there is variation in the resulting 33 virtual proportions red. Let's visualize this variation in a histogram in Figure 7.8. Note that we add `binwidth = 0.05` and `boundary = 0.4` arguments as well. Recall that setting `boundary = 0.4` ensures a binning scheme with one of the bins' boundaries at 0.4. Since the `binwidth = 0.05` is also set, this will create bins with boundaries at 0.30, 0.35, 0.45, 0.5, etc. as well.

```
ggplot(virtual_prop_red, aes(x = prop_red)) +
  geom_histogram(binwidth = 0.05, boundary = 0.4, color = "white") +
  labs(x = "Proportion of 50 balls that were red",
       title = "Distribution of 33 proportions red")
```

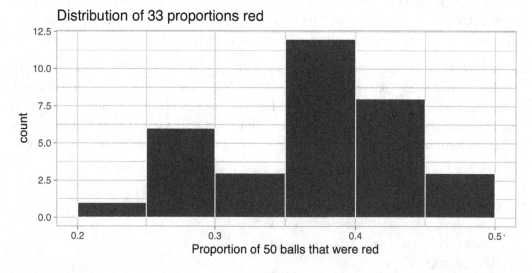

FIGURE 7.8: Distribution of 33 proportions based on 33 samples of size 50.

Observe that we occasionally obtained proportions red that are less than 30%. On the other hand, we occasionally obtained proportions that are greater than 45%. However, the most frequently occurring proportions were between 35% and 40% (for 11 out of 33 samples). Why do we have these differences in proportions red? Because of *sampling variation*.

Let's now compare our virtual results with our tactile results from the previous section in Figure 7.9. Observe that both histograms are somewhat similar in their center and variation, although not identical. These slight differences are again due to random sampling variation. Furthermore, observe that both distributions are somewhat bell-shaped.

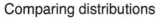

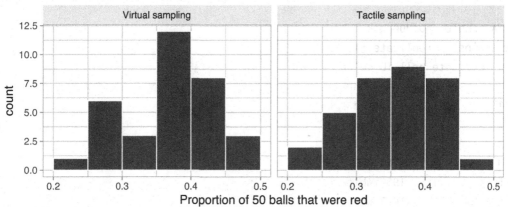

FIGURE 7.9: Comparing 33 virtual and 33 tactile proportions red.

Learning check

(LC7.3) Why couldn't we study the effects of sampling variation when we used the virtual shovel only once? Why did we need to take more than one virtual sample (in our case 33 virtual samples)?

7.2.3 Using the virtual shovel 1000 times

Now say we want to study the effects of sampling variation not for 33 samples, but rather for a larger number of samples, say 1000. We have two choices at this point. We could have our groups of friends manually take 1000 samples of 50 balls and compute the corresponding 1000 proportions. However, this would be a tedious and time-consuming task. This is where computers excel: automating long and repetitive tasks while performing them quite quickly. Thus, at this point we will abandon tactile sampling in favor of only virtual sampling. Let's once again use the `rep_sample_n()` function with sample `size` set to be 50 once again, but this time with the number of replicates `reps` set to `1000`. Be sure to scroll through the contents of `virtual_samples` in RStudio's viewer.

```
virtual_samples <- bowl %>%
  rep_sample_n(size = 50, reps = 1000)
virtual_samples
```

```
# A tibble: 50,000 x 3
# Groups:   replicate [1,000]
   replicate ball_ID color
       <int>   <int> <chr>
1          1    1236 red
2          1    1944 red
3          1    1939 white
4          1     780 white
5          1    1956 white
6          1    1003 white
7          1    2113 white
8          1    2213 white
9          1     782 white
10         1     898 white
# ... with 49,990 more rows
```

Observe that now `virtual_samples` has $1000 \cdot 50 = 50{,}000$ rows, instead of the $33 \cdot 50 = 1650$ rows from earlier. Using the same data wrangling code as earlier, let's take the data frame `virtual_samples` with $1000 \cdot 50 = 50{,}000$ rows and compute the resulting 1000 proportions of red balls.

```
virtual_prop_red <- virtual_samples %>%
  group_by(replicate) %>%
  summarize(red = sum(color == "red")) %>%
  mutate(prop_red = red / 50)
virtual_prop_red
```

```
# A tibble: 1,000 x 3
   replicate   red prop_red
       <int> <int>    <dbl>
1          1    18     0.36
2          2    19     0.38
3          3    20     0.4
4          4    15     0.3
5          5    17     0.34
6          6    16     0.32
7          7    23     0.46
```

8	8	23	0.46
9	9	15	0.3
10	10	18	0.36

`# ... with 990 more rows`

Observe that we now have 1000 replicates of `prop_red`, the proportion of 50 balls that are red. Using the same code as earlier, let's now visualize the distribution of these 1000 replicates of `prop_red` in a histogram in Figure 7.10.

```
ggplot(virtual_prop_red, aes(x = prop_red)) +
  geom_histogram(binwidth = 0.05, boundary = 0.4, color = "white") +
  labs(x = "Proportion of 50 balls that were red",
       title = "Distribution of 1000 proportions red")
```

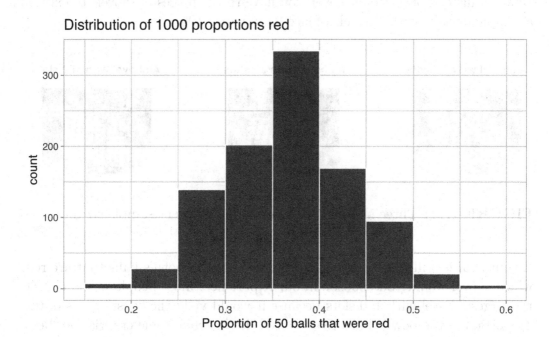

FIGURE 7.10: Distribution of 1000 proportions based on 1000 samples of size 50.

Once again, the most frequently occurring proportions of red balls occur between 35% and 40%. Every now and then, we obtain proportions as low as between 20% and 25%, and others as high as between 55% and 60%. These are rare, however. Furthermore, observe that we now have a much more symmetric and smoother bell-shaped distribution. This distribution is, in fact, approximated well by a normal distribution. At this point we recommend you read the "Normal distribution" section (Appendix A.2) for a brief discussion on the properties of the normal distribution.

Learning check

(LC7.4) Why did we not take 1000 "tactile" samples of 50 balls by hand?

(LC7.5) Looking at Figure 7.10, would you say that sampling 50 balls where 30% of them were red is likely or not? What about sampling 50 balls where 10% of them were red?

7.2.4 Using different shovels

Now say instead of just one shovel, you have three choices of shovels to extract a sample of balls with: shovels of size 25, 50, and 100.

A shovel with 25 slots   **A shovel with 50 slots**   **A shovel with 100 slots**

FIGURE 7.11: Three shovels to extract three different sample sizes.

If your goal is still to estimate the proportion of the bowl's balls that are red, which shovel would you choose? In our experience, most people would choose the largest shovel with 100 slots because it would yield the "best" guess of the proportion of the bowl's balls that are red. Let's define some criteria for "best" in this subsection.

Using our newly developed tools for virtual sampling, let's unpack the effect of having different sample sizes! In other words, let's use rep_sample_n() with size set to 25, 50, and 100, respectively, while keeping the number of repeated/replicated samples at 1000:

1. Virtually use the appropriate shovel to generate 1000 samples with size balls.
2. Compute the resulting 1000 replicates of the proportion of the shovel's balls that are red.

3. Visualize the distribution of these 1000 proportions red using a histogram.

Run each of the following code segments individually and then compare the three resulting histograms.

```
# Segment 1: sample size = 25 -------------------------------
# 1.a) Virtually use shovel 1000 times
virtual_samples_25 <- bowl %>%
  rep_sample_n(size = 25, reps = 1000)

# 1.b) Compute resulting 1000 replicates of proportion red
virtual_prop_red_25 <- virtual_samples_25 %>%
  group_by(replicate) %>%
  summarize(red = sum(color == "red")) %>%
  mutate(prop_red = red / 25)

# 1.c) Plot distribution via a histogram
ggplot(virtual_prop_red_25, aes(x = prop_red)) +
  geom_histogram(binwidth = 0.05, boundary = 0.4, color = "white") +
  labs(x = "Proportion of 25 balls that were red", title = "25")


# Segment 2: sample size = 50 -------------------------------
# 2.a) Virtually use shovel 1000 times
virtual_samples_50 <- bowl %>%
  rep_sample_n(size = 50, reps = 1000)

# 2.b) Compute resulting 1000 replicates of proportion red
virtual_prop_red_50 <- virtual_samples_50 %>%
  group_by(replicate) %>%
  summarize(red = sum(color == "red")) %>%
  mutate(prop_red = red / 50)

# 2.c) Plot distribution via a histogram
ggplot(virtual_prop_red_50, aes(x = prop_red)) +
  geom_histogram(binwidth = 0.05, boundary = 0.4, color = "white") +
  labs(x = "Proportion of 50 balls that were red", title = "50")


# Segment 3: sample size = 100 ------------------------------
```

```
# 3.a) Virtually using shovel with 100 slots 1000 times
virtual_samples_100 <- bowl %>%
  rep_sample_n(size = 100, reps = 1000)

# 3.b) Compute resulting 1000 replicates of proportion red
virtual_prop_red_100 <- virtual_samples_100 %>%
  group_by(replicate) %>%
  summarize(red = sum(color == "red")) %>%
  mutate(prop_red = red / 100)

# 3.c) Plot distribution via a histogram
ggplot(virtual_prop_red_100, aes(x = prop_red)) +
  geom_histogram(binwidth = 0.05, boundary = 0.4, color = "white") +
  labs(x = "Proportion of 100 balls that were red", title = "100")
```

For easy comparison, we present the three resulting histograms in a single row
with matching x and y axes in Figure 7.12.

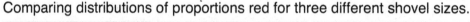

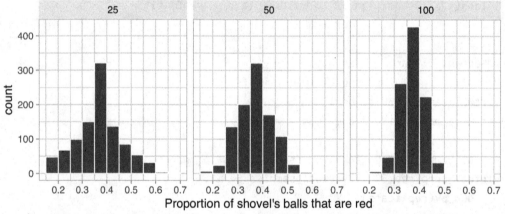

FIGURE 7.12: Comparing the distributions of proportion red for different
sample sizes.

Observe that as the sample size increases, the variation of the 1000 replicates
of the proportion of red decreases. In other words, as the sample size increases,
there are fewer differences due to sampling variation and the distribution
centers more tightly around the same value. Eyeballing Figure 7.12, all three
histograms appear to center around roughly 40%.

We can be numerically explicit about the amount of variation in our three sets of 1000 values of prop_red using the *standard deviation*. A standard deviation is a summary statistic that measures the amount of variation within a numerical variable (see Appendix A.1 for a brief discussion on the properties of the standard deviation). For all three sample sizes, let's compute the standard deviation of the 1000 proportions red by running the following data wrangling code that uses the sd() summary function.

```
# n = 25
virtual_prop_red_25 %>%
  summarize(sd = sd(prop_red))


# n = 50
virtual_prop_red_50 %>%
  summarize(sd = sd(prop_red))


# n = 100
virtual_prop_red_100 %>%
  summarize(sd = sd(prop_red))
```

Let's compare these three measures of distributional variation in Table 7.1.

TABLE 7.1: Comparing standard deviations of proportions red for three different shovels

Number of slots in shovel	Standard deviation of proportions red
25	0.094
50	0.069
100	0.045

As we observed in Figure 7.12, as the sample size increases, the variation decreases. In other words, there is less variation in the 1000 values of the proportion red. So as the sample size increases, our guesses at the true proportion of the bowl's balls that are red get more precise.

Learning check

(LC7.6) In Figure 7.12, we used shovels to take 1000 samples each, computed the resulting 1000 proportions of the shovel's balls that were red, and then visualized the distribution of these 1000 proportions in a histogram. We did this for shovels with 25, 50, and 100 slots in them. As the size of the shovels increased, the histograms got narrower. In other words, as the size of the shovels increased from 25 to 50 to 100, did the 1000 proportions

- A. vary less,
- B. vary by the same amount, or
- C. vary more?

(LC7.7) What summary statistic did we use to quantify how much the 1000 proportions red varied?

- A. The interquartile range
- B. The standard deviation
- C. The range: the largest value minus the smallest.

7.3 Sampling framework

In both our tactile and our virtual sampling activities, we used sampling for the purpose of estimation. We extracted samples in order to *estimate* the proportion of the bowl's balls that are red. We used sampling as a less time-consuming approach than performing an exhaustive count of all the balls. Our virtual sampling activity built up to the results shown in Figure 7.12 and Table 7.1: comparing 1000 proportions red based on samples of size 25, 50, and 100. This was our first attempt at understanding two key concepts relating to sampling for estimation:

1. The effect of *sampling variation* on our estimates.
2. The effect of sample size on *sampling variation*.

Let's now introduce some terminology and notation as well as statistical definitions related to sampling. Given the number of new words you'll need to learn, you will likely have to read this section a few times. Keep in mind, however, that all of the concepts underlying these terminology, notation, and definitions tie directly to the concepts underlying our tactile and virtual sampling activities. It will simply take time and practice to master them.

7.3.1 Terminology and notation

Here is a list of terminology and mathematical notation relating to sampling.

First, a **population** is a collection of individuals or observations we are interested in. This is also commonly denoted as a **study population**. We mathematically denote the population's size using upper-case N. In our sampling activities, the (study) population is the collection of $N = 2400$ identically sized red and white balls contained in the bowl.

Second, a **population parameter** is a numerical summary quantity about the population that is unknown, but you wish you knew. For example, when this quantity is a mean, the population parameter of interest is the *population mean*. This is mathematically denoted with the Greek letter μ pronounced "mu" (we'll see a sampling activity involving means in the upcoming Section 8.1). In our earlier sampling from the bowl activity, however, since we were interested in the proportion of the bowl's balls that were red, the population parameter is the *population proportion*. This is mathematically denoted with the letter p.

Third, a **census** is an exhaustive enumeration or counting of all N individuals or observations in the population in order to compute the population parameter's value *exactly*. In our sampling activity, this would correspond to counting the number of balls out of $N = 2400$ that are red and computing the *population proportion* p that are red *exactly*. When the number N of individuals or observations in our population is large as was the case with our bowl, a census can be quite expensive in terms of time, energy, and money.

Fourth, **sampling** is the act of collecting a sample from the population when we don't have the means to perform a census. We mathematically denote the sample's size using lower case n, as opposed to upper case N which denotes the population's size. Typically the sample size n is much smaller than the population size N. Thus sampling is a much cheaper alternative than performing a census. In our sampling activities, we used shovels with 25, 50, and 100 slots to extract samples of size $n = 25$, $n = 50$, and $n = 100$.

Fifth, a **point estimate (AKA sample statistic)** is a summary statistic computed from a sample that *estimates* an unknown population parameter. In our sampling activities, recall that the unknown population parameter was the population proportion and that this is mathematically denoted with p. Our point estimate is the *sample proportion*: the proportion of the shovel's balls that are red. In other words, it is our guess of the proportion of the bowl's balls balls that are red. We mathematically denote the sample proportion using $\hat{p}$. The "hat" on top of the p indicates that it is an estimate of the unknown population proportion p.

Sixth is the idea of **representative sampling**. A sample is said to be a *representative sample* if it roughly *looks like* the population. In other words, are the sample's characteristics a good representation of the population's characteristics? In our sampling activity, are the samples of n balls extracted using our shovels representative of the bowl's $N = 2400$ balls?

Seventh is the idea of **generalizability**. We say a sample is generalizable if any results based on the sample can generalize to the population. In other words, does the value of the point estimate *generalize* to the population? In our sampling activity, can we generalize the sample proportion from our shovels to the entire bowl? Using our mathematical notation, this is akin to asking if $\hat{p}$ is a "good guess" of p?

Eighth, we say **biased sampling** occurs if certain individuals or observations in a population have a higher chance of being included in a sample than others. We say a sampling procedure is *unbiased* if every observation in a population had an equal chance of being sampled. In our sampling activities, since we mixed all $N = 2400$ balls prior to each group's sampling and since each of the equally sized balls had an equal chance of being sampled, our samples were unbiased.

Ninth and lastly, the idea of **random sampling**. We say a sampling procedure is *random* if we sample randomly from the population in an unbiased fashion. In our sampling activities, this would correspond to sufficiently mixing the bowl before each use of the shovel.

Phew, that's a lot of new terminology and notation to learn! Let's put them all together to describe the paradigm of sampling.

In general:

- If the sampling of a sample of size n is done at **random**, then
- the sample is **unbiased** and **representative** of the population of size N, thus
- any result based on the sample can **generalize** to the population, thus
- the point estimate is a **"good guess"** of the unknown population parameter, thus
- instead of performing a census, we can **infer** about the population using sampling.

Specific to our sampling activity:

- If we extract a sample of $n = 50$ balls at **random**, in other words, we mix all of the equally sized balls before using the shovel, then
- the contents of the shovel are an **unbiased representation** of the contents of the bowl's 2400 balls, thus
- any result based on the shovel's balls can **generalize** to the bowl, thus
- the sample proportion $\hat{p}$ of the $n = 50$ balls in the shovel that are red is a **"good guess"** of the population proportion p of the $N = 2400$ balls that are red, thus

- instead of manually going over all 2400 balls in the bowl, we can **infer** about the bowl using the shovel.

Note that last word we wrote in bold: **infer**. The act of "inferring" means to deduce or conclude information from evidence and reasoning. In our sampling activities, we wanted to infer about the proportion of the bowl's balls that are red. *Statistical inference*[2] is the "theory, methods, and practice of forming judgments about the parameters of a population and the reliability of statistical relationships, typically on the basis of random sampling." In other words, statistical inference is the act of inference via sampling. In the upcoming Chapter 8 on confidence intervals, we'll introduce the infer package, which makes statistical inference "tidy" and transparent. It is why this third portion of the book is called "Statistical inference via infer."

Learning check

(LC7.8) In the case of our bowl activity, what is the *population parameter*? Do we know its value?

(LC7.9) What would performing a census in our bowl activity correspond to? Why did we not perform a census?

(LC7.10) What purpose do *point estimates* serve in general? What is the name of the point estimate specific to our bowl activity? What is its mathematical notation?

(LC7.11) How did we ensure that our tactile samples using the shovel were random?

(LC7.12) Why is it important that sampling be done *at random*?

(LC7.13) What are we *inferring* about the bowl based on the samples using the shovel?

7.3.2 Statistical definitions

Now, for some important statistical definitions related to sampling. As a refresher of our 1000 repeated/replicated virtual samples of size $n = 25$, $n = 50$, and $n = 100$ in Section 7.2, let's display Figure 7.12 again as Figure 7.13.

---

[2]https://en.wikipedia.org/wiki/Statistical_inference

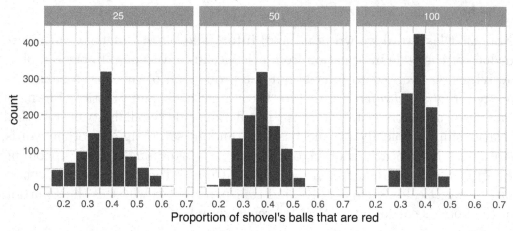

FIGURE 7.13: Previously seen three distributions of the sample proportion $\hat{p}$.

These types of distributions have a special name: **sampling distributions**; their visualization displays the effect of sampling variation on the distribution of any point estimate, in this case, the sample proportion $\hat{p}$. Using these sampling distributions, for a given sample size n, we can make statements about what values we can typically expect.

For example, observe the centers of all three sampling distributions: they are all roughly centered around $0.4 = 40\%$. Furthermore, observe that while we are somewhat likely to observe sample proportions of red balls of $0.2 = 20\%$ when using the shovel with 25 slots, we will almost never observe a proportion of 20% when using the shovel with 100 slots. Observe also the effect of sample size on the sampling variation. As the sample size n increases from 25 to 50 to 100, the variation of the sampling distribution decreases and thus the values cluster more and more tightly around the same center of around 40%. We quantified this variation using the standard deviation of our sample proportions in Table 7.1, which we display again as Table 7.2:

TABLE 7.2: Previously seen comparing standard deviations of proportions red for three different shovels

Number of slots in shovel	Standard deviation of proportions red
25	0.094
50	0.069
100	0.045

So as the sample size increases, the standard deviation of the proportion of red balls decreases. This type of standard deviation has another special name: **standard error**. Standard errors quantify the effect of sampling variation induced on our estimates. In other words, they quantify how much we can expect different proportions of a shovel's balls that are red *to vary* from one sample to another sample to another sample, and so on. As a general rule, as sample size increases, the standard error decreases.

Unfortunately, these names confuse many people who are new to statistical inference. For example, it's common for people who are new to statistical inference to call the "sampling distribution" the "sample distribution." Another additional source of confusion is the name "standard deviation" and "standard error." Remember that a standard error is merely a *kind* of standard deviation: the standard deviation of any point estimate from sampling. In other words, all standard errors are standard deviations, but not every standard deviation is necessarily a standard error.

To help reinforce these concepts, let's re-display Figure 7.12 but using our new terminology, notation, and definitions relating to sampling in Figure 7.14.

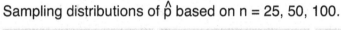

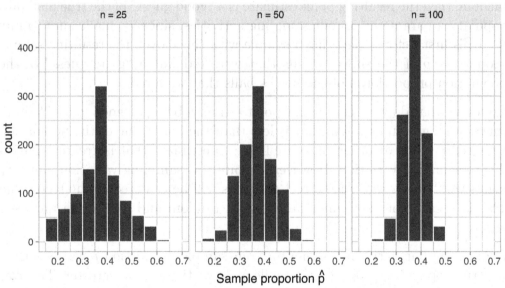

FIGURE 7.14: Three sampling distributions of the sample proportion $\hat{p}$.

Furthermore, let's re-display Table 7.1 but using our new terminology, notation, and definitions relating to sampling in Table 7.3.

TABLE 7.3: Standard errors of the sample proportion based on sample sizes of 25, 50, and 100

Sample size (n)	Standard error of $\hat{p}$
n = 25	0.094
n = 50	0.069
n = 100	0.045

Remember the key message of this last table: that as the sample size n goes up, the "typical" error of your point estimate will go down, as quantified by the *standard error*.

Learning check

(LC7.14) What purpose did the *sampling distributions* serve?

(LC7.15) What does the *standard error* of the sample proportion $\hat{p}$ quantify?

7.3.3   The moral of the story

Let's recap this section so far. We've seen that if a sample is generated at random, then the resulting point estimate is a "good guess" of the true unknown population parameter. In our sampling activities, since we made sure to mix the balls first before extracting a sample with the shovel, the resulting sample proportion $\hat{p}$ of the shovel's balls that were red was a "good guess" of the population proportion p of the bowl's balls that were red.

However, what do we mean by our point estimate being a "good guess"? Sometimes, we'll get an estimate that is less than the true value of the population parameter, while at other times we'll get an estimate that is greater. This is due to sampling variation. However, despite this sampling variation, our estimates will "on average" be correct and thus will be centered at the true value. This is because our sampling was done at random and thus in an unbiased fashion.

In our sampling activities, sometimes our sample proportion $\hat{p}$ was less than the true population proportion p, while at other times it was greater. This was due to the sampling variability. However, despite this sampling variation, our sample proportions $\hat{p}$ were "on average" correct and thus were centered at the true value of the population proportion p. This is because we mixed our bowl before taking samples and thus the sampling was done at random and thus in an unbiased fashion. This is also known as having an *accurate* estimate.

What was the value of the population proportion p of the $N = 2400$ balls in the actual bowl that were red? There were 900 red balls, for a proportion red of $900/2400 = 0.375 = 37.5\%$! How do we know this? Did the authors do an exhaustive count of all the balls? No! They were listed in the contents of the box that the bowl came in! Hence we were able to make the contents of the virtual `bowl` match the tactile bowl:

```
bowl %>%
  summarize(sum_red = sum(color == "red"),
            sum_not_red = sum(color != "red"))
```

```
# A tibble: 1 x 2
  sum_red sum_not_red
    <int>       <int>
1     900        1500
```

Let's re-display our sampling distributions from Figures 7.12 and 7.14, but now with a vertical red line marking the true population proportion p of balls that are red = 37.5% in Figure 7.15. We see that while there is a certain amount of error in the sample proportions $\hat{p}$ for all three sampling distributions, on average the $\hat{p}$ are centered at the true population proportion red p.

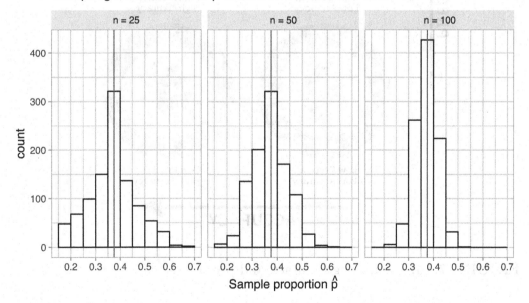

FIGURE 7.15: Three sampling distributions with population proportion p marked by vertical line.

We also saw in this section that as your sample size n increases, your point estimates will vary less and less and be more and more concentrated around the true population parameter. This variation is quantified by the decreasing *standard error*. In other words, the typical error of your point estimates will decrease. In our sampling exercise, as the sample size increased, the variation of our sample proportions $\hat{p}$ decreased. You can observe this behavior in Figure 7.15. This is also known as having a *precise* estimate.

So random sampling ensures our point estimates are *accurate*, while on the other hand having a large sample size ensures our point estimates are *precise*. While the terms "accuracy" and "precision" may sound like they mean the same thing, there is a subtle difference. Accuracy describes how "on target" our estimates are, whereas precision describes how "consistent" our estimates are. Figure 7.16 illustrates the difference.

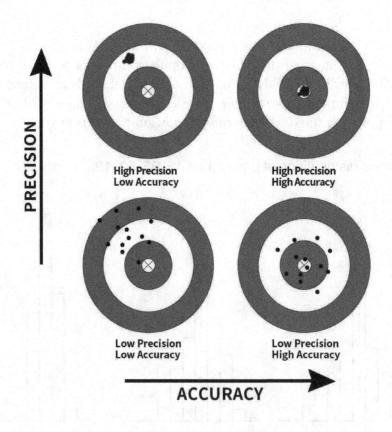

FIGURE 7.16: Comparing accuracy and precision.

At this point, you might be asking yourself: "If we already knew the true proportion of the bowl's balls that are red was 37.5%, then why did we do any sampling?". You might also be asking: "Why did we take 1000 repeated samples of size n = 25, 50, and 100? Shouldn't we be taking only *one* sample that's as large as possible?". If you did ask yourself these questions, your suspicion is merited!

The sampling activity involving the bowl is merely an *idealized version* of how sampling is done in real life. We performed this exercise only to study and understand:

1. The effect of sampling variation.
2. The effect of sample size on sampling variation.

This is not how sampling is done in real life. In a real-life scenario, we won't know what the true value of the population parameter is. Furthermore, we wouldn't take 1000 repeated/replicated samples, but rather a single sample that's as large as we can afford. In the next section, let's now study a real-life example of sampling: polls.

Learning check

(**LC7.16**) The table that follows is a version of Table 7.3 matching sample sizes n to different *standard errors* of the sample proportion $\hat{p}$, but with the rows randomly re-ordered and the sample sizes removed. Fill in the table by matching the correct sample sizes to the correct standard errors.

TABLE 7.4: Standard errors of $\hat{p}$ based on n = 25, 50, 100

Sample size	Standard error of $\hat{p}$
n =	0.094
n =	0.045
n =	0.069

For the following four *Learning checks*, let the *estimate* be the sample proportion $\hat{p}$: the proportion of a shovel's balls that were red. It estimates the population proportion p: the proportion of the bowl's balls that were red.

(**LC7.17**) What is the difference between an *accurate* and a *precise* estimate?

(**LC7.18**) How do we ensure that an estimate is *accurate*? How do we ensure that an estimate is *precise*?

(**LC7.19**) In a real-life situation, we would not take 1000 different samples to infer about a population, but rather only one. Then, what was the purpose of our exercises where we took 1000 different samples?

(LC7.20) Figure 7.16 with the targets shows four combinations of "accurate versus precise" estimates. Draw four corresponding *sampling distributions* of the sample proportion $\hat{p}$, like the one in the leftmost plot in Figure 7.15.

7.4   Case study: Polls

Let's now switch gears to a more realistic sampling scenario than our bowl activity: a poll. In practice, pollsters do not take 1000 repeated samples as we did in our previous sampling activities, but rather take only a *single sample* that's as large as possible.

On December 4, 2013, National Public Radio in the US reported on a poll of President Obama's approval rating among young Americans aged 18-29 in an article, "Poll: Support For Obama Among Young Americans Eroding."[3] The poll was conducted by the Kennedy School's Institute of Politics at Harvard University. A quote from the article:

> After voting for him in large numbers in 2008 and 2012, young Americans are souring on President Obama.
>
> According to a new Harvard University Institute of Politics poll, just 41 percent of millennials — adults ages 18-29 — approve of Obama's job performance, his lowest-ever standing among the group and an 11-point drop from April.

Let's tie elements of the real-life poll in this new article with our "tactile" and "virtual" bowl activity from Sections 7.1 and 7.2 using the terminology, notations, and definitions we learned in Section 7.3. You'll see that our sampling activity with the bowl is an idealized version of what pollsters are trying to do in real life.

First, who is the **(Study) Population** of N individuals or observations of interest?

- Bowl: $N = 2400$ identically sized red and white balls

---

[3]https://www.npr.org/sections/itsallpolitics/2013/12/04/248793753/poll-support-for-obama-among-young-americans-eroding

- Obama poll: $N = ?$ young Americans aged 18-29

Second, what is the **population parameter**?

- Bowl: The population proportion p of *all* the balls in the bowl that are red.
- Obama poll: The population proportion p of *all* young Americans who approve of Obama's job performance.

Third, what would a **census** look like?

- Bowl: Manually going over all $N = 2400$ balls and exactly computing the population proportion p of the balls that are red.
- Obama poll: Locating all N young Americans and asking them all if they approve of Obama's job performance. In this case, we don't even know what the population size N is!

Fourth, how do you perform **sampling** to obtain a sample of size n?

- Bowl: Using a shovel with n slots.
- Obama poll: One method is to get a list of phone numbers of all young Americans and pick out n phone numbers. In this poll's case, the sample size of this poll was $n = 2089$ young Americans.

Fifth, what is your **point estimate (AKA sample statistic)** of the unknown population parameter?

- Bowl: The sample proportion $\hat{p}$ of the balls in the shovel that were red.
- Obama poll: The sample proportion $\hat{p}$ of young Americans in the sample that approve of Obama's job performance. In this poll's case, $\hat{p} = 0.41 = 41\%$, the quoted percentage in the second paragraph of the article.

Sixth, is the sampling procedure **representative**?

- Bowl: Are the contents of the shovel representative of the contents of the bowl? Because we mixed the bowl before sampling, we can feel confident that they are.
- Obama poll: Is the sample of $n = 2089$ young Americans representative of *all* young Americans aged 18-29? This depends on whether the sampling was random.

Seventh, are the samples **generalizable** to the greater population?

- Bowl: Is the sample proportion $\hat{p}$ of the shovel's balls that are red a "good guess" of the population proportion p of the bowl's balls that are red? Given that the sample was representative, the answer is yes.
- Obama poll: Is the sample proportion $\hat{p} = 0.41$ of the sample of young Americans who supported Obama a "good guess" of the population proportion p of all young Americans who supported Obama at this time in 2013? In other

words, can we confidently say that roughly 41% of *all* young Americans approved of Obama at the time of the poll? Again, this depends on whether the sampling was random.

Eighth, is the sampling procedure **unbiased**? In other words, do all observations have an equal chance of being included in the sample?

- Bowl: Since each ball was equally sized and we mixed the bowl before using the shovel, each ball had an equal chance of being included in a sample and hence the sampling was unbiased.
- Obama poll: Did all young Americans have an equal chance at being represented in this poll? Again, this depends on whether the sampling was random.

Ninth and lastly, was the sampling done at **random**?

- Bowl: As long as you mixed the bowl sufficiently before sampling, your samples would be random.
- Obama poll: Was the sample conducted at random? We can't answer this question without knowing about the *sampling methodology* used by Kennedy School's Institute of Politics at Harvard University. We'll discuss this more at the end of this section.

In other words, the poll by Kennedy School's Institute of Politics at Harvard University can be thought of as *an instance* of using the shovel to sample balls from the bowl. Furthermore, if another polling company conducted a similar poll of young Americans at roughly the same time, they would likely get a different estimate than 41%. This is due to *sampling variation*.

Let's now revisit the sampling paradigm from Subsection 7.3.1:

In general:

- If the sampling of a sample of size n is done at **random**, then
- the sample is **unbiased** and **representative** of the population of size N, thus
- any result based on the sample can **generalize** to the population, thus
- the point estimate is a **"good guess"** of the unknown population parameter, thus
- instead of performing a census, we can **infer** about the population using sampling.

Specific to the bowl:

- If we extract a sample of $n = 50$ balls at **random**, in other words, we mix all of the equally sized balls before using the shovel, then

- the contents of the shovel are an **unbiased representation** of the contents of the bowl's 2400 balls, thus
- any result based on the shovel's balls can **generalize** to the bowl, thus
- the sample proportion $\hat{p}$ of the $n = 50$ balls in the shovel that are red is a **"good guess"** of the population proportion p of the $N = 2400$ balls that are red, thus
- instead of manually going over all 2400 balls in the bowl, we can **infer** about the bowl using the shovel.

Specific to the Obama poll:

- If we had a way of contacting a **randomly** chosen sample of 2089 young Americans and polling their approval of President Obama in 2013, then
- these 2089 young Americans would be an **unbiased** and **representative** sample of *all* young Americans in 2013, thus
- any results based on this sample of 2089 young Americans can **generalize** to the entire population of *all* young Americans in 2013, thus
- the reported sample approval rating of 41% of these 2089 young Americans is a **good guess** of the true approval rating among all young Americans in 2013, thus
- instead of performing an expensive census of all young Americans in 2013, we can **infer** about all young Americans in 2013 using polling.

So as you can see, it was critical for the sample obtained by Kennedy School's Institute of Politics at Harvard University to be truly random in order to infer about *all* young Americans' opinions about Obama. Was their sample truly random? It's hard to answer such questions without knowing about the *sampling methodology* they used. For example, if this poll was conducted using only mobile phone numbers, people without mobile phones would be left out and therefore not represented in the sample. What about if Kennedy School's Institute of Politics at Harvard University conducted this poll on an internet news site? Then people who don't read this particular internet news site would be left out. Ensuring that our samples were random was easy to do in our sampling bowl exercises; however, in a real-life situation like the Obama poll, this is much harder to do.

Learning check

Comment on the representativeness of the following *sampling methodologies*:

(LC7.21) The Royal Air Force wants to study how resistant all their airplanes are to bullets. They study the bullet holes on all the airplanes on the tarmac after an air battle against the Luftwaffe (German Air Force).

(LC7.22) Imagine it is 1993, a time when almost all households had landlines. You want to know the average number of people in each household in your city. You randomly pick out 500 phone numbers from the phone book and conduct a phone survey.

(LC7.23) You want to know the prevalence of illegal downloading of TV shows among students at a local college. You get the emails of 100 randomly chosen students and ask them, "How many times did you download a pirated TV show last week?".

(LC7.24) A local college administrator wants to know the average income of all graduates in the last 10 years. So they get the records of five randomly chosen graduates, contact them, and obtain their answers.

7.5   Conclusion

7.5.1   Sampling scenarios

In this chapter, we performed both tactile and virtual sampling exercises to infer about an unknown proportion. We also presented a case study of sampling in real life with polls. In each case, we used the sample proportion $\hat{p}$ to estimate the population proportion p. However, we are not just limited to scenarios related to proportions. In other words, we can use sampling to estimate other population parameters using other point estimates as well. We present four more such scenarios in Table 7.5.

TABLE 7.5: Scenarios of sampling for inference

Scenario	Population parameter	Notation	Point estimate	Symbol(s)
1	Population proportion	p	Sample proportion	$\hat{p}$
2	Population mean	μ	Sample mean	$\overline{x}$ or $\widehat{\mu}$
3	Difference in population proportions	$p_1 - p_2$	Difference in sample proportions	$\hat{p}_1 - \hat{p}_2$
4	Difference in population means	$\mu_1 - \mu_2$	Difference in sample means	$\overline{x}_1 - \overline{x}_2$
5	Population regression slope	β_1	Fitted regression slope	b_1 or $\widehat{\beta}_1$

In the rest of this book, we'll cover all the remaining scenarios as follows:

- In Chapter 8, we'll cover examples of statistical inference for
  - Scenario 2: The mean age μ of all pennies in circulation in the US.
  - Scenario 3: The difference $p_1 - p_2$ in the proportion of people who yawn *when seeing someone else yawn first* minus the proportion of people who yawn *without seeing someone else yawn first*. This is an example of *two-sample* inference.
- In Chapter 9, we'll cover an example of statistical inference for
  - Scenario 4: The difference $\mu_1 - \mu_2$ in mean IMDb ratings for action and romance movies. This is another example of *two-sample* inference.
- In Chapter 10, we'll cover an example of statistical inference for regression by revisiting the regression models for teaching score as a function of various instructor demographic variables you saw in Chapters 5 and 6.
  - Scenario 5: The slope β_1 of the population regression line.

7.5.2 Central Limit Theorem

What you visualized in Figures 7.12 and 7.14 and summarized in Tables 7.1 and 7.3 was a demonstration of a famous theorem, or mathematically proven truth, called the *Central Limit Theorem*. It loosely states that when sample means are based on larger and larger sample sizes, the sampling distribution of these sample means becomes both more and more normally shaped and more and more narrow.

In other words, their sampling distribution increasingly follows a *normal distribution* and the variation of these sampling distributions gets smaller, as quantified by their standard errors.

Shuyi Chiou, Casey Dunn, and Pathikrit Bhattacharyya created a 3-minute and 38-second video at https://youtu.be/jvoxEYmQHNM explaining this crucial statistical theorem using the average weight of wild bunny rabbits and the average wingspan of dragons as examples. Figure 7.17 shows a preview of this video.

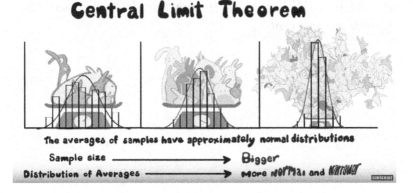

FIGURE 7.17: Preview of Central Limit Theorem video.

7.5.3 Additional resources

Solutions to all *Learning checks* can be found online in Appendix D[4].

An R script file of all R code used in this chapter is available at `https://www.moderndive.com/scripts/07-sampling.R`.

7.5.4 What's to come?

Recall in our Obama poll case study in Section 7.4 that based on this particular sample, the best guess by Kennedy School's Institute of Politics at Harvard University of the U.S. President Obama's approval rating among all young Americans was 41%. However, this isn't the end of the story. If you read the article further, it states:

> The online survey of 2,089 adults was conducted from Oct. 30 to Nov. 11, just weeks after the federal government shutdown ended and the problems surrounding the implementation of the Affordable Care Act began to take center stage. The poll's margin of error was plus or minus 2.1 percentage points.

Note the term *margin of error*, which here is "plus or minus 2.1 percentage points." Most polls won't produce an estimate that's perfectly right; there will always be a certain amount of error caused by *sampling variation*. The margin of error of plus or minus 2.1 percentage points is saying that a typical range of errors for polls of this type is about $\pm$ 2.1%, in words from about 2.1% too small to about 2.1% too big. We can restate this as the interval of $[41\% - 2.1\%, 41\% + 2.1\%] = [37.9\%, 43.1\%]$ (this notation indicates the interval contains all values between 37.9% and 43.1%, including the end points of 37.9% and 43.1%). We'll see in the next chapter that such intervals are known as *confidence intervals*.

---

[4]`https://moderndive.com/D-appendixD.html`

8

Bootstrapping and Confidence Intervals

In Chapter 7, we studied sampling. We started with a "tactile" exercise where we wanted to know the proportion of balls in the sampling bowl in Figure 7.1 that are red. While we could have performed an exhaustive count, this would have been a tedious process. So instead, we used a shovel to extract a sample of 50 balls and used the resulting proportion that were red as an *estimate*. Furthermore, we made sure to mix the bowl's contents before every use of the shovel. Because of the randomness created by the mixing, different uses of the shovel yielded different proportions red and hence different estimates of the proportion of the bowl's balls that are red.

We then mimicked this "tactile" sampling exercise with an equivalent "virtual" sampling exercise performed on the computer. Using our computer's random number generator, we quickly mimicked the above sampling procedure a large number of times. In Subsection 7.2.4, we quickly repeated this sampling procedure 1000 times, using three different "virtual" shovels with 25, 50, and 100 slots. We visualized these three sets of 1000 estimates in Figure 7.15 and saw that as the sample size increased, the variation in the estimates decreased.

In doing so, what we did was construct *sampling distributions*. The motivation for taking 1000 repeated samples and visualizing the resulting estimates was to study how these estimates varied from one sample to another; in other words, we wanted to study the effect of *sampling variation*. We quantified the variation of these estimates using their standard deviation, which has a special name: the *standard error*. In particular, we saw that as the sample size increased from 25 to 50 to 100, the standard error decreased and thus the sampling distributions narrowed. Larger sample sizes led to more *precise* estimates that varied less around the center.

We then tied these sampling exercises to terminology and mathematical notation related to sampling in Subsection 7.3.1. Our *study population* was the large bowl with $N = 2400$ balls, while the *population parameter*, the unknown quantity of interest, was the population proportion p of the bowl's balls that were red. Since performing a *census* would be expensive in terms of time and energy, we instead extracted a *sample* of size $n = 50$. The *point estimate*, also known as a *sample statistic*, used to estimate p was the sample proportion $\hat{p}$

of these 50 sampled balls that were red. Furthermore, since the sample was obtained at *random*, it can be considered as *unbiased* and *representative* of the population. Thus any results based on the sample could be *generalized* to the population. Therefore, the proportion of the shovel's balls that were red was a "good guess" of the proportion of the bowl's balls that are red. In other words, we used the sample to *infer* about the population.

However, as described in Section 7.2, both the tactile and virtual sampling exercises are not what one would do in real life; this was merely an activity used to study the effects of sampling variation. In a real-life situation, we would not take 1000 samples of size n, but rather take a *single* representative sample that's as large as possible. Additionally, we knew that the true proportion of the bowl's balls that were red was 37.5%. In a real-life situation, we will not know what this value is. Because if we did, then why would we take a sample to estimate it?

An example of a realistic sampling situation would be a poll, like the Obama poll[1] you saw in Section 7.4. Pollsters did not know the true proportion of *all* young Americans who supported President Obama in 2013, and thus they took a single sample of size $n = 2089$ young Americans to estimate this value.

So how does one quantify the effects of sampling variation when you only have a *single sample* to work with? You cannot directly study the effects of sampling variation when you only have one sample. One common method to study this is *bootstrapping resampling*, which will be the focus of the earlier sections of this chapter.

Furthermore, what if we would like not only a single estimate of the unknown population parameter, but also a *range of highly plausible* values? Going back to the Obama poll article, it stated that the pollsters' estimate of the proportion of all young Americans who supported President Obama was 41%. But in addition it stated that the poll's "margin of error was plus or minus 2.1 percentage points." This "plausible range" was [41% - 2.1%, 41% + 2.1%] = [38.9%, 43.1%]. This range of plausible values is what's known as a *confidence interval*, which will be the focus of the later sections of this chapter.

Needed packages

Let's load all the packages needed for this chapter (this assumes you've already installed them). Recall from our discussion in Section 4.4 that loading the `tidyverse` package by running `library(tidyverse)` loads the following commonly used data science packages all at once:

---

[1]https://www.npr.org/sections/itsallpolitics/2013/12/04/248793753/poll-support-for-obama-among-young-americans-eroding

- `ggplot2` for data visualization
- `dplyr` for data wrangling
- `tidyr` for converting data to tidy format
- `readr` for importing spreadsheet data into R
- As well as the more advanced `purrr`, `tibble`, `stringr`, and `forcats` packages

If needed, read Section 1.3 for information on how to install and load R packages.

```
library(tidyverse)
library(moderndive)
library(infer)
```

8.1 Pennies activity

As we did in Chapter 7, we'll begin with a hands-on tactile activity.

8.1.1 What is the average year on US pennies in 2019?

Try to imagine all the pennies being used in the United States in 2019. That's a lot of pennies! Now say we're interested in the average year of minting of *all* these pennies. One way to compute this value would be to gather up all pennies being used in the US, record the year, and compute the average. However, this would be near impossible! So instead, let's collect a *sample* of 50 pennies from a local bank in downtown Northampton, Massachusetts, USA as seen in Figure 8.1.

FIGURE 8.1: Collecting a sample of 50 US pennies from a local bank.

An image of these 50 pennies can be seen in Figure 8.2. For each of the 50 pennies starting in the top left, progressing row-by-row, and ending in the

bottom right, we assigned an "ID" identification variable and marked the year
of minting.

FIGURE 8.2: 50 US pennies labelled.

The moderndive package contains this data on our 50 sampled pennies in the
pennies_sample data frame:

```
pennies_sample
```

```
# A tibble: 50 x 2
      ID  year
   <int> <dbl>
1      1  2002
2      2  1986
3      3  2017
4      4  1988
5      5  2008
6      6  1983
7      7  2008
8      8  1996
9      9  2004
10    10  2000
# ... with 40 more rows
```

The pennies_sample data frame has 50 rows corresponding to each penny with
two variables. The first variable ID corresponds to the ID labels in Figure 8.2,

whereas the second variable year corresponds to the year of minting saved as a numeric variable, also known as a double (dbl).

Based on these 50 sampled pennies, what can we say about *all* US pennies in 2019? Let's study some properties of our sample by performing an exploratory data analysis. Let's first visualize the distribution of the year of these 50 pennies using our data visualization tools from Chapter 2. Since year is a numerical variable, we use a histogram in Figure 8.3 to visualize its distribution.

```
ggplot(pennies_sample, aes(x = year)) +
  geom_histogram(binwidth = 10, color = "white")
```

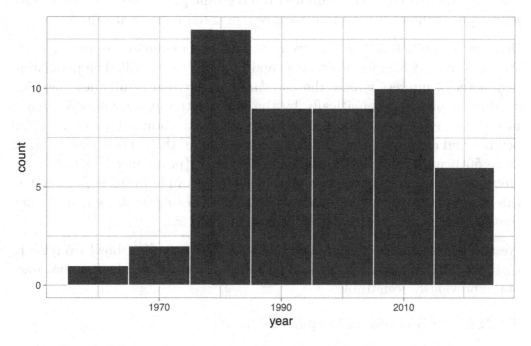

FIGURE 8.3: Distribution of year on 50 US pennies.

Observe a slightly left-skewed distribution, since most pennies fall between 1980 and 2010 with only a few pennies older than 1970. What is the average year for the 50 sampled pennies? Eyeballing the histogram it appears to be around 1990. Let's now compute this value exactly using our data wrangling tools from Chapter 3.

```
pennies_sample %>%
  summarize(mean_year = mean(year))
```

```
# A tibble: 1 x 1
```

```
mean_year
   <dbl>
1  1995.44
```

Thus, if we're willing to assume that `pennies_sample` is a representative sample from *all* US pennies, a "good guess" of the average year of minting of all US pennies would be 1995.44. In other words, around 1995. This should all start sounding similar to what we did previously in Chapter 7!

In Chapter 7, our *study population* was the bowl of $N = 2400$ balls. Our *population parameter* was the *population proportion* of these balls that were red, denoted by p. In order to estimate p, we extracted a sample of 50 balls using the shovel. We then computed the relevant *point estimate*: the *sample proportion* of these 50 balls that were red, denoted mathematically by $\hat{p}$.

Here our population is $N =$ whatever the number of pennies are being used in the US, a value which we don't know and probably never will. The population parameter of interest is now the *population mean* year of all these pennies, a value denoted mathematically by the Greek letter μ (pronounced "mu"). In order to estimate μ, we went to the bank and obtained a sample of 50 pennies and computed the relevant point estimate: the *sample mean* year of these 50 pennies, denoted mathematically by $\bar{x}$ (pronounced "x-bar"). An alternative and more intuitive notation for the sample mean is $\hat{\mu}$. However, this is unfortunately not as commonly used, so in this book we'll stick with convention and always denote the sample mean as $\bar{x}$.

We summarize the correspondence between the sampling bowl exercise in Chapter 7 and our pennies exercise in Table 8.1, which are the first two rows of the previously seen Table 7.5.

TABLE 8.1: Scenarios of sampling for inference

Scenario	Population parameter	Notation	Point estimate	Symbol(s)
1	Population proportion	p	Sample proportion	$\hat{p}$
2	Population mean	μ	Sample mean	$\bar{x}$ or $\hat{\mu}$

Going back to our 50 sampled pennies in Figure 8.2, the point estimate of interest is the sample mean $\bar{x}$ of 1995.44. This quantity is an *estimate* of the population mean year of *all* US pennies μ.

Recall that we also saw in Chapter 7 that such estimates are prone to *sampling variation*. For example, in this particular sample in Figure 8.2, we observed three pennies with the year 1999. If we sampled another 50 pennies, would we observe exactly three pennies with the year 1999 again? More than likely not. We might observe none, one, two, or maybe even all 50! The same can be said for the other 26 unique years that are represented in our sample of 50 pennies.

To study the effects of *sampling variation* in Chapter 7, we took many samples, something we could easily do with our shovel. In our case with pennies, however, how would we obtain another sample? By going to the bank and getting another roll of 50 pennies.

Say we're feeling lazy, however, and don't want to go back to the bank. How can we study the effects of sampling variation using our *single sample*? We will do so using a technique known as *bootstrap resampling with replacement*, which we now illustrate.

8.1.2 Resampling once

Step 1: Let's print out identically sized slips of paper representing our 50 pennies as seen in Figure 8.4.

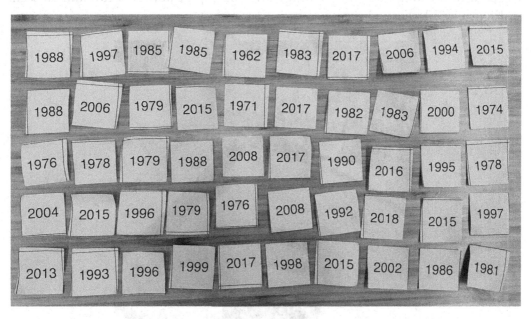

FIGURE 8.4: Step 1: 50 slips of paper representing 50 US pennies.

Step 2: Put the 50 slips of paper into a hat or tuque as seen in Figure 8.5.

FIGURE 8.5: Step 2: Putting 50 slips of paper in a hat.

Step 3: Mix the hat's contents and draw one slip of paper at random as seen in Figure 8.6. Record the year.

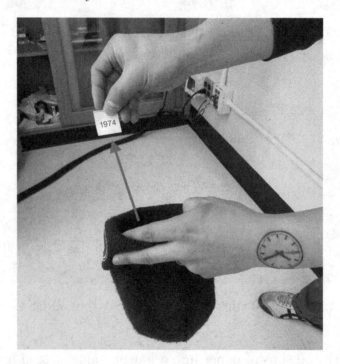

FIGURE 8.6: Step 3: Drawing one slip of paper at random.

Step 4: Put the slip of paper back in the hat! In other words, replace it as seen in Figure 8.7.

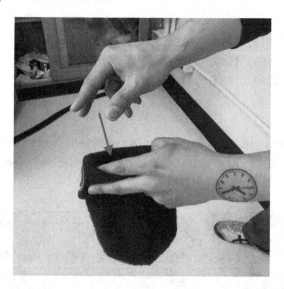

FIGURE 8.7: Step 4: Replacing slip of paper.

Step 5: Repeat Steps 3 and 4 a total of 49 more times, resulting in 50 recorded years.

What we just performed was a *resampling* of the original sample of 50 pennies. We are not sampling 50 pennies from the population of all US pennies as we did in our trip to the bank. Instead, we are mimicking this act by resampling 50 pennies from our original sample of 50 pennies.

Now ask yourselves, why did we replace our resampled slip of paper back into the hat in Step 4? Because if we left the slip of paper out of the hat each time we performed Step 4, we would end up with the same 50 original pennies! In other words, replacing the slips of paper induces *sampling variation*.

Being more precise with our terminology, we just performed a *resampling with replacement* from the original sample of 50 pennies. Had we left the slip of paper out of the hat each time we performed Step 4, this would be *resampling without replacement*.

Let's study our 50 resampled pennies via an exploratory data analysis. First, let's load the data into R by manually creating a data frame `pennies_resample` of our 50 resampled values. We'll do this using the `tibble()` command from the `dplyr` package. Note that the 50 values you resample will almost certainly not be the same as ours given the inherent randomness.

```
pennies_resample <- tibble(
  year = c(1976, 1962, 1976, 1983, 2017, 2015, 2015, 1962, 2016, 1976,
           2006, 1997, 1988, 2015, 2015, 1988, 2016, 1978, 1979, 1997,
```

```
            1974, 2013, 1978, 2015, 2008, 1982, 1986, 1979, 1981, 2004,
            2000, 1995, 1999, 2006, 1979, 2015, 1979, 1998, 1981, 2015,
            2000, 1999, 1988, 2017, 1992, 1997, 1990, 1988, 2006, 2000)
)
```

The 50 values of `year` in `pennies_resample` represent a resample of size 50 from the original sample of 50 pennies. We display the 50 resampled pennies in Figure 8.8.

FIGURE 8.8: 50 resampled US pennies labelled.

Let's compare the distribution of the numerical variable `year` of our 50 resampled pennies with the distribution of the numerical variable `year` of our original sample of 50 pennies in Figure 8.9.

```
ggplot(pennies_resample, aes(x = year)) +
  geom_histogram(binwidth = 10, color = "white") +
  labs(title = "Resample of 50 pennies")
ggplot(pennies_sample, aes(x = year)) +
  geom_histogram(binwidth = 10, color = "white") +
  labs(title = "Original sample of 50 pennies")
```

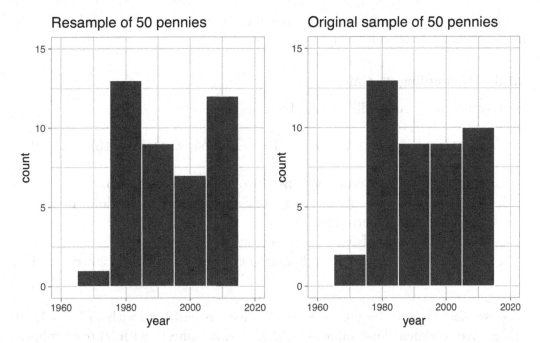

FIGURE 8.9: Comparing `year` in the resampled `pennies_resample` with the original sample `pennies_sample`.

Observe in Figure 8.9 that while the general shapes of both distributions of `year` are roughly similar, they are not identical.

Recall from the previous section that the sample mean of the original sample of 50 pennies from the bank was 1995.44. What about for our resample? Any guesses? Let's have `dplyr` help us out as before:

```
pennies_resample %>%
    summarize(mean_year = mean(year))
```

```
# A tibble: 1 x 1
  mean_year
      <dbl>
1      1996
```

We obtained a different mean year of 1996. This variation is induced by the resampling *with replacement* we performed earlier.

What if we repeated this resampling exercise many times? Would we obtain the same mean `year` each time? In other words, would our guess at the mean year of all pennies in the US in 2019 be exactly 1996 every time? Just as we did in Chapter 7, let's perform this resampling activity with the help of some of our friends: 35 friends in total.

8.1.3   Resampling 35 times

Each of our 35 friends will repeat the same five steps:

1. Start with 50 identically sized slips of paper representing the 50 pennies.
2. Put the 50 small pieces of paper into a hat or beanie cap.
3. Mix the hat's contents and draw one slip of paper at random. Record the year in a spreadsheet.
4. Replace the slip of paper back in the hat!
5. Repeat Steps 3 and 4 a total of 49 more times, resulting in 50 recorded years.

Since we had 35 of our friends perform this task, we ended up with $35 \cdot 50 = 1750$ values. We recorded these values in a shared spreadsheet[2] with 50 rows (plus a header row) and 35 columns. We display a snapshot of the first 10 rows and five columns of this shared spreadsheet in Figure 8.10.

Arianna	Artemis	Bea	Camryn	Cassandra
1988	2018	2016	2002	2015
2002	1988	1971	1997	1976
2015	1999	1986	2002	2015
1998	2015	2002	2013	1981
1979	1962	1992	1997	1988
1971	2004	1976	1979	1985
1971	2018	2015	2018	1979
2015	1988	1985	1971	1971
1988	2013	1976	1998	1978
1979	1988	1999	1996	1979
1982	2008	2013	1999	1986
2004	1983	1997	1983	1974

FIGURE 8.10: Snapshot of shared spreadsheet of resampled pennies.

For your convenience, we've taken these $35 \cdot 50 = 1750$ values and saved them in `pennies_resamples`, a "tidy" data frame included in the `moderndive` package. We saw what it means for a data frame to be "tidy" in Subsection 4.2.1.

---

[2]https://docs.google.com/spreadsheets/d/1y3kOsU_wDrDd5eiJbEtLeHT9L5SvpZb_TrzwFBsouk0/

```
pennies_resamples
```

```
# A tibble: 1,750 x 3
# Groups:   name [35]
   replicate name      year
       <int> <chr>    <dbl>
1          1 Arianna   1988
2          1 Arianna   2002
3          1 Arianna   2015
4          1 Arianna   1998
5          1 Arianna   1979
6          1 Arianna   1971
7          1 Arianna   1971
8          1 Arianna   2015
9          1 Arianna   1988
10         1 Arianna   1979
# ... with 1,740 more rows
```

What did each of our 35 friends obtain as the mean year? Once again, dplyr to
the rescue! After grouping the rows by name, we summarize each group of 50
rows by their mean year:

```
resampled_means <- pennies_resamples %>%
  group_by(name) %>%
  summarize(mean_year = mean(year))
resampled_means
```

```
# A tibble: 35 x 2
   name       mean_year
   <chr>          <dbl>
1  Arianna       1992.5
2  Artemis      1996.42
3  Bea          1996.32
4  Camryn        1996.9
5  Cassandra    1991.22
6  Cindy        1995.48
7  Claire       1995.52
8  Dahlia       1998.48
9  Dan          1993.86
10 Eindra       1993.56
# ... with 25 more rows
```

Observe that `resampled_means` has 35 rows corresponding to the 35 means based
on the 35 resamples. Furthermore, observe the variation in the 35 values in the
variable `mean_year`. Let's visualize this variation using a histogram in Figure
8.11. Recall that adding the argument `boundary = 1990` to the `geom_histogram()`
sets the binning structure so that one of the bin boundaries is at 1990 exactly.

```
ggplot(resampled_means, aes(x = mean_year)) +
  geom_histogram(binwidth = 1, color = "white", boundary = 1990) +
  labs(x = "Sampled mean year")
```

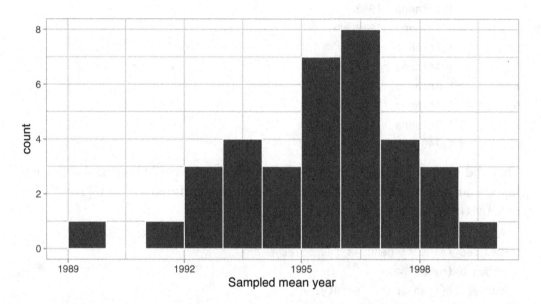

FIGURE 8.11: Distribution of 35 sample means from 35 resamples.

Observe in Figure 8.11 that the distribution looks roughly normal and that we
rarely observe sample mean years less than 1992 or greater than 2000. Also
observe how the distribution is roughly centered at 1995, which is close to the
sample mean of 1995.44 of the *original sample* of 50 pennies from the bank.

8.1.4 What did we just do?

What we just demonstrated in this activity is the statistical procedure known
as *bootstrap resampling with replacement*. We used *resampling* to mimic the
sampling variation we studied in Chapter 7 on sampling. However, in this case,
we did so using only a *single* sample from the population.

In fact, the histogram of sample means from 35 resamples in Figure 8.11
is called the *bootstrap distribution*. It is an *approximation* to the *sampling*

distribution of the sample mean, in the sense that both distributions will have a similar shape and similar spread. In fact in the upcoming Section 8.7, we'll show you that this is the case. Using this bootstrap distribution, we can study the effect of sampling variation on our estimates. In particular, we'll study the typical "error" of our estimates, known as the *standard error*.

In Section 8.2 we'll mimic our tactile resampling activity virtually on the computer, allowing us to quickly perform the resampling many more than 35 times. In Section 8.3 we'll define the statistical concept of a *confidence interval*, which builds off the concept of bootstrap distributions.

In Section 8.4, we'll construct confidence intervals using the `dplyr` package, as well as a new package: the `infer` package for "tidy" and transparent statistical inference. We'll introduce the "tidy" statistical inference framework that was the motivation for the `infer` package pipeline. The `infer` package will be the driving package throughout the rest of this book.

As we did in Chapter 7, we'll tie all these ideas together with a real-life case study in Section 8.6. This time we'll look at data from an experiment about yawning from the US television show *Mythbusters*.

8.2 Computer simulation of resampling

Let's now mimic our tactile resampling activity virtually with a computer.

8.2.1 Virtually resampling once

First, let's perform the virtual analog of resampling once. Recall that the `pennies_sample` data frame included in the `moderndive` package contains the years of our original sample of 50 pennies from the bank. Furthermore, recall in Chapter 7 on sampling that we used the `rep_sample_n()` function as a virtual shovel to sample balls from our virtual bowl of 2400 balls as follows:

```
virtual_shovel <- bowl %>%
  rep_sample_n(size = 50)
```

Let's modify this code to perform the resampling with replacement of the 50 slips of paper representing our original sample 50 pennies:

```
virtual_resample <- pennies_sample %>%
  rep_sample_n(size = 50, replace = TRUE)
```

Observe how we explicitly set the `replace` argument to `TRUE` in order to tell `rep_sample_n()` that we would like to sample pennies *with* replacement. Had we not set `replace = TRUE`, the function would've assumed the default value of `FALSE` and hence done resampling *without* replacement. Additionally, since we didn't specify the number of replicates via the `reps` argument, the function assumes the default of one replicate `reps = 1`. Lastly, observe also that the `size` argument is set to match the original sample size of 50 pennies.

Let's look at only the first 10 out of 50 rows of `virtual_resample`:

```
virtual_resample
```

```
# A tibble: 50 x 3
# Groups:    replicate [1]
   replicate    ID   year
       <int> <int>  <dbl>
1          1    37   1962
2          1     1   2002
3          1    45   1997
4          1    28   2006
5          1    50   2017
6          1    10   2000
7          1    16   2015
8          1    47   1982
9          1    23   1998
10         1    44   2015
# ... with 40 more rows
```

The replicate variable only takes on the value of 1 corresponding to us only having `reps = 1`, the `ID` variable indicates which of the 50 pennies from `pennies_sample` was resampled, and `year` denotes the year of minting. Let's now compute the mean `year` in our virtual resample of size 50 using data wrangling functions included in the `dplyr` package:

```
virtual_resample %>%
  summarize(resample_mean = mean(year))
```

```
# A tibble: 1 x 2
  replicate resample_mean
      <int>         <dbl>
1         1          1996
```

As we saw when we did our tactile resampling exercise, the resulting mean year is different than the mean year of our 50 originally sampled pennies of 1995.44.

8.2.2 Virtually resampling 35 times

Let's now perform the virtual analog of our 35 friends' resampling. Using these results, we'll be able to study the variability in the sample means from 35 resamples of size 50. Let's first add a `reps = 35` argument to `rep_sample_n()` to indicate we would like 35 replicates. Thus, we want to repeat the resampling with the replacement of 50 pennies 35 times.

```
virtual_resamples <- pennies_sample %>%
  rep_sample_n(size = 50, replace = TRUE, reps = 35)
virtual_resamples
```

```
# A tibble: 1,750 x 3
# Groups:    replicate [35]
   replicate    ID  year
       <int> <int> <dbl>
1          1    21  1981
2          1    34  1985
3          1     4  1988
4          1    11  1994
5          1    26  1979
6          1     8  1996
7          1    19  1983
8          1    21  1981
9          1    49  2006
10         1     2  1986
# ... with 1,740 more rows
```

The resulting `virtual_resamples` data frame has $35 \cdot 50 = 1750$ rows corresponding to 35 resamples of 50 pennies. Let's now compute the resulting 35 sample means using the same `dplyr` code as we did in the previous section, but this time adding a `group_by(replicate)`:

```
virtual_resampled_means <- virtual_resamples %>%
  group_by(replicate) %>%
  summarize(mean_year = mean(year))
virtual_resampled_means
```

```
# A tibble: 35 x 2
   replicate mean_year
       <int>     <dbl>
1          1   1995.58
2          2   1999.74
3          3   1993.7
4          4   1997.1
5          5   1999.42
6          6   1995.12
7          7   1994.94
8          8   1997.78
9          9   1991.26
10        10   1996.88
# ... with 25 more rows
```

Observe that `virtual_resampled_means` has 35 rows, corresponding to the 35 resampled means. Furthermore, observe that the values of `mean_year` vary. Let's visualize this variation using a histogram in Figure 8.12.

```
ggplot(virtual_resampled_means, aes(x = mean_year)) +
  geom_histogram(binwidth = 1, color = "white", boundary = 1990) +
  labs(x = "Resample mean year")
```

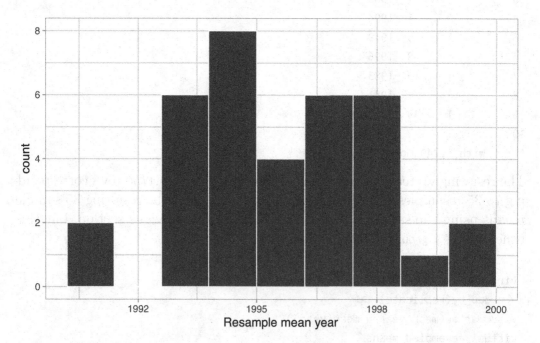

FIGURE 8.12: Distribution of 35 sample means from 35 resamples.

Let's compare our virtually constructed bootstrap distribution with the one our 35 friends constructed via our tactile resampling exercise in Figure 8.13. Observe how they are somewhat similar, but not identical.

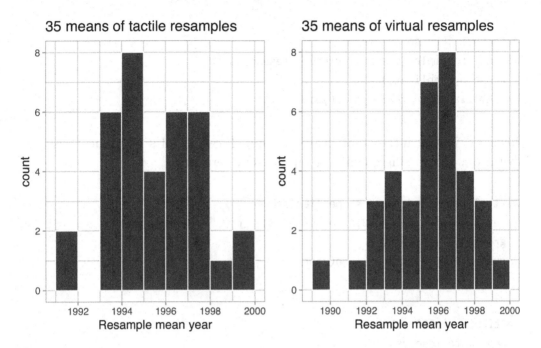

FIGURE 8.13: Comparing distributions of means from resamples.

Recall that in the "resampling with replacement" scenario we are illustrating here, both of these histograms have a special name: the *bootstrap distribution of the sample mean*. Furthermore, recall they are an approximation to the *sampling distribution* of the sample mean, a concept you saw in Chapter 7 on sampling. These distributions allow us to study the effect of sampling variation on our estimates of the true population mean, in this case the true mean year for *all* US pennies. However, unlike in Chapter 7 where we took multiple samples (something one would never do in practice), bootstrap distributions are constructed by taking multiple resamples from a *single* sample: in this case, the 50 original pennies from the bank.

8.2.3 Virtually resampling 1000 times

Remember that one of the goals of resampling with replacement is to construct the bootstrap distribution, which is an approximation of the sampling distribution. However, the bootstrap distribution in Figure 8.12 is based only on 35 resamples and hence looks a little coarse. Let's increase the number of resamples to 1000, so that we can hopefully better see the shape and the variability between different resamples.

```
# Repeat resampling 1000 times
virtual_resamples <- pennies_sample %>%
  rep_sample_n(size = 50, replace = TRUE, reps = 1000)

# Compute 1000 sample means
virtual_resampled_means <- virtual_resamples %>%
  group_by(replicate) %>%
  summarize(mean_year = mean(year))
```

However, in the interest of brevity, going forward let's combine these two operations into a single chain of pipe (%>%) operators:

```
virtual_resampled_means <- pennies_sample %>%
  rep_sample_n(size = 50, replace = TRUE, reps = 1000) %>%
  group_by(replicate) %>%
  summarize(mean_year = mean(year))
virtual_resampled_means
```

```
# A tibble: 1,000 x 2
   replicate mean_year
       <int>     <dbl>
1          1    1992.6
2          2   1994.78
3          3   1994.74
4          4   1997.88
5          5      1990
6          6   1999.48
7          7   1990.26
8          8    1993.2
9          9   1994.88
10        10    1996.3
# ... with 990 more rows
```

In Figure 8.14 let's visualize the bootstrap distribution of these 1000 means based on 1000 virtual resamples:

```
ggplot(virtual_resampled_means, aes(x = mean_year)) +
  geom_histogram(binwidth = 1, color = "white", boundary = 1990) +
  labs(x = "sample mean")
```

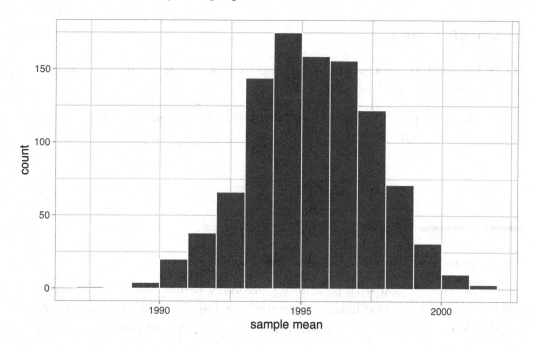

FIGURE 8.14: Bootstrap resampling distribution based on 1000 resamples.

Note here that the bell shape is starting to become much more apparent. We now have a general sense for the range of values that the sample mean may take on. But where is this histogram centered? Let's compute the mean of the 1000 resample means:

```
virtual_resampled_means %>%
  summarize(mean_of_means = mean(mean_year))
```

```
# A tibble: 1 x 1
  mean_of_means
          <dbl>
1       1995.36
```

The mean of these 1000 means is 1995.36, which is quite close to the mean of our original sample of 50 pennies of 1995.44. This is the case since each of the 1000 resamples is based on the original sample of 50 pennies.

Congratulations! You've just constructed your first bootstrap distribution! In the next section, you'll see how to use this bootstrap distribution to construct *confidence intervals*.

Learning check

(LC8.1) What is the chief difference between a bootstrap distribution and a sampling distribution?

(LC8.2) Looking at the bootstrap distribution for the sample mean in Figure 8.14, between what two values would you say *most* values lie?

8.3 Understanding confidence intervals

Let's start this section with an analogy involving fishing. Say you are trying to catch a fish. On the one hand, you could use a spear, while on the other you could use a net. Using the net will probably allow you to catch more fish!

Now think back to our pennies exercise where you are trying to estimate the true population mean year μ of *all* US pennies. Think of the value of μ as a fish.

On the one hand, we could use the appropriate *point estimate/sample statistic* to estimate μ, which we saw in Table 8.1 is the sample mean $\bar{x}$. Based on our sample of 50 pennies from the bank, the sample mean was 1995.44. Think of using this value as "fishing with a spear."

What would "fishing with a net" correspond to? Look at the bootstrap distribution in Figure 8.14 once more. Between which two years would you say that "most" sample means lie? While this question is somewhat subjective, saying that most sample means lie between 1992 and 2000 would not be unreasonable. Think of this interval as the "net."

What we've just illustrated is the concept of a *confidence interval*, which we'll abbreviate with "CI" throughout this book. As opposed to a point estimate/sample statistic that estimates the value of an unknown population parameter with a single value, a *confidence interval* gives what can be interpreted as a range of plausible values. Going back to our analogy, point estimates/sample statistics can be thought of as spears, whereas confidence intervals can be thought of as nets.

Point estimate	Confidence interval

FIGURE 8.15: Analogy of difference between point estimates and confidence intervals.

Our proposed interval of 1992 to 2000 was constructed by eye and was thus somewhat subjective. We now introduce two methods for constructing such intervals in a more exact fashion: the *percentile method* and the *standard error method*.

Both methods for confidence interval construction share some commonalities. First, they are both constructed from a bootstrap distribution, as you constructed in Subsection 8.2.3 and visualized in Figure 8.14.

Second, they both require you to specify the *confidence level*. Commonly used confidence levels include 90%, 95%, and 99%. All other things being equal, higher confidence levels correspond to wider confidence intervals, and lower confidence levels correspond to narrower confidence intervals. In this book, we'll be mostly using 95% and hence constructing "95% confidence intervals for μ" for our pennies activity.

8.3.1 Percentile method

One method to construct a confidence interval is to use the middle 95% of values of the bootstrap distribution. We can do this by computing the 2.5th and 97.5th percentiles, which are 1991.059 and 1999.283, respectively. This is known as the *percentile method* for constructing confidence intervals.

For now, let's focus only on the concepts behind a percentile method constructed confidence interval; we'll show you the code that computes these values in the next section.

Let's mark these percentiles on the bootstrap distribution with vertical lines in Figure 8.16. About 95% of the mean_year variable values in `virtual_resampled_means` fall between 1991.059 and 1999.283, with 2.5% to the left of the leftmost line and 2.5% to the right of the rightmost line.

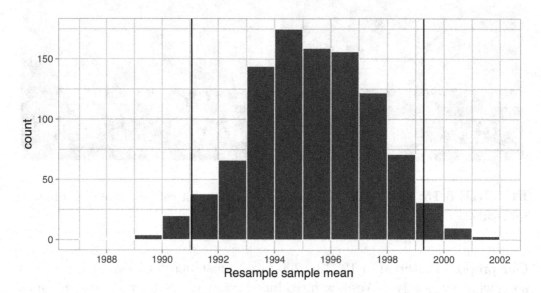

FIGURE 8.16: Percentile method 95% confidence interval. Interval endpoints marked by vertical lines.

8.3.2   Standard error method

Recall in Appendix A.2, we saw that if a numerical variable follows a normal distribution, or, in other words, the histogram of this variable is bell-shaped, then roughly 95% of values fall between ± 1.96 standard deviations of the mean. Given that our bootstrap distribution based on 1000 resamples with replacement in Figure 8.14 is normally shaped, let's use this fact about normal distributions to construct a confidence interval in a different way.

First, recall the bootstrap distribution has a mean equal to 1995.36. This value almost coincides exactly with the value of the sample mean $\bar{x}$ of our original 50 pennies of 1995.44. Second, let's compute the standard deviation of the bootstrap distribution using the values of `mean_year` in the `virtual_resampled_means` data frame:

```
virtual_resampled_means %>%
   summarize(SE = sd(mean_year))
```

```
# A tibble: 1 x 1
      SE
   <dbl>
1 2.15466
```

What is this value? Recall that the bootstrap distribution is an approximation to the sampling distribution. Recall also that the standard deviation of a sampling distribution has a special name: the *standard error*. Putting these two facts together, we can say that 2.155 is an approximation of the standard error of $\bar{x}$.

Thus, using our 95% rule of thumb about normal distributions from Appendix A.2, we can use the following formula to determine the lower and upper endpoints of a 95% confidence interval for μ:

$$\begin{aligned}
\bar{x} \pm 1.96 \cdot SE &= (\bar{x} - 1.96 \cdot SE, \bar{x} + 1.96 \cdot SE) \\
&= (1995.44 - 1.96 \cdot 2.15, 1995.44 + 1.96 \cdot 2.15) \\
&= (1991.15, 1999.73)
\end{aligned}$$

Let's now add the SE method confidence interval with dashed lines in Figure 8.17.

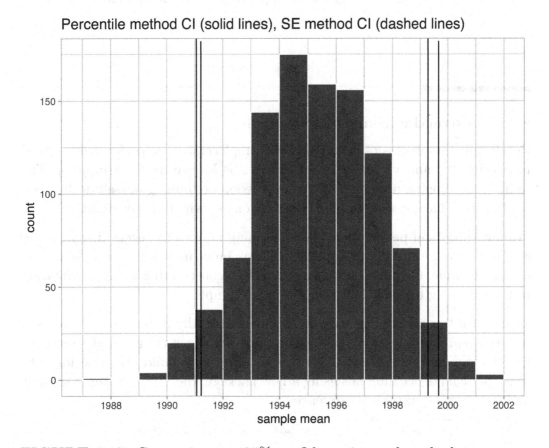

FIGURE 8.17: Comparing two 95% confidence interval methods.

We see that both methods produce nearly identical 95% confidence intervals for μ with the percentile method yielding $(1991.06, 1999.28)$ while the standard error method produces $(1991.22, 1999.66)$. However, recall that we can only use the standard error rule when the bootstrap distribution is roughly normally shaped.

Now that we've introduced the concept of confidence intervals and laid out the intuition behind two methods for constructing them, let's explore the code that allows us to construct them.

Learning check

(LC8.3) What condition about the bootstrap distribution must be met for us to be able to construct confidence intervals using the standard error method?

(LC8.4) Say we wanted to construct a 68% confidence interval instead of a 95% confidence interval for μ. Describe what changes are needed to make this happen. Hint: we suggest you look at Appendix A.2 on the normal distribution.

8.4   Constructing confidence intervals

Recall that the process of resampling with replacement we performed by hand in Section 8.1 and virtually in Section 8.2 is known as *bootstrapping*. The term bootstrapping originates in the expression of "pulling oneself up by their bootstraps," meaning to "succeed only by one's own efforts or abilities."[3]

From a statistical perspective, bootstrapping alludes to succeeding in being able to study the effects of sampling variation on estimates from the "effort" of a single sample. Or more precisely, it refers to constructing an approximation to the sampling distribution using only one sample.

To perform this resampling with replacement virtually in Section 8.2, we used the `rep_sample_n()` function, making sure that the size of the resamples matched the original sample size of 50. In this section, we'll build off these ideas to construct confidence intervals using a new package: the `infer` package for "tidy" and transparent statistical inference.

---

[3]`https://en.wiktionary.org/wiki/pull_oneself_up_by_one%27s_bootstraps`

8.4.1   Original workflow

Recall that in Section 8.2, we virtually performed bootstrap resampling with replacement to construct bootstrap distributions. Such distributions are approximations to the sampling distributions we saw in Chapter 7, but are constructed using only a single sample. Let's revisit the original workflow using the `%>%` pipe operator.

First, we used the `rep_sample_n()` function to resample `size = 50` pennies with replacement from the original sample of 50 pennies in `pennies_sample` by setting `replace = TRUE`. Furthermore, we repeated this resampling 1000 times by setting `reps = 1000`:

```
pennies_sample %>%
  rep_sample_n(size = 50, replace = TRUE, reps = 1000)
```

Second, since for each of our 1000 resamples of size 50, we wanted to compute a separate sample mean, we used the `dplyr` verb `group_by()` to group observations/rows together by the `replicate` variable...

```
pennies_sample %>%
  rep_sample_n(size = 50, replace = TRUE, reps = 1000) %>%
  group_by(replicate)
```

... followed by using `summarize()` to compute the sample `mean()` year for each replicate group:

```
pennies_sample %>%
  rep_sample_n(size = 50, replace = TRUE, reps = 1000) %>%
  group_by(replicate) %>%
  summarize(mean_year = mean(year))
```

For this simple case, we can get by with using the `rep_sample_n()` function and a couple of `dplyr` verbs to construct the bootstrap distribution. However, using only `dplyr` verbs only provides us with a limited set of tools. For more complicated situations, we'll need a little more firepower. Let's repeat this using the `infer` package.

8.4.2   `infer` package workflow

The `infer` package is an R package for statistical inference. It makes efficient use of the `%>%` pipe operator we introduced in Section 3.1 to spell out the sequence of steps necessary to perform statistical inference in a "tidy" and transparent fashion. Furthermore, just as the dplyr package provides functions with verb-like names to perform data wrangling, the infer package provides functions with intuitive verb-like names to perform statistical inference.

Let's go back to our pennies. Previously, we computed the value of the sample mean $\bar{x}$ using the dplyr function summarize():

```
pennies_sample %>%
  summarize(stat = mean(year))
```

We'll see that we can also do this using infer functions specify() and calculate():

```
pennies_sample %>%
  specify(response = year) %>%
  calculate(stat = "mean")
```

You might be asking yourself: "Isn't the infer code longer? Why would I use that code?". While not immediately apparent, you'll see that there are three chief benefits to the infer workflow as opposed to the dplyr workflow.

First, the infer verb names better align with the overall resampling framework you need to understand to construct confidence intervals and to conduct hypothesis tests (in Chapter 9). We'll see flowchart diagrams of this framework in the upcoming Figure 8.23 and in Chapter 9 with Figure 9.14.

Second, you can jump back and forth seamlessly between confidence intervals and hypothesis testing with minimal changes to your code. This will become apparent in Subsection 9.3.2 when we'll compare the infer code for both of these inferential methods.

Third, the infer workflow is much simpler for conducting inference when you have *more than one variable*. We'll see two such situations. We'll first see situations of *two-sample* inference where the sample data is collected from two groups, such as in Section 8.6 where we study the contagiousness of yawning and in Section 9.1 where we compare promotion rates of two groups at banks in the 1970s. Then in Section 10.4, we'll see situations of *inference for regression* using the regression models you fit in Chapter 5.

Let's now illustrate the sequence of verbs necessary to construct a confidence interval for μ, the population mean year of minting of all US pennies in 2019.

1. specify variables

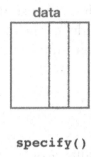

FIGURE 8.18: Diagram of the specify() verb.

As shown in Figure 8.18, the specify() function is used to choose which variables in a data frame will be the focus of our statistical inference. We do this by specifying the response argument. For example, in our pennies_sample data frame of the 50 pennies sampled from the bank, the variable of interest is year:

```
pennies_sample %>%
  specify(response = year)
```

```
Response: year (numeric)
# A tibble: 50 x 1
     year
    <dbl>
 1   2002
 2   1986
 3   2017
 4   1988
 5   2008
 6   1983
 7   2008
 8   1996
 9   2004
10   2000
# ... with 40 more rows
```

Notice how the data itself doesn't change, but the Response: year (numeric) *meta-data* does. This is similar to how the group_by() verb from dplyr doesn't change the data, but only adds "grouping" meta-data, as we saw in Section 3.4.

We can also specify which variables will be the focus of our statistical inference using a `formula = y ~ x`. This is the same formula notation you saw in Chapters 5 and 6 on regression models: the response variable `y` is separated from the explanatory variable `x` by a `~` ("tilde"). The following use of `specify()` with the `formula` argument yields the same result seen previously:

```
pennies_sample %>%
  specify(formula = year ~ NULL)
```

Since in the case of pennies we only have a response variable and no explanatory variable of interest, we set the `x` on the right-hand side of the `~` to be `NULL`.

While in the case of the pennies either specification works just fine, we'll see examples later on where the `formula` specification is simpler. In particular, this comes up in the upcoming Section 8.6 on comparing two proportions and Section 10.4 on inference for regression.

2. generate replicates

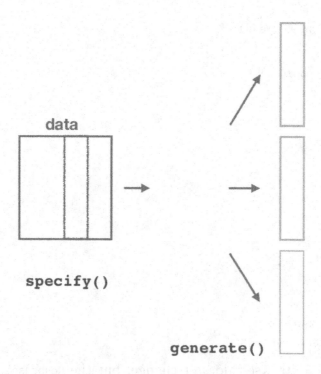

FIGURE 8.19: Diagram of generate() replicates.

After we `specify()` the variables of interest, we pipe the results into the
`generate()` function to generate replicates. Figure 8.19 shows how this is combined with `specify()` to start the pipeline. In other words, repeat the resampling
process a large number of times. Recall in Sections 8.2.2 and 8.2.3 we did this
35 and 1000 times.

The `generate()` function's first argument is `reps`, which sets the number of
replicates we would like to generate. Since we want to resample the 50 pennies
in `pennies_sample` with replacement 1000 times, we set `reps = 1000`. The second
argument `type` determines the type of computer simulation we'd like to perform.
We set this to `type = "bootstrap"` indicating that we want to perform bootstrap
resampling. You'll see different options for `type` in Chapter 9.

```
pennies_sample %>%
   specify(response = year) %>%
   generate(reps = 1000, type = "bootstrap")
```

```
Response: year (numeric)
# A tibble: 50,000 x 2
# Groups:    replicate [1,000]
   replicate  year
       <int> <dbl>
 1         1  1981
 2         1  1988
 3         1  2006
 4         1  2016
 5         1  2002
 6         1  1985
 7         1  1979
 8         1  2000
 9         1  2006
10         1  2016
# ... with 49,990 more rows
```

Observe that the resulting data frame has 50,000 rows. This is because we
performed resampling of 50 pennies with replacement 1000 times and 50,000
$= 50 \cdot 1000$.

The variable `replicate` indicates which resample each row belongs to. So it has
the value 1 50 times, the value 2 50 times, all the way through to the value
1000 50 times. The default value of the `type` argument is `"bootstrap"` in this
scenario, so if the last line was written as `generate(reps = 1000)`, we'd obtain
the same results.

Comparing with original workflow: Note that the steps of the `infer` workflow so far produce the same results as the original workflow using the `rep_sample_n()` function we saw earlier. In other words, the following two code chunks produce similar results:

```
# infer workflow:                  # Original workflow:
pennies_sample %>%                  pennies_sample %>%
  specify(response = year) %>%        rep_sample_n(size = 50, replace = TRUE,
  generate(reps = 1000)                            reps = 1000)
```

3. calculate summary statistics

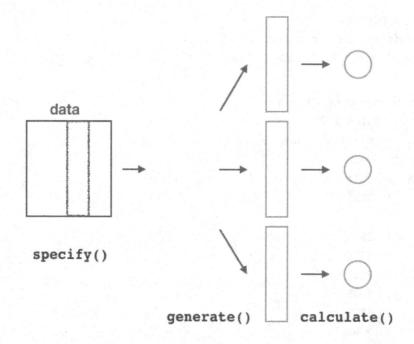

FIGURE 8.20: Diagram of calculate() summary statistics.

After we `generate()` many replicates of bootstrap resampling with replacement, we next want to summarize each of the 1000 resamples of size 50 to a single sample statistic value. As seen in the diagram, the `calculate()` function does this.

In our case, we want to calculate the mean `year` for each bootstrap resample of size 50. To do so, we set the `stat` argument to `"mean"`. You can also set the `stat` argument to a variety of other common summary statistics, like `"median"`, `"sum"`, `"sd"` (standard deviation), and `"prop"` (proportion). To see a list of all possible summary statistics you can use, type `?calculate` and read the help file.

Let's save the result in a data frame called `bootstrap_distribution` and explore its contents:

```
bootstrap_distribution <- pennies_sample %>%
  specify(response = year) %>%
  generate(reps = 1000) %>%
  calculate(stat = "mean")
bootstrap_distribution
```

```
# A tibble: 1,000 x 2
   replicate    stat
       <int>   <dbl>
1          1 1995.7
2          2 1994.04
3          3 1993.62
4          4 1994.5
5          5 1994.08
6          6 1993.6
7          7 1995.26
8          8 1996.64
9          9 1994.3
10        10 1995.94
# ... with 990 more rows
```

Observe that the resulting data frame has 1000 rows and 2 columns corresponding to the 1000 `replicate` values. It also has the mean year for each bootstrap resample saved in the variable `stat`.

Comparing with original workflow: You may have recognized at this point that the `calculate()` step in the `infer` workflow produces the same output as the `group_by() %>% summarize()` steps in the original workflow.

```
# infer workflow:               # Original workflow:
pennies_sample %>%              pennies_sample %>%
  specify(response = year) %>%    rep_sample_n(size = 50, replace = TRUE,
  generate(reps = 1000) %>%                   reps = 1000) %>%
  calculate(stat = "mean")       group_by(replicate) %>%
                                 summarize(stat = mean(year))
```

4. visualize the results

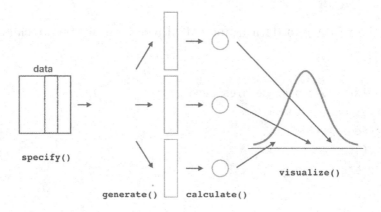

FIGURE 8.21: Diagram of visualize() results.

The visualize() verb provides a quick way to visualize the bootstrap distribution as a histogram of the numerical stat variable's values. The pipeline of the main infer verbs used for exploring bootstrap distribution results is shown in Figure 8.21.

```
visualize(bootstrap_distribution)
```

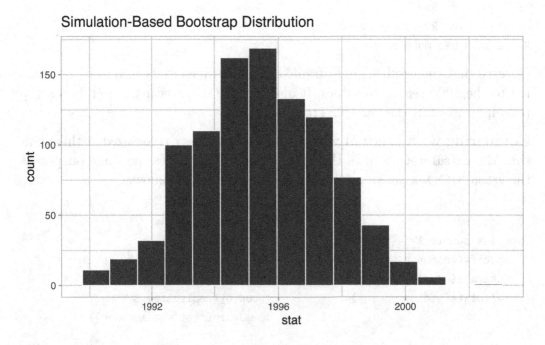

FIGURE 8.22: Bootstrap distribution.

Comparing with original workflow: In fact, `visualize()` is a *wrapper function* for the `ggplot()` function that uses a `geom_histogram()` layer. Recall that we illustrated the concept of a wrapper function in Figure 5.5 in Subsection 5.1.2.

```
# infer workflow:                    # Original workflow:
visualize(bootstrap_distribution)    ggplot(bootstrap_distribution,
                                            aes(x = stat)) +
                                        geom_histogram()
```

The `visualize()` function can take many other arguments which we'll see momentarily to customize the plot further. It also works with helper functions to do the shading of the histogram values corresponding to the confidence interval values.

Let's recap the steps of the `infer` workflow for constructing a bootstrap distribution and then visualizing it in Figure 8.23.

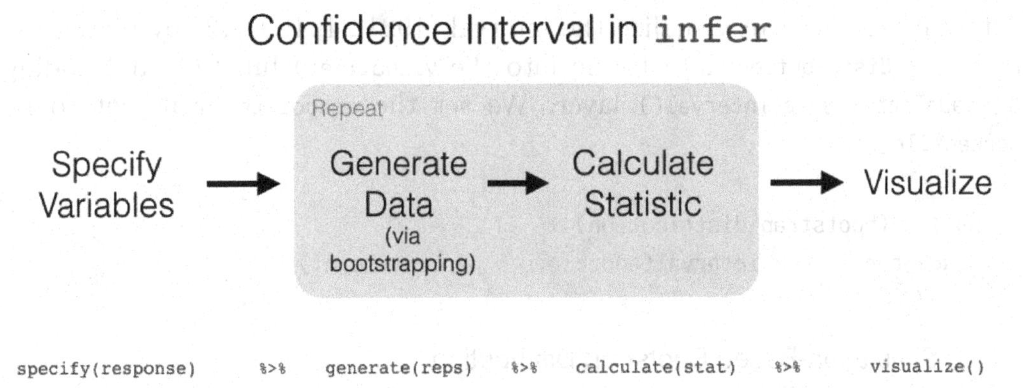

Confidence Interval in `infer`

FIGURE 8.23: infer package workflow for confidence intervals.

Recall how we introduced two different methods for constructing 95% confidence intervals for an unknown population parameter in Section 8.3: the *percentile method* and the *standard error method*. Let's now check out the `infer` package code that explicitly constructs these. There are also some additional neat functions to visualize the resulting confidence intervals built-in to the `infer` package!

8.4.3 Percentile method with `infer`

Recall the percentile method for constructing 95% confidence intervals we introduced in Subsection 8.3.1. This method sets the lower endpoint of the

confidence interval at the 2.5th percentile of the bootstrap distribution and
similarly sets the upper endpoint at the 97.5th percentile. The resulting interval
captures the middle 95% of the values of the sample mean in the bootstrap
distribution.

We can compute the 95% confidence interval by piping `bootstrap_distribution`
into the `get_confidence_interval()` function from the `infer` package, with the
confidence level set to 0.95 and the confidence interval type to be `"percentile"`.
Let's save the results in `percentile_ci`.

```
percentile_ci <- bootstrap_distribution %>%
  get_confidence_interval(level = 0.95, type = "percentile")
percentile_ci
```

```
# A tibble: 1 x 2
  `2.5%` `97.5%`
   <dbl>   <dbl>
1 1991.24 1999.42
```

Alternatively, we can visualize the interval (1991.24, 1999.42) by piping the
`bootstrap_distribution` data frame into the `visualize()` function and adding
a `shade_confidence_interval()` layer. We set the `endpoints` argument to be
`percentile_ci`.

```
visualize(bootstrap_distribution) +
  shade_confidence_interval(endpoints = percentile_ci)
```

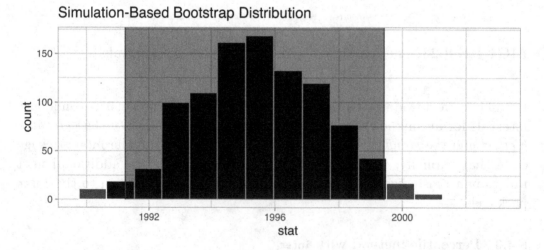

FIGURE 8.24: Percentile method 95% confidence interval shaded corresponding to potential values.

Observe in Figure 8.24 that 95% of the sample means stored in the `stat` variable in `bootstrap_distribution` fall between the two endpoints marked with the darker lines, with 2.5% of the sample means to the left of the shaded area and 2.5% of the sample means to the right. You also have the option to change the colors of the shading using the `color` and `fill` arguments.

You can also use the shorter named function `shade_ci()` and the results will be the same. This is for folks who don't want to type out all of `confidence_interval` and prefer to type out `ci` instead. Try out the following code!

```
visualize(bootstrap_distribution) +
  shade_ci(endpoints = percentile_ci, color = "hotpink", fill = "khaki")
```

8.4.4 Standard error method with `infer`

Recall the standard error method for constructing 95% confidence intervals we introduced in Subsection 8.3.2. For any distribution that is normally shaped, roughly 95% of the values lie within two standard deviations of the mean. In the case of the bootstrap distribution, the standard deviation has a special name: the *standard error*.

So in our case, 95% of values of the bootstrap distribution will lie within ± 1.96 standard errors of $\bar{x}$. Thus, a 95% confidence interval is

$$\bar{x} \pm 1.96 \cdot SE = (\bar{x} - 1.96 \cdot SE, \bar{x} + 1.96 \cdot SE).$$

Computation of the 95% confidence interval can once again be done by piping the `bootstrap_distribution` data frame we created into the `get_confidence_interval()` function. However, this time we set the first `type` argument to be `"se"`. Second, we must specify the `point_estimate` argument in order to set the center of the confidence interval. We set this to be the sample mean of the original sample of 50 pennies of 1995.44.

```
x_bar
```

```
# A tibble: 1 x 1
  mean_year
      <dbl>
1   1995.44
```

```
standard_error_ci <- bootstrap_distribution %>%
  get_confidence_interval(type = "se", point_estimate = x_bar)
standard_error_ci
```

```
# A tibble: 1 x 2
    lower    upper
    <dbl>    <dbl>
1 1991.35 1999.53
```

If we would like to visualize the interval (1991.35, 1999.53), we can once
again pipe the bootstrap_distribution data frame into the visualize() function
and add a shade_confidence_interval() layer to our plot. We set the endpoints
argument to be standard_error_ci. The resulting standard-error method based
on a 95% confidence interval for μ can be seen in Figure 8.25.

```
visualize(bootstrap_distribution) +
  shade_confidence_interval(endpoints = standard_error_ci)
```

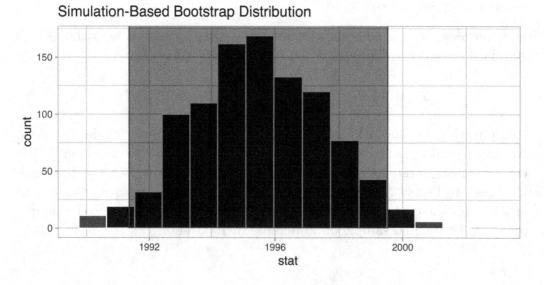

FIGURE 8.25: Standard-error-method 95% confidence interval.

As noted in Section 8.3, both methods produce similar confidence intervals:

- Percentile method: (1991.24, 1999.42)
- Standard error method: (1991.35, 1999.53)

Learning check

(LC8.5) Construct a 95% confidence interval for the *median* year of minting of *all* US pennies? Use the percentile method and, if appropriate, then use the standard-error method.

8.5   Interpreting confidence intervals

Now that we've shown you how to construct confidence intervals using a sample drawn from a population, let's now focus on how to interpret their effectiveness. The effectiveness of a confidence interval is judged by whether or not it contains the true value of the population parameter. Going back to our fishing analogy in Section 8.3, this is like asking, "Did our net capture the fish?".

So, for example, does our percentile-based confidence interval of (1991.24, 1999.42) "capture" the true mean year μ of *all* US pennies? Alas, we'll never know, because we don't know what the true value of μ is. After all, we're sampling to estimate it!

In order to interpret a confidence interval's effectiveness, we need to *know* what the value of the population parameter is. That way we can say whether or not a confidence interval "captured" this value.

Let's revisit our sampling bowl from Chapter 7. What proportion of the bowl's 2400 balls are red? Let's compute this:

```
bowl %>%
   summarize(p_red = mean(color == "red"))
```

```
# A tibble: 1 x 1
  p_red
  <dbl>
1 0.375
```

In this case, we *know* what the value of the population parameter is: we know that the population proportion p is 0.375. In other words, we know that 37.5% of the bowl's balls are red.

As we stated in Subsection 7.3.3, the sampling bowl exercise doesn't really reflect how sampling is done in real life, but rather was an *idealized* activity. In real life, we won't know what the true value of the population parameter is, hence the need for estimation.

Let's now construct confidence intervals for p using our 33 groups of friends' samples from the bowl in Chapter 7. We'll then see if the confidence intervals "captured" the true value of p, which we know to be 37.5%. That is to say, "Did the net capture the fish?".

8.5.1 Did the net capture the fish?

Recall that we had 33 groups of friends each take samples of size 50 from the bowl and then compute the sample proportion of red balls $\hat{p}$. This resulted in 33 such estimates of p. Let's focus on Ilyas and Yohan's sample, which is saved in the `bowl_sample_1` data frame in the `moderndive` package:

```
bowl_sample_1
```

```
# A tibble: 50 x 1
   color
   <chr>
 1 white
 2 white
 3 red
 4 red
 5 white
 6 white
 7 red
 8 white
 9 white
10 white
# ... with 40 more rows
```

They observed 21 red balls out of 50 and thus their sample proportion $\hat{p}$ was $21/50 = 0.42 = 42\%$. Think of this as the "spear" from our fishing analogy.

Let's now follow the `infer` package workflow from Subsection 8.4.2 to create a percentile-method-based 95% confidence interval for p using Ilyas and Yohan's sample. Think of this as the "net."

1. `specify` variables

First, we `specify()` the `response` variable of interest `color`:

```
bowl_sample_1 %>%
  specify(response = color)
```

```
Error: A level of the response variable `color` needs to be specified for the
`success` argument in `specify()`.
```

Whoops! We need to define which event is of interest! red or white balls? Since we are interested in the proportion red, let's set success to be "red":

```
bowl_sample_1 %>%
  specify(response = color, success = "red")
```

```
Response: color (factor)
# A tibble: 50 x 1
   color
   <fct>
 1 white
 2 white
 3 red
 4 red
 5 white
 6 white
 7 red
 8 white
 9 white
10 white
# ... with 40 more rows
```

2. generate replicates

Second, we generate() 1000 replicates of *bootstrap resampling with replacement* from bowl_sample_1 by setting reps = 1000 and type = "bootstrap".

```
bowl_sample_1 %>%
  specify(response = color, success = "red") %>%
  generate(reps = 1000, type = "bootstrap")
```

```
Response: color (factor)
# A tibble: 50,000 x 2
# Groups:   replicate [1,000]
   replicate color
       <int> <fct>
```

```
1          1 white
2          1 white
3          1 white
4          1 white
5          1 red
6          1 white
7          1 white
8          1 white
9          1 white
10         1 red
# ... with 49,990 more rows
```

Observe that the resulting data frame has 50,000 rows. This is because we performed resampling of 50 balls with replacement 1000 times and thus 50,000 = 50 · 1000. The variable replicate indicates which resample each row belongs to. So it has the value 1 50 times, the value 2 50 times, all the way through to the value 1000 50 times.

3. calculate summary statistics

Third, we summarize each of the 1000 resamples of size 50 with the proportion of *successes*. In other words, the proportion of the balls that are "red". We can set the summary statistic to be calculated as the proportion by setting the stat argument to be "prop". Let's save the result as sample_1_bootstrap:

```
sample_1_bootstrap <- bowl_sample_1 %>%
  specify(response = color, success = "red") %>%
  generate(reps = 1000, type = "bootstrap") %>%
  calculate(stat = "prop")
sample_1_bootstrap
```

```
# A tibble: 1,000 x 2
   replicate  stat
       <int> <dbl>
1          1  0.32
2          2  0.42
3          3  0.44
4          4  0.4
5          5  0.44
6          6  0.52
7          7  0.38
8          8  0.44
9          9  0.34
```

```
10          10  0.42
# ... with 990 more rows
```

Observe there are 1000 rows in this data frame and thus 1000 values of the variable stat. These 1000 values of stat represent our 1000 replicated values of the proportion, each based on a different resample.

4. visualize the results

Fourth and lastly, let's compute the resulting 95% confidence interval.

```
percentile_ci_1 <- sample_1_bootstrap %>%
  get_confidence_interval(level = 0.95, type = "percentile")
percentile_ci_1
```

```
# A tibble: 1 x 2
  `2.5%` `97.5%`
   <dbl>   <dbl>
1    0.3    0.56
```

Let's visualize the bootstrap distribution along with the percentile_ci_1 percentile-based 95% confidence interval for p in Figure 8.26. We'll adjust the number of bins to better see the resulting shape. Furthermore, we'll add a dashed vertical line at Ilyas and Yohan's observed $\hat{p} = 21/50 = 0.42 = 42\%$ using geom_vline().

```
sample_1_bootstrap %>%
  visualize(bins = 15) +
  shade_confidence_interval(endpoints = percentile_ci_1) +
  geom_vline(xintercept = 0.375, linetype = "dashed")
```

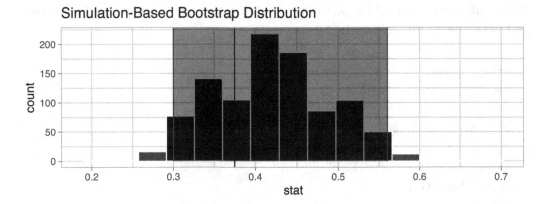

FIGURE 8.26: Bootstrap distribution.

Did Ilyas and Yohan's net capture the fish? Did their 95% confidence interval for p based on their sample contain the true value of p of 0.375? Yes! 0.375 is between the endpoints of their confidence interval (0.3, 0.56).

However, will *every* 95% confidence interval for p capture this value? In other words, if we had a different sample of 50 balls and constructed a different confidence interval, would it necessarily contain $p = 0.375$ as well? Let's see!

Let's first take a different sample from the bowl, this time using the computer as we did in Chapter 7:

```
bowl_sample_2 <- bowl %>% rep_sample_n(size = 50)
bowl_sample_2
```

```
# A tibble: 50 x 3
# Groups:   replicate [1]
   replicate ball_ID color
       <int>   <int> <chr>
1          1    1665 red
2          1    1312 red
3          1    2105 red
4          1     810 white
5          1     189 white
6          1    1429 white
7          1    2294 red
8          1    1233 white
9          1    1951 white
10         1    2061 white
# ... with 40 more rows
```

Let's reapply the same `infer` functions on `bowl_sample_2` to generate a different 95% confidence interval for p. First, we create the new bootstrap distribution and save the results in `sample_2_bootstrap`:

```
sample_2_bootstrap <- bowl_sample_2 %>%
  specify(response = color,
          success = "red") %>%
  generate(reps = 1000,
           type = "bootstrap") %>%
  calculate(stat = "prop")
sample_2_bootstrap
```

```
# A tibble: 1,000 x 2
```

```
   replicate  stat
       <int> <dbl>
1          1  0.48
2          2  0.38
3          3  0.32
4          4  0.32
5          5  0.34
6          6  0.26
7          7  0.3
8          8  0.36
9          9  0.44
10        10  0.36
# ... with 990 more rows
```

We once again compute a percentile-based 95% confidence interval for p:

```
percentile_ci_2 <- sample_2_bootstrap %>%
  get_confidence_interval(level = 0.95, type = "percentile")
percentile_ci_2
```

```
# A tibble: 1 x 2
  `2.5%` `97.5%`
   <dbl>   <dbl>
1    0.2    0.48
```

Does this new net capture the fish? In other words, does the 95% confidence interval for p based on the new sample contain the true value of p of 0.375? Yes again! 0.375 is between the endpoints of our confidence interval (0.2, 0.48).

Let's now repeat this process 100 more times: we take 100 virtual samples from the bowl and construct 100 95% confidence intervals. Let's visualize the results in Figure 8.27 where:

1. We mark the true value of $p = 0.375$ with a vertical line.
2. We mark each of the 100 95% confidence intervals with horizontal lines. These are the "nets."
3. The horizontal line is colored grey if the confidence interval "captures" the true value of p marked with the vertical line. The horizontal line is colored black otherwise.

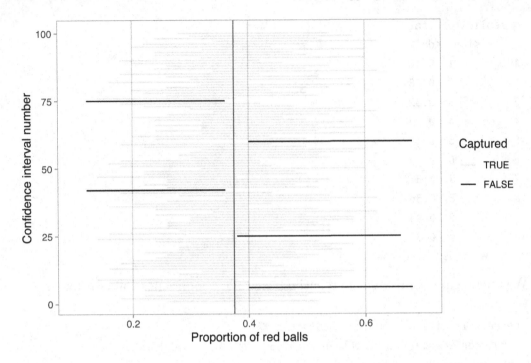

FIGURE 8.27: 100 percentile-based 95% confidence intervals for p.

Of the 100 95% confidence intervals, 95 of them captured the true value $p = 0.375$, whereas 5 of them didn't. In other words, 95 of our nets caught the fish, whereas 5 of our nets didn't.

This is where the "95% confidence level" we defined in Section 8.3 comes into play: for every 100 95% confidence intervals, we *expect* that 95 of them will capture p and that five of them won't.

Note that "expect" is a probabilistic statement referring to a long-run average. In other words, for every 100 confidence intervals, we will observe *about* 95 confidence intervals that capture p, but not necessarily exactly 95. In Figure 8.27 for example, 95 of the confidence intervals capture p.

To further accentuate our point about confidence levels, let's generate a figure similar to Figure 8.27, but this time constructing 80% standard-error method based confidence intervals instead. Let's visualize the results in Figure 8.28 with the scale on the x-axis being the same as in Figure 8.27 to make comparison easy. Furthermore, since all standard-error method 95% confidence intervals for p are centered at their respective point estimates $\hat{p}$, we mark this value on each line with dots.

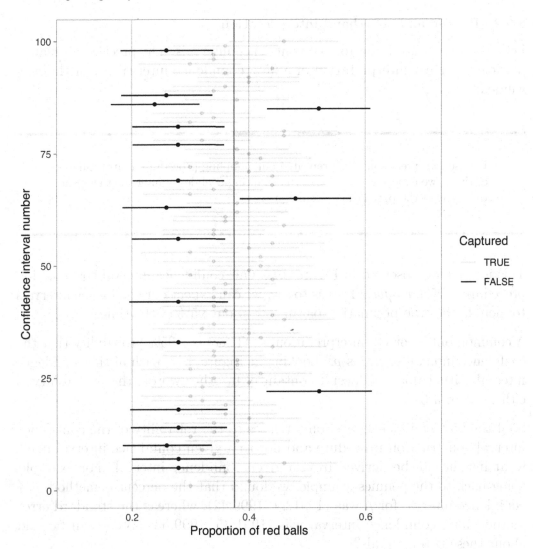

FIGURE 8.28: 100 SE-based 80% confidence intervals for p with point estimate center marked with dots.

Observe how the 80% confidence intervals are narrower than the 95% confidence intervals, reflecting our lower degree of confidence. Think of this as using a smaller "net." We'll explore other determinants of confidence interval width in the upcoming Subsection 8.5.3.

Furthermore, observe that of the 100 80% confidence intervals, 82 of them captured the population proportion $p = 0.375$, whereas 18 of them did not. Since we lowered the confidence level from 95% to 80%, we now have a much larger number of confidence intervals that failed to "catch the fish."

8.5.2   Precise and shorthand interpretation

Let's return our attention to 95% confidence intervals. The precise and mathematically correct interpretation of a 95% confidence interval is a little long-winded:

Precise interpretation: If we repeated our sampling procedure a large number of times, we expect about 95% of the resulting confidence intervals to capture the value of the population parameter.

This is what we observed in Figure 8.27. Our confidence interval construction procedure is 95% *reliable*. That is to say, we can expect our confidence intervals to include the true population parameter about 95% of the time.

A common but incorrect interpretation is: "There is a 95% probability that the confidence interval contains p." Looking at Figure 8.27, each of the confidence intervals either does or doesn't contain p. In other words, the probability is either a 1 or a 0.

So if the 95% confidence level only relates to the reliability of the confidence interval construction procedure and not to a given confidence interval itself, what insight can be derived from a given confidence interval? For example, going back to the pennies example, we found that the percentile method 95% confidence interval for μ was (1991.24, 1999.42), whereas the standard error method 95% confidence interval was (1991.35, 1999.53). What can be said about these two intervals?

Loosely speaking, we can think of these intervals as our "best guess" of a plausible range of values for the mean year μ of *all* US pennies. For the rest of this book, we'll use the following shorthand summary of the precise interpretation.

Short-hand interpretation: We are 95% "confident" that a 95% confidence interval captures the value of the population parameter.

We use quotation marks around "confident" to emphasize that while 95% relates to the reliability of our confidence interval construction procedure, ultimately a constructed confidence interval is our best guess of an interval that contains the population parameter. In other words, it's our best net.

So returning to our pennies example and focusing on the percentile method, we are 95% "confident" that the true mean year of pennies in circulation in 2019 is somewhere between 1991.24 and 1999.42.

8.5.3   Width of confidence intervals

Now that we know how to interpret confidence intervals, let's go over some factors that determine their width.

Impact of confidence level

One factor that determines confidence interval widths is the pre-specified confidence level. For example, in Figures 8.27 and 8.28, we compared the widths of 95% and 80% confidence intervals and observed that the 95% confidence intervals were wider. The quantification of the confidence level should match what many expect of the word "confident." In order to be more confident in our best guess of a range of values, we need to widen the range of values.

To elaborate on this, imagine we want to guess the forecasted high temperature in Seoul, South Korea on August 15th. Given Seoul's temperate climate with four distinct seasons, we could say somewhat confidently that the high temperature would be between 50°F - 95°F (10°C - 35°C). However, if we wanted a temperature range we were *absolutely* confident about, we would need to widen it.

We need this wider range to allow for the possibility of anomalous weather, like a freak cold spell or an extreme heat wave. So a range of temperatures we could be near certain about would be between 32°F - 110°F (0°C - 43°C). On the other hand, if we could tolerate being a little less confident, we could narrow this range to between 70°F - 85°F (21°C - 30°C).

Let's revisit our sampling bowl from Chapter 7. Let's compare $10 \cdot 3 = 30$ confidence intervals for p based on three different confidence levels: 80%, 95%, and 99%.

Specifically, we'll first take 30 different random samples of size $n = 50$ balls from the bowl. Then we'll construct 10 percentile-based confidence intervals using each of the three different confidence levels.

Finally, we'll compare the widths of these intervals. We visualize the resulting confidence intervals in Figure 8.29 along with a vertical line marking the true value of $p = 0.375$.

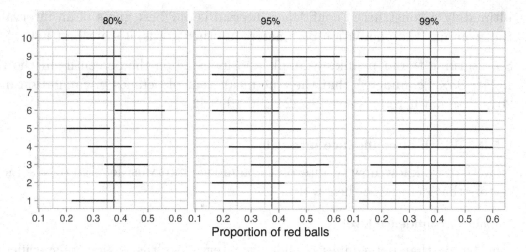

FIGURE 8.29: Ten 80, 95, and 99% confidence intervals for p based on $n = 50$.

Observe that as the confidence level increases from 80% to 95% to 99%, the confidence intervals tend to get wider as seen in Table 8.2 where we compare their average widths.

TABLE 8.2: Average width of 80, 95, and 99% confidence intervals

Confidence level	Mean width
80%	0.162
95%	0.262
99%	0.338

So in order to have a higher confidence level, our confidence intervals must be wider. Ideally, we would have both a high confidence level and narrow confidence intervals. However, we cannot have it both ways. If we want to *be more confident*, we need to allow for wider intervals. Conversely, if we would like a narrow interval, we must tolerate a lower confidence level.

The moral of the story is: **Higher confidence levels tend to produce wider confidence intervals.** When looking at Figure 8.29 it is important to keep in mind that we kept the sample size fixed at $n = 50$. Thus, all $10 \cdot 3 = 30$ random samples from the bowl had the same sample size. What happens if instead we took samples of different sizes? Recall that we did this in Subsection 7.2.4 using virtual shovels with 25, 50, and 100 slots.

Impact of sample size

This time, let's fix the confidence level at 95%, but consider three different sample sizes for n: 25, 50, and 100. Specifically, we'll first take 10 different random samples of size 25, 10 different random samples of size 50, and 10 different random samples of size 100. We'll then construct 95% percentile-based confidence intervals for each sample. Finally, we'll compare the widths of these intervals. We visualize the resulting 30 confidence intervals in Figure 8.30. Note also the vertical line marking the true value of $p = 0.375$.

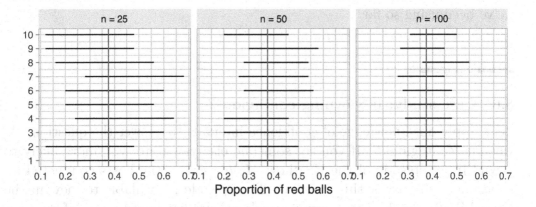

FIGURE 8.30: Ten 95% confidence intervals for p with $n = 25, 50$, and 100.

Observe that as the confidence intervals are constructed from larger and larger sample sizes, they tend to get narrower. Let's compare the average widths in Table 8.3.

TABLE 8.3: Average width of 95% confidence intervals based on $n = 25$, 50, and 100

Sample size	Mean width
n = 25	0.380
n = 50	0.268
n = 100	0.189

The moral of the story is: **Larger sample sizes tend to produce narrower confidence intervals.** Recall that this was a key message in Subsection 7.3.3. As we used larger and larger shovels for our samples, the sample proportions red $\hat{p}$ tended to vary less. In other words, our estimates got more and more *precise*.

Recall that we visualized these results in Figure 7.15, where we compared the *sampling distributions* for $\hat{p}$ based on samples of size n equal 25, 50, and 100. We also quantified the sampling variation of these sampling distributions using their standard deviation, which has that special name: the *standard error*. So as the sample size increases, the standard error decreases.

In fact, the standard error is another related factor in determining confidence interval width. We'll explore this fact in Subsection 8.7.2 when we discuss theory-based methods for constructing confidence intervals using mathematical formulas. Such methods are an alternative to the computer-based methods we've been using so far.

8.6 Case study: Is yawning contagious?

Let's apply our knowledge of confidence intervals to answer the question: "Is yawning contagious?". If you see someone else yawn, are you more likely to yawn? In an episode of the US show *Mythbusters*[4], the hosts conducted an experiment to answer this question. The episode is available to view in the United States on the Discovery Network website here[5] and more information about the episode is also available on IMDb[6].

8.6.1 *Mythbusters* study data

Fifty adult participants who thought they were being considered for an appearance on the show were interviewed by a show recruiter. In the interview, the recruiter either yawned or did not. Participants then sat by themselves in a large van and were asked to wait. While in the van, the *Mythbusters* team watched the participants using a hidden camera to see if they yawned. The data frame containing the results of their experiment is available in the mythbusters_yawn data frame included in the moderndive package:

```
mythbusters_yawn
```

```
# A tibble: 50 x 3
    subj group   yawn
   <int> <chr>   <chr>
 1     1 seed    yes
```

[4]http://www.discovery.com/tv-shows/mythbusters/mythbusters-database/yawning-contagious/
[5]https://www.discovery.com/tv-shows/mythbusters/videos/is-yawning-contagious
[6]https://www.imdb.com/title/tt0768479/

```
 2       2 control yes
 3       3 seed    no
 4       4 seed    yes
 5       5 seed    no
 6       6 control no
 7       7 seed    yes
 8       8 control no
 9       9 control no
10      10 seed    no
# ... with 40 more rows
```

The variables are:

- subj: The participant ID with values 1 through 50.
- group: A binary *treatment* variable indicating whether the participant was exposed to yawning. "seed" indicates the participant was exposed to yawning while "control" indicates the participant was not.
- yawn: A binary *response* variable indicating whether the participant ultimately yawned.

Recall that you learned about treatment and response variables in Subsection 5.3.1 in our discussion on confounding variables.

Let's use some data wrangling to obtain counts of the four possible outcomes:

```
mythbusters_yawn %>%
  group_by(group, yawn) %>%
  summarize(count = n())
```

```
# A tibble: 4 x 3
# Groups:   group [2]
  group   yawn  count
  <chr>   <chr> <int>
1 control no       12
2 control yes       4
3 seed    no       24
4 seed    yes      10
```

Let's first focus on the "control" group participants who were not exposed to yawning. 12 such participants did not yawn, while 4 such participants did. So out of the 16 people who were not exposed to yawning, $4/16 = 0.25 = 25\%$ did yawn.

Let's now focus on the "seed" group participants who were exposed to yawning where 24 such participants did not yawn, while 10 such participants did yawn.

So out of the 34 people who were exposed to yawning, $10/34 = 0.294 = 29.4\%$ did yawn. Comparing these two percentages, the participants who were exposed to yawning yawned 29.4% - $25\% = 4.4\%$ more often than those who were not.

8.6.2   Sampling scenario

Let's review the terminology and notation related to sampling we studied in Subsection 7.3.1. In Chapter 7 our *study population* was the bowl of $N = 2400$ balls. Our *population parameter* of interest was the *population proportion* of these balls that were red, denoted mathematically by p. In order to estimate p, we extracted a sample of 50 balls using the shovel and computed the relevant *point estimate*: the *sample proportion* that were red, denoted mathematically by $\hat{p}$.

Who is the study population here? All humans? All the people who watch the show *Mythbusters*? It's hard to say! This question can only be answered if we know how the show's hosts recruited participants! In other words, what was the *sampling methodology* used by the *Mythbusters* to recruit participants? We alas are not provided with this information. Only for the purposes of this case study, however, we'll *assume* that the 50 participants are a representative sample of all Americans given the popularity of this show. Thus, we'll be assuming that any results of this experiment will generalize to all $N = 327$ million Americans (2018 population).

Just like with our sampling bowl, the population parameter here will involve proportions. However, in this case it will be the *difference in population proportions* $p_{seed} - p_{control}$, where p_{seed} is the proportion of *all* Americans who if exposed to yawning will yawn themselves, and $p_{control}$ is the proportion of *all* Americans who if not exposed to yawning still yawn themselves. Correspondingly, the point estimate/sample statistic based the *Mythbusters'* sample of participants will be the *difference in sample proportions* $\hat{p}_{seed} - \hat{p}_{control}$. Let's extend Table 7.5 of scenarios of sampling for inference to include our latest scenario.

TABLE 8.4: Scenarios of sampling for inference

Scenario	Population parameter	Notation	Point estimate	Symbol(s)
1	Population proportion	p	Sample proportion	$\hat{p}$
2	Population mean	μ	Sample mean	$\bar{x}$ or $\hat{\mu}$
3	Difference in population proportions	$p_1 - p_2$	Difference in sample proportions	$\hat{p}_1 - \hat{p}_2$

This is known as a *two-sample* inference situation since we have two separate samples. Based on their two-samples of size $n_{seed} = 34$ and $n_{control} = 16$, the point estimate is

$$\hat{p}_{seed} - \hat{p}_{control} = \frac{24}{34} - \frac{12}{16} = 0.04411765 \approx 4.4\%$$

However, say the *Mythbusters* repeated this experiment. In other words, say they recruited 50 new participants and exposed 34 of them to yawning and 16 not. Would they obtain the exact same estimated difference of 4.4%? Probably not, again, because of *sampling variation*.

How does this sampling variation affect their estimate of 4.4%? In other words, what would be a plausible range of values for this difference that accounts for this sampling variation? We can answer this question with confidence intervals! Furthermore, since the *Mythbusters* only have a single two-sample of 50 participants, they would have to construct a 95% confidence interval for $p_{seed} - p_{control}$ using *bootstrap resampling with replacement*.

We make a couple of important notes. First, for the comparison between the "seed" and "control" groups to make sense, however, both groups need to be *independent* from each other. Otherwise, they could influence each other's results. This means that a participant being selected for the "seed" or "control" group has no influence on another participant being assigned to one of the two groups. As an example, if there were a mother and her child as participants in the study, they wouldn't necessarily be in the same group. They would each be assigned randomly to one of the two groups of the explanatory variable.

Second, the order of the subtraction in the difference doesn't matter so long as you are consistent and tailor your interpretations accordingly. In other words, using a point estimate of $\hat{p}_{seed} - \hat{p}_{control}$ or $\hat{p}_{control} - \hat{p}_{seed}$ does not make a material difference, you just need to stay consistent and interpret your results accordingly.

8.6.3 Constructing the confidence interval

As we did in Subsection 8.4.2, let's first construct the bootstrap distribution for $\hat{p}_{seed} - \hat{p}_{control}$ and then use this to construct 95% confidence intervals for $p_{seed} - p_{control}$. We'll do this using the infer workflow again. However, since the difference in proportions is a new scenario for inference, we'll need to use some new arguments in the infer functions along the way.

1. specify variables

Let's take our mythbusters_yawn data frame and specify() which variables are of interest using the y ~ x formula interface where:

- Our response variable is yawn: whether or not a participant yawned. It has levels "yes" and "no".
- The explanatory variable is group: whether or not a participant was exposed to yawning. It has levels "seed" (exposed to yawning) and "control" (not exposed to yawning).

```
mythbusters_yawn %>%
  specify(formula = yawn ~ group)
```

```
Error: A level of the response variable `yawn` needs to be
specified for the `success` argument in `specify()`.
```

Alas, we got an error message similar to the one from Subsection 8.5.1: infer is telling us that one of the levels of the categorical variable yawn needs to be defined as the success. Recall that we define success to be the event of interest we are trying to count and compute proportions of. Are we interested in those participants who "yes" yawned or those who "no" didn't yawn? This isn't clear to R or someone just picking up the code and results for the first time, so we need to set the success argument to "yes" as follows to improve the transparency of the code:

```
mythbusters_yawn %>%
  specify(formula = yawn ~ group, success = "yes")
```

```
Response: yawn (factor)
Explanatory: group (factor)
# A tibble: 50 x 2
   yawn  group
   <fct> <fct>
 1 yes   seed
 2 yes   control
 3 no    seed
 4 yes   seed
 5 no    seed
 6 no    control
 7 yes   seed
 8 no    control
 9 no    control
```

```
10 no     seed
# ... with 40 more rows
```

2. generate replicates

Our next step is to perform *bootstrap resampling with replacement* like we did with the slips of paper in our pennies activity in Section 8.1. We saw how it works with both a single variable in computing bootstrap means in Section 8.4 and in computing bootstrap proportions in Section 8.5, but we haven't yet worked with bootstrapping involving multiple variables.

In the infer package, bootstrapping with multiple variables means that each *row* is potentially resampled. Let's investigate this by focusing only on the first six rows of mythbusters_yawn:

```
first_six_rows <- head(mythbusters_yawn)
first_six_rows
```

```
# A tibble: 6 x 3
   subj group   yawn
  <int> <chr>   <chr>
1     1 seed    yes
2     2 control yes
3     3 seed    no
4     4 seed    yes
5     5 seed    no
6     6 control no
```

When we bootstrap this data, we are potentially pulling the subject's readings multiple times. Thus, we could see the entries of "seed" for group and "no" for yawn together in a new row in a bootstrap sample. This is further seen by exploring the sample_n() function in dplyr on this smaller 6-row data frame comprised of head(mythbusters_yawn). The sample_n() function can perform this bootstrapping procedure and is similar to the rep_sample_n() function in infer, except that it is not repeated, but rather only performs one sample with or without replacement.

```
first_six_rows %>%
  sample_n(size = 6, replace = TRUE)
```

```
# A tibble: 6 x 3
   subj group   yawn
  <int> <chr>   <chr>
```

```
1      1 seed    yes
2      6 control no
3      1 seed    yes
4      5 seed    no
5      4 seed    yes
6      4 seed    yes
```

We can see that in this bootstrap sample generated from the first six rows
of mythbusters_yawn, we have some rows repeated. The same is true when we
perform the generate() step in infer as done in what follows. Using this fact,
we generate 1000 replicates, or, in other words, we bootstrap resample the 50
participants with replacement 1000 times.

```
mythbusters_yawn %>%
  specify(formula = yawn ~ group, success = "yes") %>%
  generate(reps = 1000, type = "bootstrap")
```

```
Response: yawn (factor)
Explanatory: group (factor)
# A tibble: 50,000 x 3
# Groups:    replicate [1,000]
   replicate yawn  group
       <int> <fct> <fct>
1          1 yes   seed
2          1 yes   control
3          1 no    control
4          1 no    control
5          1 yes   seed
6          1 yes   seed
7          1 yes   seed
8          1 yes   seed
9          1 no    seed
10         1 yes   seed
# ... with 49,990 more rows
```

Observe that the resulting data frame has 50,000 rows. This is because we
performed resampling of 50 participants with replacement 1000 times and
$50,000 = 1000 \cdot 50$. The variable replicate indicates which resample each row
belongs to. So it has the value 1 50 times, the value 2 50 times, all the way
through to the value 1000 50 times.

3. calculate summary statistics

After we `generate()` many replicates of bootstrap resampling with replacement, we next want to summarize the bootstrap resamples of size 50 with a single summary statistic, the difference in proportions. We do this by setting the `stat` argument to `"diff in props"`:

```
mythbusters_yawn %>%
  specify(formula = yawn ~ group, success = "yes") %>%
  generate(reps = 1000, type = "bootstrap") %>%
  calculate(stat = "diff in props")
```

```
Error: Statistic is based on a difference; specify the `order` in which to
subtract the levels of the explanatory variable.
```

We see another error here. We need to specify the order of the subtraction. Is it $\hat{p}_{seed} - \hat{p}_{control}$ or $\hat{p}_{control} - \hat{p}_{seed}$. We specify it to be $\hat{p}_{seed} - \hat{p}_{control}$ by setting `order = c("seed", "control")`. Note that you could've also set `order = c("control", "seed")`. As we stated earlier, the order of the subtraction does not matter, so long as you stay consistent throughout your analysis and tailor your interpretations accordingly.

Let's save the output in a data frame `bootstrap_distribution_yawning`:

```
bootstrap_distribution_yawning <- mythbusters_yawn %>%
  specify(formula = yawn ~ group, success = "yes") %>%
  generate(reps = 1000, type = "bootstrap") %>%
  calculate(stat = "diff in props", order = c("seed", "control"))
bootstrap_distribution_yawning
```

```
# A tibble: 1,000 x 2
   replicate         stat
       <int>        <dbl>
1          1   0.0357143
2          2   0.229167
3          3   0.00952381
4          4   0.0106952
5          5   0.00483092
6          6   0.00793651
7          7  -0.0845588
8          8  -0.00466200
9          9   0.164686
10        10   0.124777
```

```
# ... with 990 more rows
```

Observe that the resulting data frame has 1000 rows and 2 columns corresponding to the 1000 `replicate` ID's and the 1000 differences in proportions for each bootstrap resample in `stat`.

4. `visualize` the results

In Figure 8.31 we `visualize()` the resulting bootstrap resampling distribution. Let's also add a vertical line at 0 by adding a `geom_vline()` layer.

```
visualize(bootstrap_distribution_yawning) +
  geom_vline(xintercept = 0)
```

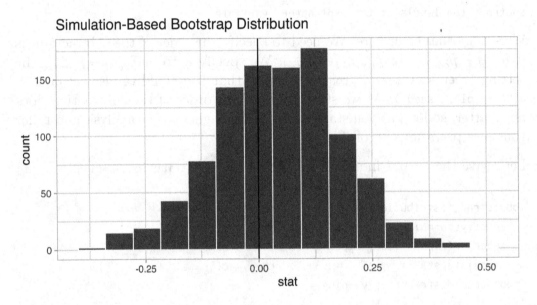

FIGURE 8.31: Bootstrap distribution.

First, let's compute the 95% confidence interval for $p_{seed} - p_{control}$ using the percentile method, in other words, by identifying the 2.5th and 97.5th percentiles which include the middle 95% of values. Recall that this method does not require the bootstrap distribution to be normally shaped.

```
bootstrap_distribution_yawning %>%
  get_confidence_interval(type = "percentile", level = 0.95)
```

```
# A tibble: 1 x 2
    `2.5%`   `97.5%`
```

```
      <dbl>    <dbl>
1 -0.238276 0.302464
```

Second, since the bootstrap distribution is roughly bell-shaped, we can construct a confidence interval using the standard error method as well. Recall that to construct a confidence interval using the standard error method, we need to specify the center of the interval using the point_estimate argument. In our case, we need to set it to be the difference in sample proportions of 4.4% that the *Mythbusters* observed.

We can also use the infer workflow to compute this value by excluding the generate() 1000 bootstrap replicates step. In other words, do not generate replicates, but rather use only the original sample data. We can achieve this by commenting out the generate() line, telling R to ignore it:

```
obs_diff_in_props <- mythbusters_yawn %>%
  specify(formula = yawn ~ group, success = "yes") %>%
  # generate(reps = 1000, type = "bootstrap") %>%
  calculate(stat = "diff in props", order = c("seed", "control"))
obs_diff_in_props
```

```
# A tibble: 1 x 1
      stat
     <dbl>
1 0.0441176
```

We thus plug this value in as the point_estimate argument.

```
myth_ci_se <- bootstrap_distribution_yawning %>%
  get_confidence_interval(type = "se", point_estimate = obs_diff_in_props)
myth_ci_se
```

```
# A tibble: 1 x 2
     lower    upper
     <dbl>    <dbl>
1 -0.227291 0.315526
```

Let's visualize both confidence intervals in Figure 8.32, with the percentile-method interval marked with black lines and the standard-error-method marked with grey lines. Observe that they are both similar to each other.

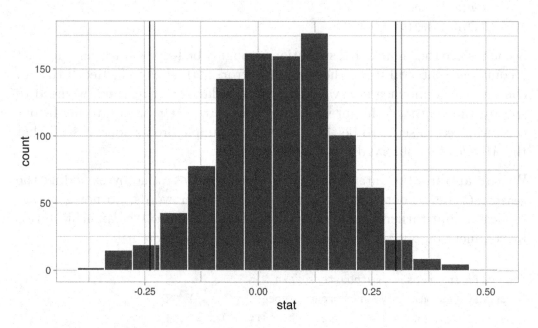

FIGURE 8.32: Two 95% confidence intervals: percentile method (black) and standard error method (grey).

8.6.4   Interpreting the confidence interval

Given that both confidence intervals are quite similar, let's focus our interpretation to only the percentile-method confidence interval of (-0.238, 0.302). Recall from Subsection 8.5.2 that the precise statistical interpretation of a 95% confidence interval is: if this construction procedure is repeated 100 times, then we expect about 95 of the confidence intervals to capture the true value of $p_{seed} - p_{control}$. In other words, if we gathered 100 samples of $n = 50$ participants from a similar pool of people and constructed 100 confidence intervals each based on each of the 100 samples, about 95 of them will contain the true value of $p_{seed} - p_{control}$ while about five won't. Given that this is a little long winded, we use the shorthand interpretation: we're 95% "confident" that the true difference in proportions $p_{seed} - p_{control}$ is between (-0.238, 0.302).

There is one value of particular interest that this 95% confidence interval contains: zero. If $p_{seed} - p_{control}$ were equal to 0, then there would be no difference in proportion yawning between the two groups. This would suggest that there is no associated effect of being exposed to a yawning recruiter on whether you yawn yourself.

In our case, since the 95% confidence interval includes 0, we cannot conclusively say if either proportion is larger. Of our 1000 bootstrap resamples with replacement, sometimes $\hat{p}_{seed}$ was higher and thus those exposed to yawning yawned themselves more often. At other times, the reverse happened.

Say, on the other hand, the 95% confidence interval was entirely above zero. This would suggest that $p_{seed} - p_{control} > 0$, or, in other words $p_{seed} > p_{control}$, and thus we'd have evidence suggesting those exposed to yawning do yawn more often.

8.7   Conclusion

8.7.1   Comparing bootstrap and sampling distributions

Let's talk more about the relationship between *sampling distributions* and *bootstrap distributions*.

Recall back in Subsection 7.2.3, we took 1000 virtual samples from the bowl using a virtual shovel, computed 1000 values of the sample proportion red $\hat{p}$, then visualized their distribution in a histogram. Recall that this distribution is called the *sampling distribution of $\hat{p}$* . Furthermore, the standard deviation of the sampling distribution has a special name: the *standard error*.

We also mentioned that this sampling activity does not reflect how sampling is done in real life. Rather, it was an *idealized version* of sampling so that we could study the effects of sampling variation on estimates, like the proportion of the shovel's balls that are red. In real life, however, one would take a single sample that's as large as possible, much like in the Obama poll we saw in Section 7.4. But how can we get a sense of the effect of sampling variation on estimates if we only have one sample and thus only one estimate? Don't we need many samples and hence many estimates?

The workaround to having a *single* sample was to perform *bootstrap resampling with replacement* from the single sample. We did this in the resampling activity in Section 8.1 where we focused on the mean year of minting of pennies. We used pieces of paper representing the original sample of 50 pennies from the bank and resampled them with replacement from a hat. We had 35 of our friends perform this activity and visualized the resulting 35 sample means $\bar{x}$ in a histogram in Figure 8.11.

This distribution was called the *bootstrap distribution* of $\bar{x}$. We stated at the time that the bootstrap distribution is an *approximation* to the sampling distribution of $\bar{x}$ in the sense that both distributions will have a similar shape and similar spread. Thus the *standard error* of the bootstrap distribution can be used as an approximation to the *standard error* of the sampling distribution.

Let's show you that this is the case by now comparing these two types of distributions. Specifically, we'll compare

1. the sampling distribution of $\hat{p}$ based on 1000 virtual samples from the `bowl` from Subsection 7.2.3 to
2. the bootstrap distribution of $\hat{p}$ based on 1000 virtual resamples with replacement from Ilyas and Yohan's single sample `bowl_sample_1` from Subsection 8.5.1.

Sampling distribution

Here is the code you saw in Subsection 7.2.3 to construct the sampling distribution of $\hat{p}$ shown again in Figure 8.33, with some changes to incorporate the statistical terminology relating to sampling from Subsection 7.3.1.

```
# Take 1000 virtual samples of size 50 from the bowl:
virtual_samples <- bowl %>%
  rep_sample_n(size = 50, reps = 1000)
# Compute the sampling distribution of 1000 values of p-hat
sampling_distribution <- virtual_samples %>%
  group_by(replicate) %>%
  summarize(red = sum(color == "red")) %>%
  mutate(prop_red = red / 50)
# Visualize sampling distribution of p-hat
ggplot(sampling_distribution, aes(x = prop_red)) +
  geom_histogram(binwidth = 0.05, boundary = 0.4, color = "white") +
  labs(x = "Proportion of 50 balls that were red",
       title = "Sampling distribution")
```

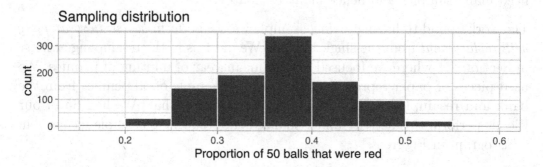

FIGURE 8.33: Previously seen sampling distribution of sample proportion red for $n = 1000$.

An important thing to keep in mind is the default value for `replace` is `FALSE` when using `rep_sample_n()`. This is because when sampling 50 balls with a shovel, we are extracting 50 balls one-by-one *without* replacing them. This is in contrast to bootstrap resampling *with* replacement, where we resample a ball and put it back, and repeat this process 50 times.

Let's quantify the variability in this sampling distribution by calculating the standard deviation of the `prop_red` variable representing 1000 values of the sample proportion $\hat{p}$. Remember that the standard deviation of the sampling distribution is the *standard error*, frequently denoted as `se`.

```
sampling_distribution %>% summarize(se = sd(prop_red))
```

```
# A tibble: 1 x 1
       se
    <dbl>
1 0.0673987
```

Bootstrap distribution

Here is the code you previously saw in Subsection 8.5.1 to construct the bootstrap distribution of $\hat{p}$ based on Ilyas and Yohan's original sample of 50 balls saved in `bowl_sample_1`.

```
bootstrap_distribution <- bowl_sample_1 %>%
  specify(response = color, success = "red") %>%
  generate(reps = 1000, type = "bootstrap") %>%
  calculate(stat = "prop")
```

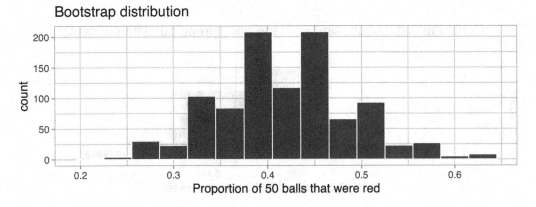

FIGURE 8.34: Bootstrap distribution of proportion red for $n = 1000$.

```
bootstrap_distribution %>% summarize(se = sd(stat))
```

```
# A tibble: 1 x 1
        se
     <dbl>
1 0.0712212
```

Comparison

Now that we have computed both the sampling distribution and the bootstrap distributions, let's compare them side-by-side in Figure 8.35. We'll make both histograms have matching scales on the x- and y-axes to make them more comparable. Furthermore, we'll add:

1. To the sampling distribution on the top: a solid line denoting the proportion of the bowl's balls that are red $p = 0.375$.
2. To the bootstrap distribution on the bottom: a dashed line at the sample proportion $\hat{p} = 21/50 = 0.42 = 42\%$ that Ilyas and Yohan observed.

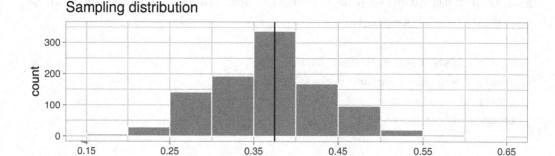

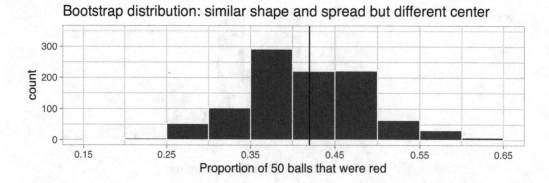

FIGURE 8.35: Comparing the sampling and bootstrap distributions of $\hat{p}$.

There is a lot going on in Figure 8.35, so let's break down all the comparisons slowly. First, observe how the sampling distribution on top is centered at p = 0.375. This is because the sampling is done at random and in an unbiased fashion. So the estimates $\hat{p}$ are centered at the true value of p.

However, this is not the case with the following bootstrap distribution. The bootstrap distribution is centered at 0.42, which is the proportion red of Ilyas and Yohan's 50 sampled balls. This is because we are resampling from the same sample over and over again. Since the bootstrap distribution is centered at the original sample's proportion, it doesn't necessarily provide a better estimate of p = 0.375. This leads us to our first lesson about bootstrapping:

The bootstrap distribution will likely not have the same center as the sampling distribution. In other words, bootstrapping cannot improve the quality of an estimate.

Second, let's now compare the spread of the two distributions: they are somewhat similar. In the previous code, we computed the standard deviations of both distributions as well. Recall that such standard deviations have a special name: *standard errors*. Let's compare them in Table 8.5.

TABLE 8.5: Comparing standard errors

Distribution type	Standard error
Sampling distribution	0.067
Bootstrap distribution	0.071

Notice that the bootstrap distribution's standard error is a rather good *approximation* to the sampling distribution's standard error. This leads us to our second lesson about bootstrapping:

Even if the bootstrap distribution might not have the same center as the sampling distribution, it will likely have very similar shape and spread. In other words, bootstrapping will give you a good estimate of the *standard error*.

Thus, using the fact that the bootstrap distribution and sampling distributions have similar spreads, we can build confidence intervals using bootstrapping as we've done all throughout this chapter!

8.7.2 Theory-based confidence intervals

So far in this chapter, we've constructed confidence intervals using two methods: the percentile method and the standard error method. Recall also from Subsection 8.3.2 that we can only use the standard-error method if the bootstrap distribution is bell-shaped (i.e., normally distributed).

In a similar vein, if the sampling distribution is normally shaped, there is another method for constructing confidence intervals that does not involve using your computer. You can use a *theory-based method* involving a mathematical formulas!

The formula uses the rule of thumb we saw in Appendix A.2 that 95% of values in a normal distribution are within ± 1.96 standard deviations of the mean. In the case of sampling and bootstrap distributions, recall that the standard deviation has a special name: the *standard error*.

Theory-based method for computing standard errors

There exists in many cases a formula that approximates the standard error! In the case of our bowl where we used the sample proportion red $\hat{p}$ to estimate the proportion of the bowl's balls that are red, the formula that approximates the standard error is:

$$\text{SE}_{\hat{p}} \approx \sqrt{\frac{\hat{p}(1-\hat{p})}{n}}$$

For example, recall from bowl_sample_1 that Yohan and Ilyas sampled $n = 50$ balls and observed a sample proportion $\hat{p}$ of $21/50 = 0.42$. So, using the formula, an approximation of the standard error of $\hat{p}$ is

$$\text{SE}_{\hat{p}} \approx \sqrt{\frac{0.42(1-0.42)}{50}} = \sqrt{0.004872} = 0.0698 \approx 0.070$$

The key observation to make here is that there is an n in the denominator. So as the sample size n increases, the standard error decreases. We've demonstrated this fact using our virtual shovels in Subsection 7.3.3. If you don't recall this demonstration, we highly recommend you go back and read that subsection.

Let's compare this theory-based standard error to the standard error of the sampling and bootstrap distributions you computed previously in Subsection 8.7.1 in Table 8.6. Notice how they are all similar!

TABLE 8.6: Comparing standard errors

Distribution type	Standard error
Sampling distribution	0.067
Bootstrap distribution	0.071
Formula approximation	0.070

Going back to Yohan and Ilyas' sample proportion of $\hat{p}$ of $21/50 = 0.42$, say this were based on a sample of size $n = 100$ instead of 50. Then the standard error would be:

$$\text{SE}_{\hat{p}} \approx \sqrt{\frac{0.42(1 - 0.42)}{100}} = \sqrt{0.002436} = 0.0494$$

Observe that the standard error has gone down from 0.0698 to 0.0494. In other words, the "typical" error of our estimates using $n = 100$ will go down and hence be more *precise*. Recall that we illustrated the difference between accuracy and precision of estimates in Figure 7.16.

Why is this formula true? Unfortunately, we don't have the tools at this point to prove this; you'll need to take a more advanced course in probability and statistics. (It is related to the concepts of Bernoulli and Binomial Distributions. You can read more about its derivation here[7] if you like.)

Theory-based method for constructing confidence intervals

Using these theory-based standard errors, let's present a theory-based method for constructing 95% confidence intervals that does not involve using a computer, but rather mathematical formulas. Note that this theory-based method only holds if the sampling distribution is normally shaped, so that we can use the 95% rule of thumb about normal distributions discussed in Appendix A.2.

1. Collect a single representative sample of size n that's as large as possible.
2. Compute the *point estimate*: the *sample proportion* $\hat{p}$. Think of this as the center of your "net."
3. Compute the approximation to the standard error

---

[7]http://onlinestatbook.com/2/sampling_distributions/samp_dist_p.html

$$\mathrm{SE}_{\hat{p}} \approx \sqrt{\frac{\hat{p}(1-\hat{p})}{n}}$$

4. Compute a quantity known as the *margin of error* (more on this later after we list the five steps):

$$\mathrm{MoE}_{\hat{p}} = 1.96 \cdot \mathrm{SE}_{\hat{p}} = 1.96 \cdot \sqrt{\frac{\hat{p}(1-\hat{p})}{n}}$$

5. Compute both endpoints of the confidence interval.
   - The lower end-point. Think of this as the left end-point of the net:

$$\hat{p} - \mathrm{MoE}_{\hat{p}} = \hat{p} - 1.96 \cdot \mathrm{SE}_{\hat{p}} = \hat{p} - 1.96 \cdot \sqrt{\frac{\hat{p}(1-\hat{p})}{n}}$$

   - The upper endpoint. Think of this as the right end-point of the net:

$$\hat{p} + \mathrm{MoE}_{\hat{p}} = \hat{p} + 1.96 \cdot \mathrm{SE}_{\hat{p}} = \hat{p} + 1.96 \cdot \sqrt{\frac{\hat{p}(1-\hat{p})}{n}}$$

   - Alternatively, you can succinctly summarize a 95% confidence interval for p using the $\pm$ symbol:

$$\hat{p} \pm \mathrm{MoE}_{\hat{p}} = \hat{p} \pm (1.96 \cdot \mathrm{SE}_{\hat{p}}) = \hat{p} \pm \left(1.96 \cdot \sqrt{\frac{\hat{p}(1-\hat{p})}{n}} \right)$$

So going back to Yohan and Ilyas' sample of $n = 50$ balls that had 21 red balls, the 95% confidence interval for p is

$$
\begin{aligned}
0.41 \pm 1.96 \cdot 0.0698 &= 0.41 \pm 0.137 \\
&= (0.41 - 0.137, \, 0.41 + 0.137) \\
&= (0.273, 0.547).
\end{aligned}
$$

Yohan and Ilyas are 95% "confident" that the true proportion red of the bowl's balls is between 28.3% and 55.7%. Given that the true population proportion p was 0.375, in this case they successfully captured the fish.

In Step 4, we defined a statistical quantity known as the *margin of error*. You can think of this quantity as how much the net extends to the left and to the right of the center of our net. The 1.96 multiplier is rooted in the 95% rule of thumb we introduced earlier and the fact that we want the confidence level to be 95%. The value of the margin of error entirely determines the width of the confidence interval. Recall from Subsection 8.5.3 that confidence interval widths are determined by an interplay of the confidence level, the sample size n, and the standard error.

Let's revisit the poll of President Obama's approval rating among young Americans aged 18-29 which we introduced in Section 7.4. Pollsters found that based on a representative sample of $n = 2089$ young Americans, $\hat{p} = 0.41 = 41\%$ supported President Obama.

If you look towards the end of the article, it also states: "The poll's margin of error was plus or minus 2.1 percentage points." This is precisely the MoE:

$$\text{MoE} = 1.96 \cdot \text{SE} = 1.96 \cdot \sqrt{\frac{\hat{p}(1-\hat{p})}{n}} = 1.96 \cdot \sqrt{\frac{0.41(1-0.41)}{2089}}$$
$$= 1.96 \cdot 0.0108 = 0.021 = 2.1\%$$

Their poll results are based on a confidence level of 95% and the resulting 95% confidence interval for the proportion of all young Americans who support Obama is:

$$\hat{p} \pm \text{MoE} = 0.41 \pm 0.021 = (0.389, 0.431) = (38.9\%, 43.1\%).$$

Confidence intervals based on 33 tactile samples

Let's revisit our 33 friends' samples from the bowl from Subsection 7.1.3. We'll use their 33 samples to construct 33 theory-based 95% confidence intervals for p. Recall this data was saved in the tactile_prop_red data frame included in the moderndive package:

1. rename() the variable prop_red to p_hat, the statistical name of the sample proportion $\hat{p}$.
2. mutate() a new variable n making explicit the sample size of 50.
3. mutate() other new variables computing:
   - The standard error SE for $\hat{p}$ using the previous formula.
   - The margin of error MoE by multiplying the SE by 1.96
   - The left endpoint of the confidence interval lower_ci
   - The right endpoint of the confidence interval upper_ci

```
conf_ints <- tactile_prop_red %>%
  rename(p_hat = prop_red) %>%
  mutate(
    n = 50,
    SE = sqrt(p_hat * (1 - p_hat) / n),
```

```
    MoE = 1.96 * SE,
    lower_ci = p_hat - MoE,
    upper_ci = p_hat + MoE
)
```

In Figure 8.36, let's plot the 33 confidence intervals for p saved in `conf_ints` along with a vertical line at $p = 0.375$ indicating the true proportion of the bowl's balls that are red. Furthermore, let's mark the sample proportions $\hat{p}$ with dots since they represent the centers of these confidence intervals.

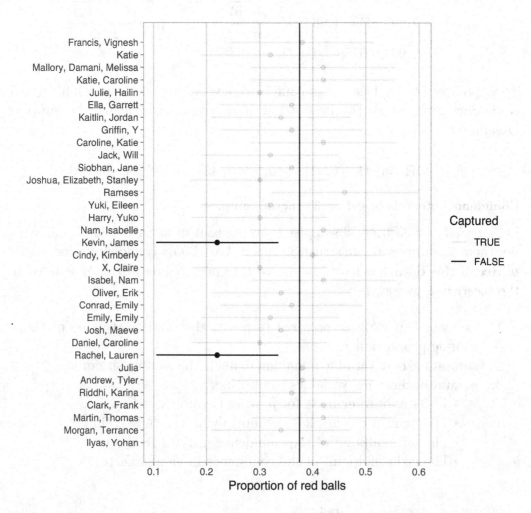

FIGURE 8.36: 33 confidence intervals at the 95% level based on 33 tactile samples of size $n = 50$.

Observe that 31 of the 33 confidence intervals "captured" the true value of p, for a success rate of 31 / 33 = 93.94%. While this is not quite 95%, recall that we *expect* about 95% of such confidence intervals to capture p. The actual observed success rate will vary slightly.

Theory-based methods like this have largely been used in the past because we didn't have the computing power to perform simulation-based methods such as bootstrapping. They are still commonly used, however, and if the sampling distribution is normally distributed, we have access to an alternative method for constructing confidence intervals as well as performing hypothesis tests as we will see in Chapter 9.

The kind of computer-based statistical inference we've seen so far has a particular name in the field of statistics: *simulation-based inference*. This is because we are performing statistical inference using computer simulations. In our opinion, two large benefits of simulation-based methods over theory-based methods are that (1) they are easier for people new to statistical inference to understand and (2) they also work in situations where theory-based methods and mathematical formulas don't exist.

8.7.3 Additional resources

Solutions to all *Learning checks* can be found online in Appendix D[8].

An R script file of all R code used in this chapter is available at `https://www.moderndive.com/scripts/08-confidence-intervals.R`.

If you want more examples of the `infer` workflow to construct confidence intervals, we suggest you check out the `infer` package homepage, in particular, a series of example analyses available at `https://infer.netlify.com/articles/`.

8.7.4 What's to come?

Now that we've equipped ourselves with confidence intervals, in Chapter 9 we'll cover the other common tool for statistical inference: hypothesis testing. Just like confidence intervals, hypothesis tests are used to infer about a population using a sample. However, we'll see that the framework for making such inferences is slightly different.

---

[8]`https://moderndive.com/D-appendixD.html`

9

Hypothesis Testing

Now that we've studied confidence intervals in Chapter 8, let's study another commonly used method for statistical inference: hypothesis testing. Hypothesis tests allow us to take a sample of data from a population and infer about the plausibility of competing hypotheses. For example, in the upcoming "promotions" activity in Section 9.1, you'll study the data collected from a psychology study in the 1970s to investigate whether gender-based discrimination in promotion rates existed in the banking industry at the time of the study.

The good news is we've already covered many of the necessary concepts to understand hypothesis testing in Chapters 7 and 8. We will expand further on these ideas here and also provide a general framework for understanding hypothesis tests. By understanding this general framework, you'll be able to adapt it to many different scenarios.

The same can be said for confidence intervals. There was one general framework that applies to *all* confidence intervals and the infer package was designed around this framework. While the specifics may change slightly for different types of confidence intervals, the general framework stays the same.

We believe that this approach is much better for long-term learning than focusing on specific details for specific confidence intervals using theory-based approaches. As you'll now see, we prefer this general framework for hypothesis tests as well.

Needed packages

Let's load all the packages needed for this chapter (this assumes you've already installed them). Recall from our discussion in Section 4.4 that loading the tidyverse package by running library(tidyverse) loads the following commonly used data science packages all at once:

- ggplot2 for data visualization
- dplyr for data wrangling
- tidyr for converting data to "tidy" format
- readr for importing spreadsheet data into R
- As well as the more advanced purrr, tibble, stringr, and forcats packages

If needed, read Section 1.3 for information on how to install and load R packages.

```
library(tidyverse)
library(infer)
library(moderndive)
library(nycflights13)
library(ggplot2movies)
```

9.1 Promotions activity

Let's start with an activity studying the effect of gender on promotions at a bank.

9.1.1 Does gender affect promotions at a bank?

Say you are working at a bank in the 1970s and you are submitting your résumé to apply for a promotion. Will your gender affect your chances of getting promoted? To answer this question, we'll focus on data from a study published in the *Journal of Applied Psychology* in 1974. This data is also used in the *OpenIntro*[1] series of statistics textbooks.

To begin the study, 48 bank supervisors were asked to assume the role of a hypothetical director of a bank with multiple branches. Every one of the bank supervisors was given a résumé and asked whether or not the candidate on the résumé was fit to be promoted to a new position in one of their branches.

However, each of these 48 résumés were identical in all respects except one: the name of the applicant at the top of the résumé. Of the supervisors, 24 were randomly given résumés with stereotypically "male" names, while 24 of the supervisors were randomly given résumés with stereotypically "female" names. Since only (binary) gender varied from résumé to résumé, researchers could isolate the effect of this variable in promotion rates.

While many people today (including us, the authors) disagree with such binary views of gender, it is important to remember that this study was conducted at a time where more nuanced views of gender were not as prevalent. Despite this imperfection, we decided to still use this example as we feel it presents ideas still relevant today about how we could study discrimination in the workplace.

---

[1] https://www.openintro.org/

The `moderndive` package contains the data on the 48 applicants in the `promotions` data frame. Let's explore this data by looking at six randomly selected rows:

```
promotions %>%
  sample_n(size = 6) %>%
  arrange(id)
```

```
# A tibble: 6 x 3
     id decision gender
  <int> <fct>    <fct>
1    11 promoted male
2    26 promoted female
3    28 promoted female
4    36 not      male
5    37 not      male
6    46 not      female
```

The variable `id` acts as an identification variable for all 48 rows, the `decision` variable indicates whether the applicant was selected for promotion or not, while the `gender` variable indicates the gender of the name used on the résumé. Recall that this data does not pertain to 24 actual men and 24 actual women, but rather 48 identical résumés of which 24 were assigned stereotypically "male" names and 24 were assigned stereotypically "female" names.

Let's perform an exploratory data analysis of the relationship between the two categorical variables `decision` and `gender`. Recall that we saw in Subsection 2.8.3 that one way we can visualize such a relationship is by using a stacked barplot.

```
ggplot(promotions, aes(x = gender, fill = decision)) +
  geom_bar() +
  labs(x = "Gender of name on résumé")
```

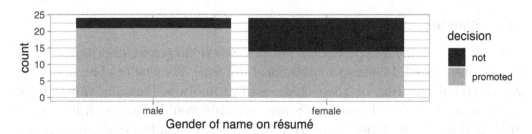

FIGURE 9.1: Barplot relating gender to promotion decision.

Observe in Figure 9.1 that it appears that résumés with female names were much less likely to be accepted for promotion. Let's quantify these promotion rates by computing the proportion of résumés accepted for promotion for each group using the `dplyr` package for data wrangling. Note the use of the `tally()` function here which is a shortcut for `summarize(n = n())` to get counts.

```
promotions %>%
  group_by(gender, decision) %>%
  tally()
```

```
# A tibble: 4 x 3
# Groups:   gender [2]
  gender decision      n
  <fct>  <fct>     <int>
1 male   not           3
2 male   promoted     21
3 female not          10
4 female promoted     14
```

So of the 24 résumés with male names, 21 were selected for promotion, for a proportion of $21/24 = 0.875 = 87.5\%$. On the other hand, of the 24 résumés with female names, 14 were selected for promotion, for a proportion of $14/24 = 0.583 = 58.3\%$. Comparing these two rates of promotion, it appears that résumés with male names were selected for promotion at a rate 0.875 - 0.583 $= 0.292 = 29.2\%$ higher than résumés with female names. This is suggestive of an advantage for résumés with a male name on it.

The question is, however, does this provide *conclusive* evidence that there is gender discrimination in promotions at banks? Could a difference in promotion rates of 29.2% still occur by chance, even in a hypothetical world where no gender-based discrimination existed? In other words, what is the role of *sampling variation* in this hypothesized world? To answer this question, we'll again rely on a computer to run *simulations*.

9.1.2 Shuffling once

First, try to imagine a hypothetical universe where no gender discrimination in promotions existed. In such a hypothetical universe, the gender of an applicant would have no bearing on their chances of promotion. Bringing things back to our `promotions` data frame, the `gender` variable would thus be an irrelevant label. If these `gender` labels were irrelevant, then we could randomly reassign them by "shuffling" them to no consequence!

To illustrate this idea, let's narrow our focus to 6 arbitrarily chosen résumés of the 48 in Table 9.1. The decision column shows that 3 résumés resulted in promotion while 3 didn't. The gender column shows what the original gender of the résumé name was.

However, in our hypothesized universe of no gender discrimination, gender is irrelevant and thus it is of no consequence to randomly "shuffle" the values of gender. The shuffled_gender column shows one such possible random shuffling. Observe in the fourth column how the number of male and female names remains the same at 3 each, but they are now listed in a different order.

TABLE 9.1: One example of shuffling gender variable

résumé number	decision	gender	shuffled gender
1	not	male	male
2	not	female	male
3	not	female	female
4	promoted	male	female
5	promoted	male	female
6	promoted	female	male

Again, such random shuffling of the gender label only makes sense in our hypothesized universe of no gender discrimination. How could we extend this shuffling of the gender variable to all 48 résumés by hand? One way would be by using standard deck of 52 playing cards, which we display in Figure 9.2.

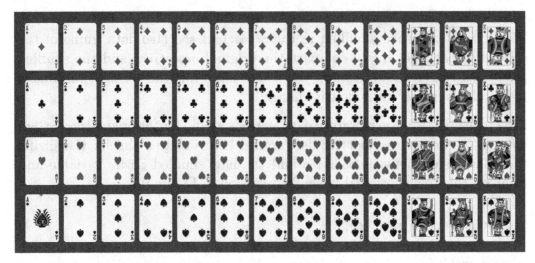

FIGURE 9.2: Standard deck of 52 playing cards.

Since half the cards are red (diamonds and hearts) and the other half are black (spades and clubs), by removing two red cards and two black cards, we would end up with 24 red cards and 24 black cards. After shuffling these 48 cards as seen in Figure 9.3, we can flip the cards over one-by-one, assigning "male" for each red card and "female" for each black card.

FIGURE 9.3: Shuffling a deck of cards.

We've saved one such shuffling in the promotions_shuffled data frame of the moderndive package. If you compare the original promotions and the shuffled promotions_shuffled data frames, you'll see that while the decision variable is identical, the gender variable has changed.

Let's repeat the same exploratory data analysis we did for the original promotions data on our promotions_shuffled data frame. Let's create a barplot visualizing the relationship between decision and the new shuffled gender variable and compare this to the original unshuffled version in Figure 9.4.

```
ggplot(promotions_shuffled,
       aes(x = gender, fill = decision)) +
  geom_bar() +
  labs(x = "Gender of résumé name")
```

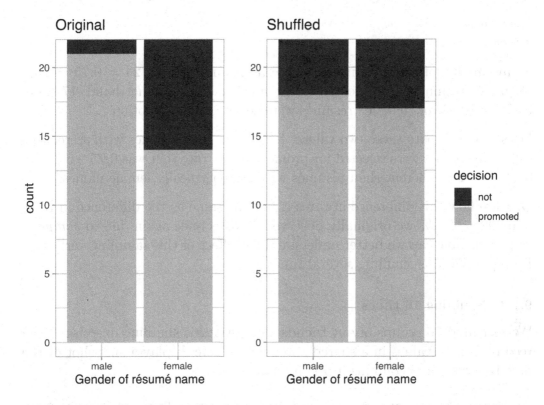

FIGURE 9.4: Barplots of relationship of promotion with gender (left) and shuffled gender (right).

It appears the difference in "male names" versus "female names" promotion rates is now different. Compared to the original data in the left barplot, the new "shuffled" data in the right barplot has promotion rates that are much more similar.

Let's also compute the proportion of résumés accepted for promotion for each group:

```
promotions_shuffled %>%
  group_by(gender, decision) %>%
  tally() # Same as summarize(n = n())
```

```
# A tibble: 4 x 3
# Groups:   gender [2]
  gender decision       n
  <fct>  <fct>      <int>
1 male   not            6
2 male   promoted      18
```

```
3 female not        7
4 female promoted   17
```

So in this hypothetical universe of no discrimination, $18/24 = 0.75 = 75\%$ of "male" résumés were selected for promotion. On the other hand, $17/24 = 0.708 = 70.8\%$ of "female" résumés were selected for promotion.

Let's next compare these two values. It appears that résumés with stereotypically male names were selected for promotion at a rate that was $0.75 - 0.708 = 0.042 = 4.2\%$ different than résumés with stereotypically female names.

Observe how this difference in rates is not the same as the difference in rates of $0.292 = 29.2\%$ we originally observed. This is once again due to *sampling variation*. How can we better understand the effect of this sampling variation? By repeating this shuffling several times!

9.1.3 Shuffling 16 times

We recruited 16 groups of our friends to repeat this shuffling exercise. They recorded these values in a shared spreadsheet[2]; we display a snapshot of the first 10 rows and 5 columns in Figure 9.5.

id	decision	Cassandra, Nox	Priya, Jenny, Eindra	Maddie, Grace, Stephanie	Dahlia, Sarah	Claire, Cindy, Danna
1	not	m	m	m	m	m
2	not	m	m	f	m	m
3	not	m	f	m	m	f
4	not	f	f	f	f	f
5	not	f	m	f	f	f
6	not	m	m	m	f	f
7	not	f	f	m	f	m
8	not	m	f	f	m	f
9	not	m	f	f	m	f
10	not	m	f	m	f	f

FIGURE 9.5: Snapshot of shared spreadsheet of shuffling results (m for male, f for female).

For each of these 16 columns of *shuffles*, we computed the difference in promotion rates, and in Figure 9.6 we display their distribution in a histogram. We also mark the observed difference in promotion rate that occurred in real life of $0.292 = 29.2\%$ with a dark line.

---

[2]https://docs.google.com/spreadsheets/d/1Q-ENy3o5IrpJshJ7gn3hJ5A0TOWV2AZrKNHMsshQtiE/

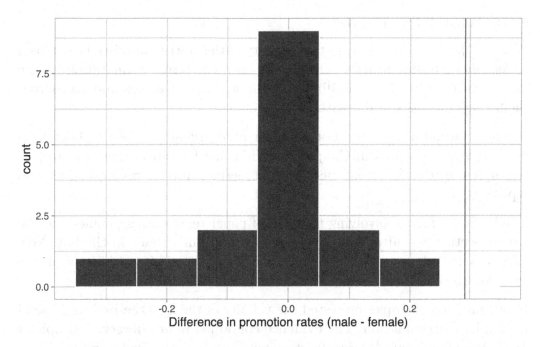

FIGURE 9.6: Distribution of shuffled differences in promotions.

Before we discuss the distribution of the histogram, we emphasize the key thing to remember: this histogram represents differences in promotion rates that one would observe in our *hypothesized universe* of no gender discrimination.

Observe first that the histogram is roughly centered at 0. Saying that the difference in promotion rates is 0 is equivalent to saying that both genders had the same promotion rate. In other words, the center of these 16 values is consistent with what we would expect in our hypothesized universe of no gender discrimination.

However, while the values are centered at 0, there is variation about 0. This is because even in a hypothesized universe of no gender discrimination, you will still likely observe small differences in promotion rates because of chance *sampling variation*. Looking at the histogram in Figure 9.6, such differences could even be as extreme as -0.292 or 0.208.

Turning our attention to what we observed in real life: the difference of 0.292 = 29.2% is marked with a vertical dark line. Ask yourself: in a hypothesized world of no gender discrimination, how likely would it be that we observe this difference? While opinions here may differ, in our opinion not often! Now ask yourself: what do these results say about our hypothesized universe of no gender discrimination?

9.1.4   What did we just do?

What we just demonstrated in this activity is the statistical procedure known as *hypothesis testing* using a *permutation test*. The term "permutation" is the mathematical term for "shuffling": taking a series of values and reordering them randomly, as you did with the playing cards.

In fact, permutations are another form of *resampling*, like the bootstrap method you performed in Chapter 8. While the bootstrap method involves resampling *with* replacement, permutation methods involve resampling *without* replacement.

Think of our exercise involving the slips of paper representing pennies and the hat in Section 8.1: after sampling a penny, you put it back in the hat. Now think of our deck of cards. After drawing a card, you laid it out in front of you, recorded the color, and then you *did not* put it back in the deck.

In our previous example, we tested the validity of the hypothesized universe of no gender discrimination. The evidence contained in our observed sample of 48 résumés was somewhat inconsistent with our hypothesized universe. Thus, we would be inclined to *reject* this hypothesized universe and declare that the evidence suggests there is gender discrimination.

Recall our case study on whether yawning is contagious from Section 8.6. The previous example involves inference about an unknown difference of population proportions as well. This time, it will be $p_m - p_f$, where p_m is the population proportion of résumés with male names being recommended for promotion and p_f is the equivalent for résumés with female names. Recall that this is one of the scenarios for inference we've seen so far in Table 9.2.

TABLE 9.2: Scenarios of sampling for inference

Scenario	Population parameter	Notation	Point estimate	Symbol(s)
1	Population proportion	p	Sample proportion	$\hat{p}$
2	Population mean	μ	Sample mean	$\bar{x}$ or $\hat{\mu}$
3	Difference in population proportions	$p_1 - p_2$	Difference in sample proportions	$\hat{p}_1 - \hat{p}_2$

So, based on our sample of $n_m = 24$ "male" applicants and $n_w = 24$ "female" applicants, the *point estimate* for $p_m - p_f$ is the *difference in sample proportions*

$\hat{p}_m - \hat{p}_f = 0.875$ - $0.583 = 0.292 = 29.2\%$. This difference in favor of "male" résumés of 0.292 is greater than 0, suggesting discrimination in favor of men.

However, the question we asked ourselves was "is this difference meaningfully greater than 0?". In other words, is that difference indicative of true discrimination, or can we just attribute it to *sampling variation*? Hypothesis testing allows us to make such distinctions.

9.2 Understanding hypothesis tests

Much like the terminology, notation, and definitions relating to sampling you saw in Section 7.3, there are a lot of terminology, notation, and definitions related to hypothesis testing as well. Learning these may seem like a very daunting task at first. However, with practice, practice, and more practice, anyone can master them.

First, a **hypothesis** is a statement about the value of an unknown population parameter. In our résumé activity, our population parameter of interest is the difference in population proportions $p_m - p_f$. Hypothesis tests can involve any of the population parameters in Table 7.5 of the five inference scenarios we'll cover in this book and also more advanced types we won't cover here.

Second, a **hypothesis test** consists of a test between two competing hypotheses: (1) a **null hypothesis** H_0 (pronounced "H-naught") versus (2) an **alternative hypothesis** H_A (also denoted H_1).

Generally the null hypothesis is a claim that there is "no effect" or "no difference of interest." In many cases, the null hypothesis represents the status quo or a situation that nothing interesting is happening. Furthermore, generally the alternative hypothesis is the claim the experimenter or researcher wants to establish or find evidence to support. It is viewed as a "challenger" hypothesis to the null hypothesis H_0. In our résumé activity, an appropriate hypothesis test would be:

$$H_0 : \text{men and women are promoted at the same rate}$$
$$\text{vs } H_A : \text{men are promoted at a higher rate than women}$$

Note some of the choices we have made. First, we set the null hypothesis H_0 to be that there is no difference in promotion rate and the "challenger" alternative hypothesis H_A to be that there is a difference. While it would not be wrong in principle to reverse the two, it is a convention in statistical inference that the null hypothesis is set to reflect a "null" situation where "nothing is going on." As we discussed earlier, in this case, H_0 corresponds to there being no difference in promotion rates. Furthermore, we set H_A to be that men are promoted at a *higher* rate, a subjective choice reflecting a prior suspicion we have that this is the case. We call such alternative hypotheses *one-sided alternatives*. If someone else however does not share such suspicions and only wants to investigate that there is a difference, whether higher or lower, they would set what is known as a *two-sided alternative*.

We can re-express the formulation of our hypothesis test using the mathematical notation for our population parameter of interest, the difference in population proportions $p_m - p_f$:

$$H_0 : p_m - p_f = 0$$
$$\text{vs } H_A : p_m - p_f > 0$$

Observe how the alternative hypothesis H_A is one-sided with $p_m - p_f > 0$. Had we opted for a two-sided alternative, we would have set $p_m - p_f \neq 0$. To keep things simple for now, we'll stick with the simpler one-sided alternative. We'll present an example of a two-sided alternative in Section 9.5.

Third, a **test statistic** is a *point estimate/sample statistic* formula used for hypothesis testing. Note that a sample statistic is merely a summary statistic based on a sample of observations. Recall we saw in Section 3.3 that a summary statistic takes in many values and returns only one. Here, the samples would be the $n_m = 24$ résumés with male names and the $n_f = 24$ résumés with female names. Hence, the point estimate of interest is the difference in sample proportions $\hat{p}_m - \hat{p}_f$.

Fourth, the **observed test statistic** is the value of the test statistic that we observed in real life. In our case, we computed this value using the data saved in the promotions data frame. It was the observed difference of $\hat{p}_m - \hat{p}_f = 0.875 - 0.583 = 0.292 = 29.2\%$ in favor of résumés with male names.

Fifth, the **null distribution** is the sampling distribution of the test statistic *assuming the null hypothesis H_0 is true*. Ooof! That's a long one! Let's unpack it slowly. The key to understanding the null distribution is that the null hypothesis H_0 is *assumed* to be true. We're not saying that H_0 is true at this point, we're only assuming it to be true for hypothesis testing purposes. In our case, this corresponds to our hypothesized universe of no gender discrimination in promotion rates. Assuming the null hypothesis H_0, also stated as "Under H_0," how does the test statistic vary due to sampling variation? In our case, how will the difference in sample proportions $\hat{p}_m - \hat{p}_f$ vary due to sampling under H_0? Recall from Subsection 7.3.2 that distributions displaying how point estimates vary due to sampling variation are called *sampling* *distributions.* The only additional thing to keep in mind about null distributions is that they are sampling distributions *assuming the null hypothesis H_0 is true.*

In our case, we previously visualized a null distribution in Figure 9.6, which we re-display in Figure 9.7 using our new notation and terminology. It is the distribution of the 16 differences in sample proportions our friends computed *assuming* a hypothetical universe of no gender discrimination. We also mark the value of the observed test statistic of 0.292 with a vertical line.

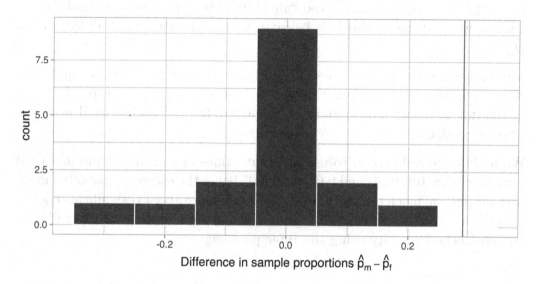

FIGURE 9.7: Null distribution and observed test statistic.

Sixth, the *p*-**value** is the probability of obtaining a test statistic just as extreme or more extreme than the observed test statistic *assuming the null hypothesis H_0 is true.* Double ooof! Let's unpack this slowly as well. You can think of the *p*-value as a quantification of "surprise": assuming H_0 is true, how surprised are we with what we observed? Or in our case, in our hypothesized universe of no gender discrimination, how surprised are we that we observed a difference in promotion rates of 0.292 from our collected samples assuming H_0 is true? Very surprised? Somewhat surprised?

The *p*-value quantifies this probability, or in the case of our 16 differences in sample proportions in Figure 9.7, what proportion had a more "extreme" result? Here, extreme is defined in terms of the alternative hypothesis H_A that "male" applicants are promoted at a higher rate than "female" applicants. In other words, how often was the discrimination in favor of men *even more* pronounced than $0.875 - 0.583 = 0.292 = 29.2\%$?

In this case, 0 times out of 16, we obtained a difference in proportion greater than or equal to the observed difference of $0.292 = 29.2\%$. A very rare (in fact, not occurring) outcome! Given the rarity of such a pronounced difference in promotion rates in our hypothesized universe of no gender discrimination, we're inclined to *reject* our hypothesized universe. Instead, we favor the hypothesis stating there is discrimination in favor of the "male" applicants. In other words, we reject H_0 in favor of H_A.

Seventh and lastly, in many hypothesis testing procedures, it is commonly recommended to set the **significance level** of the test beforehand. It is denoted by the Greek letter α (pronounced "alpha"). This value acts as a cutoff on the p-value, where if the p-value falls below α, we would "reject the null hypothesis H_0."

Alternatively, if the p-value does not fall below α, we would "fail to reject H_0." Note the latter statement is not quite the same as saying we "accept H_0." This distinction is rather subtle and not immediately obvious. So we'll revisit it later in Section 9.4.

While different fields tend to use different values of α, some commonly used values for α are 0.1, 0.01, and 0.05; with 0.05 being the choice people often make without putting much thought into it. We'll talk more about α significance levels in Section 9.4, but first let's fully conduct the hypothesis test corresponding to our promotions activity using the `infer` package.

9.3   Conducting hypothesis tests

In Section 8.4, we showed you how to construct confidence intervals. We first illustrated how to do this using `dplyr` data wrangling verbs and the `rep_sample_n()` function from Subsection 7.2.3 which we used as a virtual shovel. In particular, we constructed confidence intervals by resampling with replacement by setting the `replace = TRUE` argument to the `rep_sample_n()` function.

We then showed you how to perform the same task using the `infer` package workflow. While both workflows resulted in the same bootstrap distribution from which we can construct confidence intervals, the `infer` package workflow emphasizes each of the steps in the overall process in Figure 9.8. It does so using function names that are intuitively named with verbs:

1. `specify()` the variables of interest in your data frame.
2. `generate()` replicates of bootstrap resamples with replacement.
3. `calculate()` the summary statistic of interest.

4. `visualize()` the resulting bootstrap distribution and confidence interval.

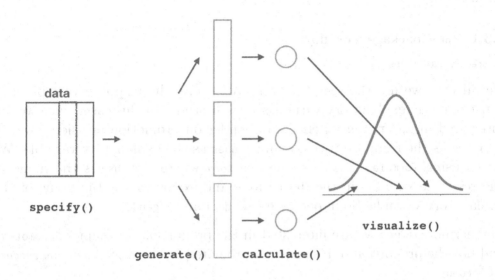

FIGURE 9.8: Confidence intervals with the infer package.

In this section, we'll now show you how to seamlessly modify the previously seen `infer` code for constructing confidence intervals to conduct hypothesis tests. You'll notice that the basic outline of the workflow is almost identical, except for an additional `hypothesize()` step between the `specify()` and `generate()` steps, as can be seen in Figure 9.9.

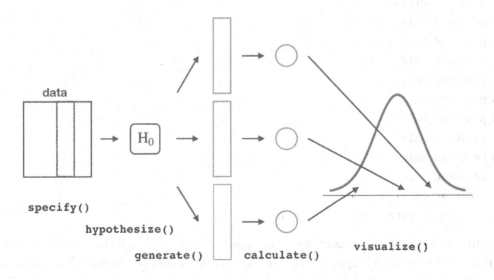

FIGURE 9.9: Hypothesis testing with the infer package.

Furthermore, we'll use a pre-specified significance level $\alpha = 0.05$ for this
hypothesis test. Let's leave discussion on the choice of this α value until later
on in Section 9.4.

9.3.1 `infer` package workflow

1. `specify variables`

Recall that we use the `specify()` verb to specify the response variable and,
if needed, any explanatory variables for our study. In this case, since we are
interested in any potential effects of gender on promotion decisions, we set
`decision` as the response variable and `gender` as the explanatory variable. We
do so using `formula = response ~ explanatory` where `response` is the name of
the response variable in the data frame and `explanatory` is the name of the
explanatory variable. So in our case it is `decision ~ gender`.

Furthermore, since we are interested in the proportion of résumés `"promoted"`,
and not the proportion of résumés `not` promoted, we set the argument `success`
to `"promoted"`.

```
promotions %>%
  specify(formula = decision ~ gender, success = "promoted")
```

```
Response: decision (factor)
Explanatory: gender (factor)
# A tibble: 48 x 2
   decision gender
   <fct>    <fct>
 1 promoted male
 2 promoted male
 3 promoted male
 4 promoted male
 5 promoted male
 6 promoted male
 7 promoted male
 8 promoted male
 9 promoted male
10 promoted male
# ... with 38 more rows
```

Again, notice how the `promotions` data itself doesn't change, but the `Response:`
`decision (factor)` and `Explanatory: gender (factor)` *meta-data* do. This is similar
to how the `group_by()` verb from `dplyr` doesn't change the data, but only adds
"grouping" meta-data, as we saw in Section 3.4.

2. `hypothesize` the null

In order to conduct hypothesis tests using the `infer` workflow, we need a new step not present for confidence intervals: `hypothesize()`. Recall from Section 9.2 that our hypothesis test was

$$H_0 : p_m - p_f = 0$$
$$\text{vs. } H_A : p_m - p_f > 0$$

In other words, the null hypothesis H_0 corresponding to our "hypothesized universe" stated that there was no difference in gender-based discrimination rates. We set this null hypothesis H_0 in our `infer` workflow using the `null` argument of the `hypothesize()` function to either:

- `"point"` for hypotheses involving a single sample or
- `"independence"` for hypotheses involving two samples.

In our case, since we have two samples (the résumés with "male" and "female" names), we set `null = "independence"`.

```
promotions %>%
  specify(formula = decision ~ gender, success = "promoted") %>%
  hypothesize(null = "independence")
```

```
Response: decision (factor)
Explanatory: gender (factor)
Null Hypothesis: independence
# A tibble: 48 x 2
   decision gender
   <fct>    <fct>
 1 promoted male
 2 promoted male
 3 promoted male
 4 promoted male
 5 promoted male
 6 promoted male
 7 promoted male
 8 promoted male
 9 promoted male
10 promoted male
# ... with 38 more rows
```

Again, the data has not changed yet. This will occur at the upcoming `generate()` step; we're merely setting meta-data for now.

Where do the terms `"point"` and `"independence"` come from? These are two technical statistical terms. The term "point" relates from the fact that for a single group of observations, you will test the value of a single point. Going back to the pennies example from Chapter 8, say we wanted to test if the mean year of all US pennies was equal to 1993 or not. We would be testing the value of a "point" μ, the mean year of *all* US pennies, as follows

$$H_0 : \mu = 1993$$
$$\text{vs } H_A : \mu \neq 1993$$

The term "independence" relates to the fact that for two groups of observations, you are testing whether or not the response variable is *independent* of the explanatory variable that assigns the groups. In our case, we are testing whether the `decision` response variable is "independent" of the explanatory variable `gender` that assigns each résumé to either of the two groups.

3. generate replicates

After we `hypothesize()` the null hypothesis, we `generate()` replicates of "shuffled" datasets assuming the null hypothesis is true. We do this by repeating the shuffling exercise you performed in Section 9.1 several times. Instead of merely doing it 16 times as our groups of friends did, let's use the computer to repeat this 1000 times by setting `reps = 1000` in the `generate()` function. However, unlike for confidence intervals where we generated replicates using `type = "bootstrap"` resampling with replacement, we'll now perform shuffles/permutations by setting `type = "permute"`. Recall that shuffles/permutations are a kind of resampling, but unlike the bootstrap method, they involve resampling *without* replacement.

```
promotions_generate <- promotions %>%
  specify(formula = decision ~ gender, success = "promoted") %>%
  hypothesize(null = "independence") %>%
  generate(reps = 1000, type = "permute")
nrow(promotions_generate)
```

```
[1] 48000
```

Observe that the resulting data frame has 48,000 rows. This is because we performed shuffles/permutations for each of the 48 rows 1000 times and $48,000 = 1000 \cdot 48$. If you explore the `promotions_generate` data frame with `View()`, you'll notice that the variable `replicate` indicates which resample each row belongs to. So it has the value `1` 48 times, the value `2` 48 times, all the way through to the value `1000` 48 times.

4. calculate summary statistics

Now that we have generated 1000 replicates of "shuffles" assuming the null hypothesis is true, let's calculate() the appropriate summary statistic for each of our 1000 shuffles. From Section 9.2, point estimates related to hypothesis testing have a specific name: *test statistics*. Since the unknown population parameter of interest is the difference in population proportions $p_m - p_f$, the test statistic here is the difference in sample proportions $\hat{p}_m - \hat{p}_f$.

For each of our 1000 shuffles, we can calculate this test statistic by setting stat = "diff in props". Furthermore, since we are interested in $\hat{p}_m - \hat{p}_f$ we set order = c("male", "female"). As we stated earlier, the order of the subtraction does not matter, so long as you stay consistent throughout your analysis and tailor your interpretations accordingly.

Let's save the result in a data frame called null_distribution:

```
null_distribution <- promotions %>%
    specify(formula = decision ~ gender, success = "promoted") %>%
    hypothesize(null = "independence") %>%
    generate(reps = 1000, type = "permute") %>%
    calculate(stat = "diff in props", order = c("male", "female"))
null_distribution
```

```
# A tibble: 1,000 x 2
    replicate         stat
        <int>        <dbl>
1           1  -0.0416667
2           2  -0.125
3           3  -0.125
4           4  -0.0416667
5           5  -0.0416667
6           6  -0.125
7           7  -0.125
8           8  -0.125
9           9  -0.0416667
10         10  -0.0416667
# ... with 990 more rows
```

Observe that we have 1000 values of stat, each representing one instance of $\hat{p}_m - \hat{p}_f$ in a hypothesized world of no gender discrimination. Observe as well that we chose the name of this data frame carefully: null_distribution. Recall once again from Section 9.2 that sampling distributions when the null hypothesis H_0 is assumed to be true have a special name: the *null distribution*.

What was the *observed* difference in promotion rates? In other words, what was the *observed test statistic* $\hat{p}_m - \hat{p}_f$? Recall from Section 9.1 that we computed this observed difference by hand to be 0.875 - 0.583 = 0.292 = 29.2%. We can also compute this value using the previous `infer` code but with the `hypothesize()` and `generate()` steps removed. Let's save this in `obs_diff_prop`:

```
obs_diff_prop <- promotions %>%
  specify(decision ~ gender, success = "promoted") %>%
  calculate(stat = "diff in props", order = c("male", "female"))
obs_diff_prop
```

```
# A tibble: 1 x 1
     stat
    <dbl>
1 0.291667
```

5. visualize the p-value

The final step is to measure how surprised we are by a promotion difference of 29.2% in a hypothesized universe of no gender discrimination. If the observed difference of 0.292 is highly unlikely, then we would be inclined to reject the validity of our hypothesized universe.

We start by visualizing the *null distribution* of our 1000 values of $\hat{p}_m - \hat{p}_f$ using `visualize()` in Figure 9.10. Recall that these are values of the difference in promotion rates assuming H_0 is true. This corresponds to being in our hypothesized universe of no gender discrimination.

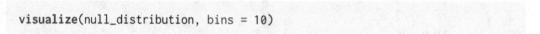

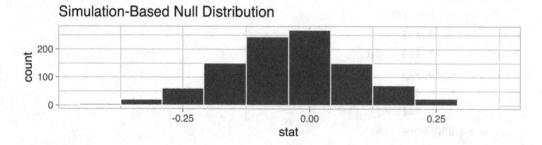

FIGURE 9.10: Null distribution.

Let's now add what happened in real life to Figure 9.10, the observed difference in promotion rates of 0.875 - 0.583 = 0.292 = 29.2%. However, instead of merely adding a vertical line using geom_vline(), let's use the shade_p_value() function with obs_stat set to the observed test statistic value we saved in obs_diff_prop.

Furthermore, we'll set the direction = "right" reflecting our alternative hypothesis $H_A : p_m - p_f > 0$. Recall our alternative hypothesis H_A is that $p_m - p_f > 0$, stating that there is a difference in promotion rates in favor of résumés with male names. "More extreme" here corresponds to differences that are "bigger" or "more positive" or "more to the right." Hence we set the direction argument of shade_p_value() to be "right".

On the other hand, had our alternative hypothesis H_A been the other possible one-sided alternative $p_m - p_f < 0$, suggesting discrimination in favor of résumés with female names, we would've set direction = "left". Had our alternative hypothesis H_A been two-sided $p_m - p_f \neq 0$, suggesting discrimination in either direction, we would've set direction = "both".

```
visualize(null_distribution, bins = 10) +
  shade_p_value(obs_stat = obs_diff_prop, direction = "right")
```

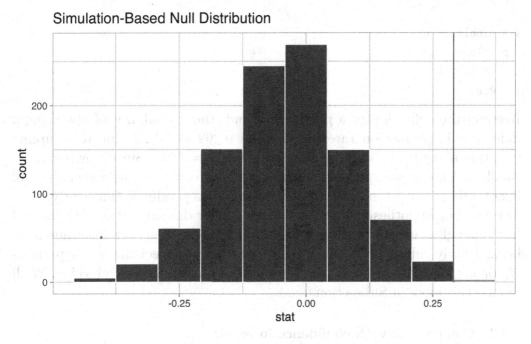

FIGURE 9.11: Shaded histogram to show *p*-value.

In the resulting Figure 9.11, the solid dark line marks $0.292 = 29.2\%$. However, what does the shaded-region correspond to? This is the *p-value*. Recall the definition of the *p*-value from Section 9.2:

A *p*-value is the probability of obtaining a test statistic just as or more extreme
than the observed test statistic *assuming the null hypothesis H_0 is true.*

So judging by the shaded region in Figure 9.11, it seems we would somewhat
rarely observe differences in promotion rates of $0.292 = 29.2\%$ or more in a
hypothesized universe of no gender discrimination. In other words, the *p*-value
is somewhat small. Hence, we would be inclined to reject this hypothesized
universe, or using statistical language we would "reject H_0."

What fraction of the null distribution is shaded? In other words, what is the
exact value of the *p*-value? We can compute it using the `get_p_value()` function
with the same arguments as the previous `shade_p_value()` code:

```
null_distribution %>%
  get_p_value(obs_stat = obs_diff_prop, direction = "right")
```

```
# A tibble: 1 x 1
  p_value
    <dbl>
1   0.027
```

Keeping the definition of a *p*-value in mind, the probability of observing a
difference in promotion rates as large as $0.292 = 29.2\%$ due to sampling
variation alone in the null distribution is $0.027 = 2.7\%$. Since this *p*-value is
smaller than our pre-specified significance level $\alpha = 0.05$, we reject the null
hypothesis $H_0 : p_m - p_f = 0$. In other words, this *p*-value is sufficiently small
to reject our hypothesized universe of no gender discrimination. We instead
have enough evidence to change our mind in favor of gender discrimination
being a likely culprit here. Observe that whether we reject the null hypothesis
H_0 or not depends in large part on our choice of significance level α. We'll
discuss this more in Subsection 9.4.3.

9.3.2 Comparison with confidence intervals

One of the great things about the `infer` package is that we can jump seamlessly
between conducting hypothesis tests and constructing confidence intervals with
minimal changes! Recall the code from the previous section that creates the
null distribution, which in turn is needed to compute the *p*-value:

```
null_distribution <- promotions %>%
  specify(formula = decision ~ gender, success = "promoted") %>%
  hypothesize(null = "independence") %>%
  generate(reps = 1000, type = "permute") %>%
  calculate(stat = "diff in props", order = c("male", "female"))
```

To create the corresponding bootstrap distribution needed to construct a 95% confidence interval for $p_m - p_f$, we only need to make two changes. First, we remove the hypothesize() step since we are no longer assuming a null hypothesis H_0 is true. We can do this by deleting or commenting out the hypothesize() line of code. Second, we switch the type of resampling in the generate() step to be "bootstrap" instead of "permute".

```
bootstrap_distribution <- promotions %>%
  specify(formula = decision ~ gender, success = "promoted") %>%
  # Change 1 - Remove hypothesize():
  # hypothesize(null = "independence") %>%
  # Change 2 - Switch type from "permute" to "bootstrap":
  generate(reps = 1000, type = "bootstrap") %>%
  calculate(stat = "diff in props", order = c("male", "female"))
```

Using this bootstrap_distribution, let's first compute the percentile-based confidence intervals, as we did in Section 8.4:

```
percentile_ci <- bootstrap_distribution %>%
  get_confidence_interval(level = 0.95, type = "percentile")
percentile_ci
```

```
# A tibble: 1 x 2
    `2.5%`   `97.5%`
     <dbl>     <dbl>
1 0.0444444 0.538542
```

Using our shorthand interpretation for 95% confidence intervals from Subsection 8.5.2, we are 95% "confident" that the true difference in population proportions $p_m - p_f$ is between (0.044, 0.539). Let's visualize bootstrap_distribution and this percentile-based 95% confidence interval for $p_m - p_f$ in Figure 9.12.

```
visualize(bootstrap_distribution) +
  shade_confidence_interval(endpoints = percentile_ci)
```

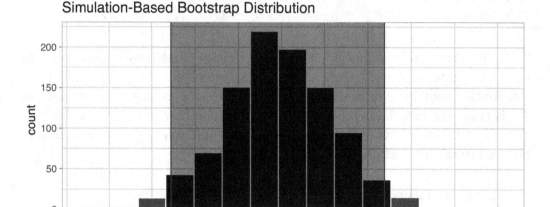

FIGURE 9.12: Percentile-based 95% confidence interval.

Notice a key value that is not included in the 95% confidence interval for $p_m - p_f$: the value 0. In other words, a difference of 0 is not included in our net, suggesting that p_m and p_f are truly different! Furthermore, observe how the entirety of the 95% confidence interval for $p_m - p_f$ lies above 0, suggesting that this difference is in favor of men.

Since the bootstrap distribution appears to be roughly normally shaped, we can also use the standard error method as we did in Section 8.4. In this case, we must specify the point_estimate argument as the observed difference in promotion rates $0.292 = 29.2\%$ saved in obs_diff_prop. This value acts as the center of the confidence interval.

```
se_ci <- bootstrap_distribution %>%
  get_confidence_interval(level = 0.95, type = "se",
                          point_estimate = obs_diff_prop)
  se_ci
```

```
# A tibble: 1 x 2
     lower    upper
     <dbl>    <dbl>
1 0.0514129 0.531920
```

Let's visualize `bootstrap_distribution` again, but now the standard error based 95% confidence interval for $p_m - p_f$ in Figure 9.13. Again, notice how the value 0 is not included in our confidence interval, again suggesting that p_m and p_f are truly different!

```
visualize(bootstrap_distribution) +
  shade_confidence_interval(endpoints = se_ci)
```

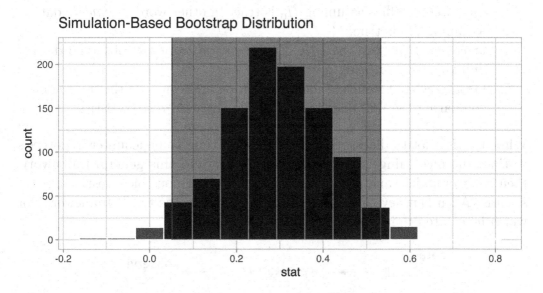

FIGURE 9.13: Standard error-based 95% confidence interval.

Learning check

(LC9.1) Conduct the same hypothesis test and confidence interval analysis comparing male and female promotion rates using the median rating instead of the mean rating. What was different and what was the same?

(LC9.2) Why are we relatively confident that the distributions of the sample proportions will be good approximations of the population distributions of promotion proportions for the two genders?

(LC9.3) Using the definition of *p-value*, write in words what the *p*-value represents for the hypothesis test comparing the promotion rates for males and females.

9.3.3 "There is only one test"

Let's recap the steps necessary to conduct a hypothesis test using the terminology, notation, and definitions related to sampling you saw in Section 9.2 and the `infer` workflow from Subsection 9.3.1:

1. `specify()` the variables of interest in your data frame.
2. `hypothesize()` the null hypothesis H_0. In other words, set a "model for the universe" assuming H_0 is true.
3. `generate()` shuffles assuming H_0 is true. In other words, *simulate* data assuming H_0 is true.
4. `calculate()` the *test statistic* of interest, both for the observed data and your *simulated* data.
5. `visualize()` the resulting *null distribution* and compute the *p-value* by comparing the null distribution to the observed test statistic.

While this is a lot to digest, especially the first time you encounter hypothesis testing, the nice thing is that once you understand this general framework, then you can understand *any* hypothesis test. In a famous blog post, computer scientist Allen Downey called this the "There is only one test"[3] framework, for which he created the flowchart displayed in Figure 9.14.

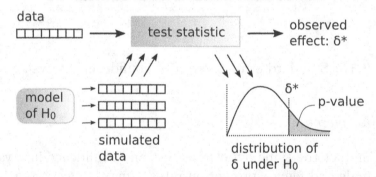

FIGURE 9.14: Allen Downey's hypothesis testing framework.

Notice its similarity with the "hypothesis testing with `infer`" diagram you saw in Figure 9.9. That's because the `infer` package was explicitly designed to match the "There is only one test" framework. So if you can understand the framework, you can easily generalize these ideas for all hypothesis testing scenarios. Whether for population proportions p, population means μ, differences in population proportions $p_1 - p_2$, differences in population means $\mu_1 - \mu_2$, and as you'll see in Chapter 10 on inference for regression, population regression

---

[3]http://allendowney.blogspot.com/2016/06/there-is-still-only-one-test.html

slopes β_1 as well. In fact, it applies more generally even than just these examples to more complicated hypothesis tests and test statistics as well.

Learning check

(LC9.4) Describe in a paragraph how we used Allen Downey's diagram to conclude if a statistical difference existed between the promotion rate of males and females using this study.

9.4 Interpreting hypothesis tests

Interpreting the results of hypothesis tests is one of the more challenging aspects of this method for statistical inference. In this section, we'll focus on ways to help with deciphering the process and address some common misconceptions.

9.4.1 Two possible outcomes

In Section 9.2, we mentioned that given a pre-specified significance level α there are two possible outcomes of a hypothesis test:

- If the p-value is less than α, then we *reject* the null hypothesis H_0 in favor of H_A.
- If the p-value is greater than or equal to α, we *fail to reject* the null hypothesis H_0.

Unfortunately, the latter result is often misinterpreted as "accepting the null hypothesis H_0." While at first glance it may seem that the statements "failing to reject H_0" and "accepting H_0" are equivalent, there actually is a subtle difference. Saying that we "accept the null hypothesis H_0" is equivalent to stating that "we think the null hypothesis H_0 is true." However, saying that we "fail to reject the null hypothesis H_0" is saying something else: "While H_0 might still be false, we don't have enough evidence to say so." In other words, there is an absence of enough proof. However, the absence of proof is not proof of absence.

To further shed light on this distinction, let's use the United States criminal justice system as an analogy. A criminal trial in the United States is a similar situation to hypothesis tests whereby a choice between two contradictory claims must be made about a defendant who is on trial:

1. The defendant is truly either "innocent" or "guilty."
2. The defendant is presumed "innocent until proven guilty."
3. The defendant is found guilty only if there is *strong evidence* that the defendant is guilty. The phrase "beyond a reasonable doubt" is often used as a guideline for determining a cutoff for when enough evidence exists to find the defendant guilty.
4. The defendant is found to be either "not guilty" or "guilty" in the ultimate verdict.

In other words, *not guilty* verdicts are not suggesting the defendant is *innocent*, but instead that "while the defendant may still actually be guilty, there wasn't enough evidence to prove this fact." Now let's make the connection with hypothesis tests:

1. Either the null hypothesis H_0 or the alternative hypothesis H_A is true.
2. Hypothesis tests are conducted assuming the null hypothesis H_0 is true.
3. We reject the null hypothesis H_0 in favor of H_A only if the evidence found in the sample suggests that H_A is true. The significance level α is used as a guideline to set the threshold on just how strong of evidence we require.
4. We ultimately decide to either "fail to reject H_0" or "reject H_0."

So while gut instinct may suggest "failing to reject H_0" and "accepting H_0" are equivalent statements, they are not. "Accepting H_0" is equivalent to finding a defendant innocent. However, courts do not find defendants "innocent," but rather they find them "not guilty." Putting things differently, defense attorneys do not need to prove that their clients are innocent, rather they only need to prove that clients are not "guilty beyond a reasonable doubt".

So going back to our résumés activity in Section 9.3, recall that our hypothesis test was $H_0 : p_m - p_f = 0$ versus $H_A : p_m - p_f > 0$ and that we used a pre-specified significance level of $\alpha = 0.05$. We found a p-value of 0.027. Since the p-value was smaller than $\alpha = 0.05$, we rejected H_0. In other words, we found needed levels of evidence in this particular sample to say that H_0 is false at the $\alpha = 0.05$ significance level. We also state this conclusion using non-statistical language: we found enough evidence in this data to suggest that there was gender discrimination at play.

9.4.2 Types of errors

Unfortunately, there is some chance a jury or a judge can make an incorrect decision in a criminal trial by reaching the wrong verdict. For example, finding a truly innocent defendant "guilty". Or on the other hand, finding a truly guilty defendant "not guilty." This can often stem from the fact that prosecutors don't have access to all the relevant evidence, but instead are limited to whatever evidence the police can find.

The same holds for hypothesis tests. We can make incorrect decisions about a population parameter because we only have a sample of data from the population and thus sampling variation can lead us to incorrect conclusions.

There are two possible erroneous conclusions in a criminal trial: either (1) a truly innocent person is found guilty or (2) a truly guilty person is found not guilty. Similarly, there are two possible errors in a hypothesis test: either (1) rejecting H_0 when in fact H_0 is true, called a **Type I error** or (2) failing to reject H_0 when in fact H_0 is false, called a **Type II error**. Another term used for "Type I error" is "false positive," while another term for "Type II error" is "false negative."

This risk of error is the price researchers pay for basing inference on a sample instead of performing a census on the entire population. But as we've seen in our numerous examples and activities so far, censuses are often very expensive and other times impossible, and thus researchers have no choice but to use a sample. Thus in any hypothesis test based on a sample, we have no choice but to tolerate some chance that a Type I error will be made and some chance that a Type II error will occur.

To help understand the concepts of Type I error and Type II errors, we apply these terms to our criminal justice analogy in Figure 9.15.

	Truly not guilty	Truly guilty
Verdict		
Not guilty verdict	Correct	Type II error
Guilty verdict	Type I error	Correct

FIGURE 9.15: Type I and Type II errors in criminal trials.

Thus a Type I error corresponds to incorrectly putting a truly innocent person in jail, whereas a Type II error corresponds to letting a truly guilty person go free. Let's show the corresponding table in Figure 9.16 for hypothesis tests.

	H0 true	HA true
Verdict		
Fail to reject H0	Correct	Type II error
Reject H0	Type I error	Correct

FIGURE 9.16: Type I and Type II errors in hypothesis tests.

9.4.3 How do we choose alpha?

If we are using a sample to make inferences about a population, we run the risk of making errors. For confidence intervals, a corresponding "error" would be constructing a confidence interval that does not contain the true value of the population parameter. For hypothesis tests, this would be making either a Type I or Type II error. Obviously, we want to minimize the probability of either error; we want a small probability of making an incorrect conclusion:

- The probability of a Type I Error occurring is denoted by α. The value of α is called the *significance level* of the hypothesis test, which we defined in Section 9.2.
- The probability of a Type II Error is denoted by β. The value of $1 - \beta$ is known as the *power* of the hypothesis test.

In other words, α corresponds to the probability of incorrectly rejecting H_0 when in fact H_0 is true. On the other hand, β corresponds to the probability of incorrectly failing to reject H_0 when in fact H_0 is false.

Ideally, we want $\alpha = 0$ and $\beta = 0$, meaning that the chance of making either error is 0. However, this can never be the case in any situation where we are sampling for inference. There will always be the possibility of making either error when we use sample data. Furthermore, these two error probabilities are inversely related. As the probability of a Type I error goes down, the probability of a Type II error goes up.

What is typically done in practice is to fix the probability of a Type I error by pre-specifying a significance level α and then try to minimize β. In other words, we will tolerate a certain fraction of incorrect rejections of the null hypothesis H_0, and then try to minimize the fraction of incorrect non-rejections of H_0.

So for example if we used $\alpha = 0.01$, we would be using a hypothesis testing procedure that in the long run would incorrectly reject the null hypothesis H_0 one percent of the time. This is analogous to setting the confidence level of a confidence interval.

So what value should you use for α? Different fields have different conventions, but some commonly used values include 0.10, 0.05, 0.01, and 0.001. However, it is important to keep in mind that if you use a relatively small value of α, then all things being equal, p-values will have a harder time being less than α. Thus we would reject the null hypothesis less often. In other words, we would reject the null hypothesis H_0 only if we have *very strong* evidence to do so. This is known as a "conservative" test.

On the other hand, if we used a relatively large value of α, then all things being equal, p-values will have an easier time being less than α. Thus we would reject the null hypothesis more often. In other words, we would reject the null hypothesis H_0 even if we only have *mild* evidence to do so. This is known as a "liberal" test.

Learning check

(LC9.5) What is wrong about saying, "The defendant is innocent." based on the US system of criminal trials?

(LC9.6) What is the purpose of hypothesis testing?

(LC9.7) What are some flaws with hypothesis testing? How could we alleviate them?

(LC9.8) Consider two α significance levels of 0.1 and 0.01. Of the two, which would lead to a more *liberal* hypothesis testing procedure? In other words, one that will, all things being equal, lead to more rejections of the null hypothesis H_0.

9.5 Case study: Are action or romance movies rated higher?

Let's apply our knowledge of hypothesis testing to answer the question: "Are action or romance movies rated higher on IMDb?". IMDb[4] is a database on the internet providing information on movie and television show casts, plot summaries, trivia, and ratings. We'll investigate if, on average, action or romance movies get higher ratings on IMDb.

---

[4]https://www.imdb.com/

9.5.1   IMDb ratings data

The `movies` dataset in the `ggplot2movies` package contains information on 58,788 movies that have been rated by users of IMDb.com.

```
movies
```

```
# A tibble: 58,788 x 24
     title  year length budget rating votes    r1    r2    r3    r4    r5
     <chr> <int>  <int>  <int>  <dbl> <int> <dbl> <dbl> <dbl> <dbl> <dbl>
 1 $         1971    121     NA    6.4   348   4.5   4.5   4.5   4.5  14.5
 2 $100~     1939     71     NA    6      20   0    14.5   4.5  24.5  14.5
 3 $21 ~     1941      7     NA    8.200    5   0     0     0     0     0
 4 $40,~     1996     70     NA    8.200    6  14.5   0     0     0     0
 5 $50,~     1975     71     NA    3.4    17  24.5   4.5   0    14.5  14.5
 6 $pent     2000     91     NA    4.3    45   4.5   4.5   4.5  14.5  14.5
 7 $win~     2002     93     NA    5.3   200   4.5   0     4.5   4.5  24.5
 8 '15'      2002     25     NA    6.7    24   4.5   4.5   4.5   4.5   4.5
 9 '38       1987     97     NA    6.6    18   4.5   4.5   4.5   0     0
10 '49-~     1917     61     NA    6      51   4.5   0     4.5   4.5   4.5
# ... with 58,778 more rows, and 13 more variables: r6 <dbl>, r7 <dbl>,
#   r8 <dbl>, r9 <dbl>, r10 <dbl>, mpaa <chr>, Action <int>,
#   Animation <int>, Comedy <int>, Drama <int>, Documentary <int>,
#   Romance <int>, Short <int>
```

We'll focus on a random sample of 68 movies that are classified as either "action" or "romance" movies but not both. We disregard movies that are classified as both so that we can assign all 68 movies into either category. Furthermore, since the original `movies` dataset was a little messy, we provide a pre-wrangled version of our data in the `movies_sample` data frame included in the `moderndive` package. If you're curious, you can look at the necessary data wrangling code to do this on GitHub[5].

```
movies_sample
```

```
# A tibble: 68 x 4
    title            year rating genre
    <chr>           <int>  <dbl> <chr>
 1 Underworld       1985    3.1 Action
 2 Love Affair      1932    6.3 Romance
 3 Junglee          1961    6.8 Romance
```

[5]https://github.com/moderndive/moderndive/blob/master/data-raw/process_data_sets.R

```
 4 Eversmile, New Jersey    1989   5    Romance
 5 Search and Destroy       1979   4    Action
 6 Secreto de Romelia, El   1988   4.9  Romance
 7 Amants du Pont-Neuf, Les 1991   7.4  Romance
 8 Illicit Dreams           1995   3.5  Action
 9 Kabhi Kabhie             1976   7.7  Romance
10 Electric Horseman, The   1979   5.8  Romance
# ... with 58 more rows
```

The variables include the title and year the movie was filmed. Furthermore, we have a numerical variable rating, which is the IMDb rating out of 10 stars, and a binary categorical variable genre indicating if the movie was an Action or Romance movie. We are interested in whether Action or Romance movies got a higher rating on average.

Let's perform an exploratory data analysis of this data. Recall from Subsection 2.7.1 that a boxplot is a visualization we can use to show the relationship between a numerical and a categorical variable. Another option you saw in Section 2.6 would be to use a faceted histogram. However, in the interest of brevity, let's only present the boxplot in Figure 9.17.

```
ggplot(data = movies_sample, aes(x = genre, y = rating)) +
  geom_boxplot() +
  labs(y = "IMDb rating")
```

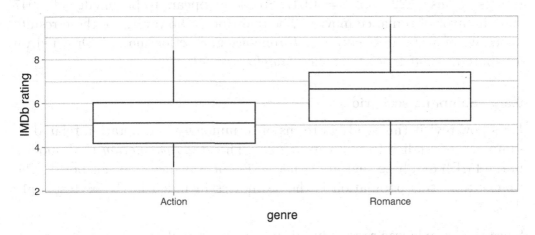

FIGURE 9.17: Boxplot of IMDb rating vs. genre.

Eyeballing Figure 9.17, romance movies have a higher median rating. Do we have reason to believe, however, that there is a *significant* difference between the mean rating for action movies compared to romance movies? It's hard to say just based on this plot. The boxplot does show that the median sample rating is higher for romance movies.

However, there is a large amount of overlap between the boxes. Recall that the median isn't necessarily the same as the mean either, depending on whether the distribution is skewed.

Let's calculate some summary statistics split by the binary categorical variable genre: the number of movies, the mean rating, and the standard deviation split by genre. We'll do this using dplyr data wrangling verbs. Notice in particular how we count the number of each type of movie using the n() summary function.

```
movies_sample %>%
  group_by(genre) %>%
  summarize(n = n(), mean_rating = mean(rating), std_dev = sd(rating))
```

```
# A tibble: 2 x 4
  genre        n mean_rating std_dev
  <chr>    <int>       <dbl>   <dbl>
1 Action      32       5.275 1.36121
2 Romance     36     6.32222 1.60963
```

Observe that we have 36 movies with an average rating of 6.322 stars and 32 movies with an average rating of 5.275 stars. The difference in these average ratings is thus 6.322 - 5.275 = 1.047. So there appears to be an edge of 1.047 stars in favor of romance movies. The question is, however, are these results indicative of a true difference for *all* romance and action movies? Or could we attribute this difference to chance *sampling variation*?

9.5.2 Sampling scenario

Let's now revisit this study in terms of terminology and notation related to sampling we studied in Subsection 7.3.1. The *study population* is all movies in the IMDb database that are either action or romance (but not both). The *sample* from this population is the 68 movies included in the movies_sample dataset.

Since this sample was randomly taken from the population movies, it is representative of all romance and action movies on IMDb. Thus, any analysis and results based on movies_sample can generalize to the entire population. What are the relevant *population parameter* and *point estimates*? We introduce the fourth sampling scenario in Table 9.3.

TABLE 9.3: Scenarios of sampling for inference

Scenario	Population parameter	Notation	Point estimate	Symbol(s)
1	Population proportion	p	Sample proportion	$\hat{p}$
2	Population mean	μ	Sample mean	$\bar{x}$ or $\hat{\mu}$
3	Difference in population proportions	$p_1 - p_2$	Difference in sample proportions	$\hat{p}_1 - \hat{p}_2$
4	Difference in population means	$\mu_1 - \mu_2$	Difference in sample means	$\bar{x}_1 - \bar{x}_2$

So, whereas the sampling bowl exercise in Section 7.1 concerned *proportions*, the pennies exercise in Section 8.1 concerned *means*, the case study on whether yawning is contagious in Section 8.6 and the promotions activity in Section 9.1 concerned *differences in proportions*, we are now concerned with *differences in means*.

In other words, the population parameter of interest is the difference in population mean ratings $\mu_a - \mu_r$, where μ_a is the mean rating of all action movies on IMDb and similarly μ_r is the mean rating of all romance movies. Additionally the point estimate/sample statistic of interest is the difference in sample means $\bar{x}_a - \bar{x}_r$, where $\bar{x}_a$ is the mean rating of the $n_a = 32$ movies in our sample and $\bar{x}_r$ is the mean rating of the $n_r = 36$ in our sample. Based on our earlier exploratory data analysis, our estimate $\bar{x}_a - \bar{x}_r$ is $5.275 - 6.322 = -1.047$.

So there appears to be a slight difference of -1.047 in favor of romance movies. The question is, however, could this difference of -1.047 be merely due to chance and sampling variation? Or are these results indicative of a true difference in mean ratings for *all* romance and action movies on IMDb? To answer this question, we'll use hypothesis testing.

9.5.3 Conducting the hypothesis test

We'll be testing:

$$H_0 : \mu_a - \mu_r = 0$$
$$\text{vs } H_A : \mu_a - \mu_r \neq 0$$

In other words, the null hypothesis H_0 suggests that both romance and action movies have the same mean rating. This is the "hypothesized universe" we'll *assume* is true. On the other hand, the alternative hypothesis H_A suggests that there is a difference. Unlike the one-sided alternative we used in the promotions exercise $H_a : p_m - p_f > 0$, we are now considering a two-sided alternative of $H_A : \mu_a - \mu_r \neq 0$.

Furthermore, we'll pre-specify a low significance level of $\alpha = 0.001$. By setting this value low, all things being equal, there is a lower chance that the *p*-value will be less than α. Thus, there is a lower chance that we'll reject the null hypothesis H_0 in favor of the alternative hypothesis H_A. In other words, we'll reject the hypothesis that there is no difference in mean ratings for all action and romance movies, only if we have quite strong evidence. This is known as a "conservative" hypothesis testing procedure.

1. specify variables

Let's now perform all the steps of the `infer` workflow. We first `specify()` the variables of interest in the `movies_sample` data frame using the formula `rating ~ genre`. This tells `infer` that the numerical variable `rating` is the outcome variable, while the binary variable `genre` is the explanatory variable. Note that unlike previously when we were interested in proportions, since we are now interested in the mean of a numerical variable, we do not need to set the success argument.

```
movies_sample %>%
    specify(formula = rating ~ genre)
```

```
Response: rating (numeric)
Explanatory: genre (factor)
# A tibble: 68 x 2
    rating genre
     <dbl> <fct>
 1     3.1 Action
 2     6.3 Romance
 3     6.8 Romance
 4     5   Romance
 5     4   Action
 6     4.9 Romance
 7     7.4 Romance
 8     3.5 Action
 9     7.7 Romance
10     5.8 Romance
# ... with 58 more rows
```

Observe at this point that the data in `movies_sample` has not changed. The only change so far is the newly defined Response: rating (numeric) and Explanatory: genre (factor) *meta-data.*

2. hypothesize the null

We set the null hypothesis $H_0 : \mu_a - \mu_r = 0$ by using the `hypothesize()` function. Since we have two samples, action and romance movies, we set `null` to be `"independence"` as we described in Section 9.3.

```
movies_sample %>%
   specify(formula = rating ~ genre) %>%
   hypothesize(null = "independence")
```

```
Response: rating (numeric)
Explanatory: genre (factor)
Null Hypothesis: independence
# A tibble: 68 x 2
    rating genre
     <dbl> <fct>
 1     3.1 Action
 2     6.3 Romance
 3     6.8 Romance
 4     5   Romance
 5     4   Action
 6     4.9 Romance
 7     7.4 Romance
 8     3.5 Action
 9     7.7 Romance
10     5.8 Romance
# ... with 58 more rows
```

3. generate replicates

After we have set the null hypothesis, we generate "shuffled" replicates assuming the null hypothesis is true by repeating the shuffling/permutation exercise you performed in Section 9.1.

We'll repeat this resampling without replacement of `type = "permute"` a total of `reps = 1000` times. Feel free to run the code below to check out what the `generate()` step produces.

```
movies_sample %>%
  specify(formula = rating ~ genre) %>%
  hypothesize(null = "independence") %>%
  generate(reps = 1000, type = "permute") %>%
  View()
```

4. calculate summary statistics

Now that we have 1000 replicated "shuffles" assuming the null hypothesis H_0 that both `Action` and `Romance` movies on average have the same ratings on IMDb, let's `calculate()` the appropriate summary statistic for these 1000 replicated shuffles. From Section 9.2, summary statistics relating to hypothesis testing have a specific name: *test statistics*. Since the unknown population parameter of interest is the difference in population means $\mu_a - \mu_r$, the test statistic of interest here is the difference in sample means $\overline{x}_a - \overline{x}_r$.

For each of our 1000 shuffles, we can calculate this test statistic by setting `stat = "diff in means"`. Furthermore, since we are interested in $\overline{x}_a - \overline{x}_r$, we set `order = c("Action", "Romance")`. Let's save the results in a data frame called `null_distribution_movies`:

```
null_distribution_movies <- movies_sample %>%
  specify(formula = rating ~ genre) %>%
  hypothesize(null = "independence") %>%
  generate(reps = 1000, type = "permute") %>%
  calculate(stat = "diff in means", order = c("Action", "Romance"))
null_distribution_movies
```

```
# A tibble: 1,000 x 2
   replicate      stat
       <int>     <dbl>
1          1  0.511111
2          2  0.345833
3          3 -0.327083
4          4 -0.209028
5          5 -0.433333
6          6 -0.102778
7          7  0.387153
8          8  0.16875
9          9  0.257292
10        10  0.334028
# ... with 990 more rows
```

Observe that we have 1000 values of stat, each representing one instance of $\overline{x}_a - \overline{x}_r$. The 1000 values form the *null distribution*, which is the technical term for the sampling distribution of the difference in sample means $\overline{x}_a - \overline{x}_r$ assuming H_0 is true. What happened in real life? What was the observed difference in promotion rates? What was the *observed test statistic* $\overline{x}_a - \overline{x}_r$? Recall from our earlier data wrangling, this observed difference in means was $5.275 - 6.322 = -1.047$. We can also achieve this using the code that constructed the null distribution null_distribution_movies but with the hypothesize() and generate() steps removed. Let's save this in obs_diff_means:

```
obs_diff_means <- movies_sample %>%
  specify(formula = rating ~ genre) %>%
  calculate(stat = "diff in means", order = c("Action", "Romance"))
obs_diff_means
```

```
# A tibble: 1 x 1
     stat
    <dbl>
1 -1.04722
```

5. visualize the p-value

Lastly, in order to compute the *p*-value, we have to assess how "extreme" the observed difference in means of -1.047 is. We do this by comparing -1.047 to our null distribution, which was constructed in a hypothesized universe of no true difference in movie ratings. Let's visualize both the null distribution and the *p*-value in Figure 9.18. Unlike our example in Subsection 9.3.1 involving promotions, since we have a two-sided $H_A : \mu_a - \mu_r \neq 0$, we have to allow for both possibilities for *more extreme*, so we set direction = "both".

```
visualize(null_distribution_movies, bins = 10) +
  shade_p_value(obs_stat = obs_diff_means, direction = "both")
```

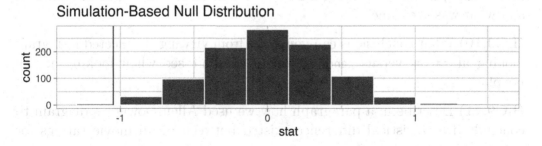

FIGURE 9.18: Null distribution, observed test statistic, and *p*-value.

Let's go over the elements of this plot. First, the histogram is the *null distribution*. Second, the solid line is the *observed test statistic*, or the difference in sample means we observed in real life of $5.275 - 6.322 = -1.047$. Third, the two shaded areas of the histogram form the *p-value*, or the probability of obtaining a test statistic just as or more extreme than the observed test statistic *assuming the null hypothesis H_0 is true.*

What proportion of the null distribution is shaded? In other words, what is the numerical value of the *p*-value? We use the `get_p_value()` function to compute this value:

```
null_distribution_movies %>%
  get_p_value(obs_stat = obs_diff_means, direction = "both")
```

```
# A tibble: 1 x 1
  p_value
    <dbl>
1   0.004
```

This *p*-value of 0.004 is very small. In other words, there is a very small chance that we'd observe a difference of 5.275 - 6.322 = -1.047 in a hypothesized universe where there was truly no difference in ratings.

But this *p*-value is larger than our (even smaller) pre-specified α significance level of 0.001. Thus, we are inclined to fail to reject the null hypothesis $H_0 : \mu_a - \mu_r = 0$. In non-statistical language, the conclusion is: we do not have the evidence needed in this sample of data to suggest that we should reject the hypothesis that there is no difference in mean IMDb ratings between romance and action movies. We, thus, cannot say that a difference exists in romance and action movie ratings, on average, for all IMDb movies.

Learning check

(LC9.9) Conduct the same analysis comparing action movies versus romantic movies using the median rating instead of the mean rating. What was different and what was the same?

(LC9.10) What conclusions can you make from viewing the faceted histogram looking at `rating` versus `genre` that you couldn't see when looking at the boxplot?

(LC9.11) Describe in a paragraph how we used Allen Downey's diagram to conclude if a statistical difference existed between mean movie ratings for action and romance movies.

(LC9.12) Why are we relatively confident that the distributions of the sample ratings will be good approximations of the population distributions of ratings for the two genres?

(LC9.13) Using the definition of p-value, write in words what the p-value represents for the hypothesis test comparing the mean rating of romance to action movies.

(LC9.14) What is the value of the p-value for the hypothesis test comparing the mean rating of romance to action movies?

(LC9.15) Test your data wrangling knowledge and EDA skills:

- Use `dplyr` and `tidyr` to create the necessary data frame focused on only action and romance movies (but not both) from the `movies` data frame in the `ggplot2movies` package.
- Make a boxplot and a faceted histogram of this population data comparing ratings of action and romance movies from IMDb.
- Discuss how these plots compare to the similar plots produced for the `movies_sample` data.

9.6 Conclusion

9.6.1 Theory-based hypothesis tests

Much as we did in Subsection 8.7.2 when we showed you a theory-based method for constructing confidence intervals that involved mathematical formulas, we now present an example of a traditional theory-based method to conduct hypothesis tests. This method relies on probability models, probability distributions, and a few assumptions to construct the null distribution. This is in contrast to the approach we've been using throughout this book where we relied on computer simulations to construct the null distribution.

These traditional theory-based methods have been used for decades mostly because researchers didn't have access to computers that could run thousands of calculations quickly and efficiently. Now that computing power is much cheaper and more accessible, simulation-based methods are much more feasible. However, researchers in many fields continue to use theory-based methods. Hence, we make it a point to include an example here.

As we'll show in this section, any theory-based method is ultimately an approximation to the simulation-based method. The theory-based method we'll focus on is known as the *two-sample t-test* for testing differences in sample means. However, the test statistic we'll use won't be the difference in sample means $\overline{x}_1 - \overline{x}_2$, but rather the related *two-sample t-statistic*. The data we'll use will once again be the movies_sample data of action and romance movies from Section 9.5.

Two-sample t-statistic

A common task in statistics is the process of "standardizing a variable." By standardizing different variables, we make them more comparable. For example, say you are interested in studying the distribution of temperature recordings from Portland, Oregon, USA and comparing it to that of the temperature recordings in Montreal, Quebec, Canada. Given that US temperatures are generally recorded in degrees Fahrenheit and Canadian temperatures are generally recorded in degrees Celsius, how can we make them comparable? One approach would be to convert degrees Fahrenheit into Celsius, or vice versa. Another approach would be to convert them both to a common "standardized" scale, like degrees Kelvin.

One common method for standardizing a variable from probability and statistics theory is to compute the z-score:

$$z = \frac{x - \mu}{\sigma}$$

where x represents one value of a variable, μ represents the mean of that variable, and σ represents the standard deviation of that variable. You first subtract the mean μ from each value of x and then divide $x - \mu$ by the standard deviation σ. These operations will have the effect of *re-centering* your variable around 0 and *re-scaling* your variable x so that they have what are known as "standard units." Thus for every value that your variable can take, it has a corresponding z-score that gives how many standard units away that value is from the mean μ. z-scores are normally distributed with mean 0 and standard deviation 1. This curve is called a "z-distribution" or "standard normal" curve and has the common, bell-shaped pattern from Figure 9.19 discussed in Appendix A.2.

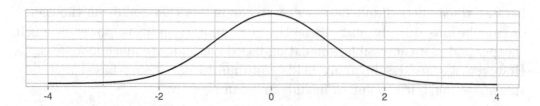

FIGURE 9.19: Standard normal z curve.

Bringing these back to the difference of sample mean ratings $\bar{x}_a - \bar{x}_r$ of action versus romance movies, how would we standardize this variable? By once again subtracting its mean and dividing by its standard deviation. Recall two facts from Subsection 7.3.3. First, if the sampling was done in a representative fashion, then the sampling distribution of $\bar{x}_a - \bar{x}_r$ will be centered at the true population parameter $\mu_a - \mu_r$. Second, the standard deviation of point estimates like $\bar{x}_a - \bar{x}_r$ has a special name: the standard error.

Applying these ideas, we present the *two-sample t-statistic*:

$$t = \frac{(\bar{x}_a - \bar{x}_r) - (\mu_a - \mu_r)}{SE_{\bar{x}_a - \bar{x}_r}} = \frac{(\bar{x}_a - \bar{x}_r) - (\mu_a - \mu_r)}{\sqrt{\dfrac{s_a{}^2}{n_a} + \dfrac{s_r{}^2}{n_r}}}$$

Oofda! There is a lot to try to unpack here! Let's go slowly. In the numerator, $\bar{x}_a - \bar{x}_r$ is the difference in sample means, while $\mu_a - \mu_r$ is the difference in population means. In the denominator, s_a and s_r are the *sample standard deviations* of the action and romance movies in our sample movies_sample. Lastly, n_a and n_r are the sample sizes of the action and romance movies. Putting this together under the square root gives us the standard error $SE_{\bar{x}_a - \bar{x}_r}$.

Observe that the formula for $SE_{\bar{x}_a - \bar{x}_r}$ has the sample sizes n_a and n_r in them. So as the sample sizes increase, the standard error goes down. We've seen this concept numerous times now, in particular in our simulations using the three virtual shovels with $n = 25$, 50, and 100 slots in Figure 7.15 and in Subsection 8.5.3 where we studied the effect of using larger sample sizes on the widths of confidence intervals.

So how can we use the two-sample t-statistic as a test statistic in our hypothesis test? First, assuming the null hypothesis $H_0 : \mu_a - \mu_r = 0$ is true, the right-hand side of the numerator (to the right of the $-$ sign), $\mu_a - \mu_r$, becomes 0.

Second, similarly to how the Central Limit Theorem from Subsection 7.5.2 states that sample means follow a normal distribution, it can be mathematically proven that the two-sample t-statistic follows a t *distribution with degrees of freedom* "roughly equal" to $df = n_a + n_r - 2$. To better understand this concept of *degrees of freedom*, we next display three examples of t-distributions in Figure 9.20 along with the standard normal z curve.

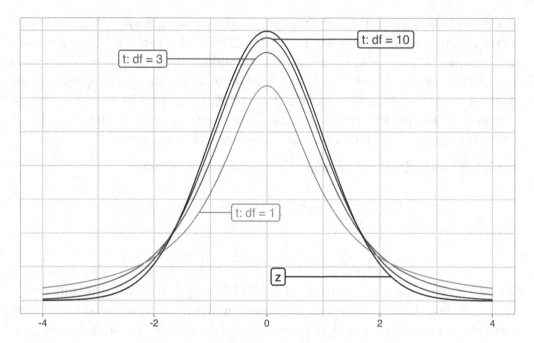

FIGURE 9.20: Examples of t-distributions and the z curve.

Begin by looking at the center of the plot at 0 on the horizontal axis. As you move up from the value of 0, follow along with the labels and note that the bottom curve corresponds to 1 degree of freedom, the curve above it is for 3 degrees of freedom, the curve above that is for 10 degrees of freedom, and lastly the dotted curve is the standard normal z curve.

Observe that all four curves have a bell shape, are centered at 0, and that as the degrees of freedom increase, the t-distribution more and more resembles the standard normal z curve. The "degrees of freedom" measures how different the t distribution will be from a normal distribution. t-distributions tend to have more values in the tails of their distributions than the standard normal z curve.

This "roughly equal" statement indicates that the equation $df = n_a + n_r - 2$ is a "good enough" approximation to the true degrees of freedom. The true formula[6] is a bit more complicated than this simple expression, but we've found the formula to be beyond the reach of those new to statistical inference and it does little to build the intuition of the t-test.

The message to retain, however, is that small sample sizes lead to small degrees of freedom and thus small sample sizes lead to t-distributions that are different

---

[6]https://en.wikipedia.org/wiki/Student%27s_t-test#Equal_or_unequal_sample_sizes,_unequal_
variances

than the z curve. On the other hand, large sample sizes correspond to large degrees of freedom and thus produce t distributions that closely align with the standard normal z-curve.

So, assuming the null hypothesis H_0 is true, our formula for the test statistic simplifies a bit:

$$t = \frac{(\bar{x}_a - \bar{x}_r) - 0}{\sqrt{\dfrac{s_a{}^2}{n_a} + \dfrac{s_r{}^2}{n_r}}} = \frac{\bar{x}_a - \bar{x}_r}{\sqrt{\dfrac{s_a{}^2}{n_a} + \dfrac{s_r{}^2}{n_r}}}$$

Let's compute the values necessary for this two-sample t-statistic. Recall the summary statistics we computed during our exploratory data analysis in Section 9.5.1.

```
movies_sample %>%
  group_by(genre) %>%
  summarize(n = n(), mean_rating = mean(rating), std_dev = sd(rating))
```

```
# A tibble: 2 x 4
  genre        n mean_rating std_dev
  <chr>    <int>       <dbl>   <dbl>
1 Action      32       5.275 1.36121
2 Romance     36     6.32222 1.60963
```

Using these values, the observed two-sample t-test statistic is

$$\frac{\bar{x}_a - \bar{x}_r}{\sqrt{\dfrac{s_a{}^2}{n_a} + \dfrac{s_r{}^2}{n_r}}} = \frac{5.28 - 6.32}{\sqrt{\dfrac{1.36^2}{32} + \dfrac{1.61^2}{36}}} = -2.906$$

Great! How can we compute the p-value using this theory-based test statistic? We need to compare it to a null distribution, which we construct next.

Null distribution

Let's revisit the null distribution for the test statistic $\bar{x}_a - \bar{x}_r$ we constructed in Section 9.5. Let's visualize this in the left-hand plot of Figure 9.21.

```
# Construct null distribution of xbar_a - xbar_m:
null_distribution_movies <- movies_sample %>%
  specify(formula = rating ~ genre) %>%
```

```
  hypothesize(null = "independence") %>%
  generate(reps = 1000, type = "permute") %>%
  calculate(stat = "diff in means", order = c("Action", "Romance"))
visualize(null_distribution_movies, bins = 10)
```

The `infer` package also includes some built-in theory-based test statistics as well. So instead of calculating the test statistic of interest as the `"diff in means"` $\bar{x}_a - \bar{x}_r$, we can calculate this defined two-sample t-statistic by setting `stat = "t"`. Let's visualize this in the right-hand plot of Figure 9.21.

```
# Construct null distribution of t:
null_distribution_movies_t <- movies_sample %>%
  specify(formula = rating ~ genre) %>%
  hypothesize(null = "independence") %>%
  generate(reps = 1000, type = "permute") %>%
  # Notice we switched stat from "diff in means" to "t"
  calculate(stat = "t", order = c("Action", "Romance"))
visualize(null_distribution_movies_t, bins = 10)
```

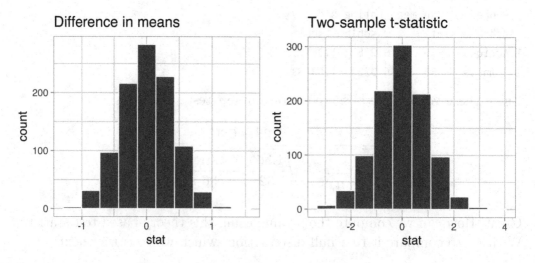

FIGURE 9.21: Comparing the null distributions of two test statistics.

Observe that while the shape of the null distributions of both the difference in means $\bar{x}_a - \bar{x}_r$ and the two-sample t-statistics are similar, the scales on the x-axis are different. The two-sample t-statistic values are spread out over a larger range.

However, a traditional theory-based t-test doesn't look at the simulated histogram in `null_distribution_movies_t`, but instead it looks at the t-distribution curve with degrees of freedom equal to roughly 65.85. This calculation is based on the complicated formula referenced previously, which we approximated with $df = n_a + n_r - 2 = 32 + 36 - 2 = 66$. Let's overlay this t-distribution curve over the top of our simulated two-sample t-statistics using the `method = "both"` argument in `visualize()`.

```
visualize(null_distribution_movies_t, bins = 10, method = "both")
```

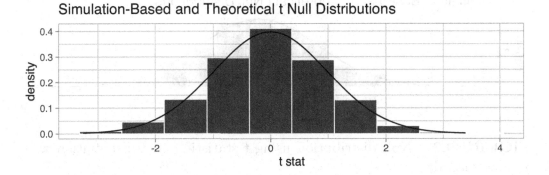

FIGURE 9.22: Null distribution using t-statistic and t-distribution.

Observe that the curve does a good job of approximating the histogram here. To calculate the p-value in this case, we need to figure out how much of the total area under the t-distribution curve is at or "more extreme" than our observed two-sample t-statistic. Since $H_A : \mu_a - \mu_r \neq 0$ is a two-sided alternative, we need to add up the areas in both tails.

We first compute the observed two-sample t-statistic using `infer` verbs. This shortcut calculation further assumes that the null hypothesis is true: that the population of action and romance movies have an equal average rating.

```
obs_two_sample_t <- movies_sample %>%
  specify(formula = rating ~ genre) %>%
  calculate(stat = "t", order = c("Action", "Romance"))
obs_two_sample_t
```

```
# A tibble: 1 x 1
     stat
    <dbl>
1 -2.90589
```

We want to find the percentage of values that are at or above `obs_two_sample_t` $= -2.906$ or at or below `-obs_two_sample_t` $= 2.906$. We use the `shade_p_value()` function with the `direction` argument set to `"both"` to do this:

```
visualize(null_distribution_movies_t, method = "both") +
  shade_p_value(obs_stat = obs_two_sample_t, direction = "both")
```

```
Warning: Check to make sure the conditions have been met for the
theoretical method. {infer} currently does not check these for you.
```

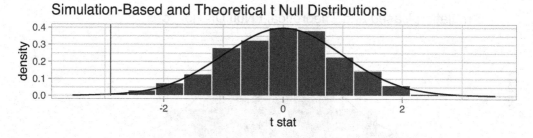

FIGURE 9.23: Null distribution using t-statistic and t-distribution with p-value shaded.

(We'll discuss this warning message shortly.) What is the p-value? We apply `get_p_value()` to our null distribution saved in `null_distribution_movies_t`:

```
null_distribution_movies_t %>%
  get_p_value(obs_stat = obs_two_sample_t, direction = "both")
```

```
# A tibble: 1 x 1
  p_value
    <dbl>
1   0.002
```

We have a very small p-value, and thus it is very unlikely that these results are due to *sampling variation*. Thus, we are inclined to reject H_0.

Let's come back to that earlier warning message: `Check to make sure the conditions have been met for the theoretical method. {infer} currently does not check these for you.` To be able to use the t-test and other such theoretical methods, there are always a few conditions to check. The `infer` package does not automatically check these conditions, hence the warning message we received. These conditions are necessary so that the underlying mathematical theory holds. In order for the results of our two-sample t-test to be valid, three conditions must be met:

1. Nearly normal populations or large sample sizes. A general rule of thumb that works in many (but not all) situations is that the sample size n should be greater than 30.
2. Both samples are selected independently of each other.
3. All observations are independent from each other.

Let's see if these conditions hold for our `movies_sample` data:

1. This is met since $n_a = 32$ and $n_r = 36$ are both larger than 30, satisfying our rule of thumb.
2. This is met since we sampled the action and romance movies at random and in an unbiased fashion from the database of all IMDb movies.
3. Unfortunately, we don't know how IMDb computes the ratings. For example, if the same person rated multiple movies, then those observations would be related and hence not independent.

Assuming all three conditions are roughly met, we can be reasonably certain that the theory-based t-test results are valid. If any of the conditions were clearly not met, we couldn't put as much trust into any conclusions reached. On the other hand, in most scenarios, the only assumption that needs to be met in the simulation-based method is that the sample is selected at random. Thus, in our experience, we prefer simulation-based methods as they have fewer assumptions, are conceptually easier to understand, and since computing power has recently become easily accessible, they can be run quickly. That being said since much of the world's research still relies on traditional theory-based methods, we also believe it is important to understand them.

You may be wondering why we chose `reps = 1000` for these simulation-based methods. We've noticed that after around 1000 replicates for the null distribution and the bootstrap distribution for most problems you can start to get a general sense for how the statistic behaves. You can change this value to something like 10,000 though for `reps` if you would like even finer detail but this will take more time to compute. Feel free to iterate on this as you like to get an even better idea about the shape of the null and bootstrap distributions as you wish.

9.6.2   When inference is not needed

We've now walked through several different examples of how to use the `infer` package to perform statistical inference: constructing confidence intervals and conducting hypothesis tests. For each of these examples, we made it a point to always perform an exploratory data analysis (EDA) first; specifically, by looking at the raw data values, by using data visualization with `ggplot2`, and by data wrangling with `dplyr` beforehand. We *highly* encourage you to always do the same. As a beginner to statistics, EDA helps you develop intuition as to what statistical methods like confidence intervals and hypothesis tests can tell us. Even as a seasoned practitioner of statistics, EDA helps guide your statistical investigations. In particular, is statistical inference even needed?

Let's consider an example. Say we're interested in the following question: Of *all* flights leaving a New York City airport, are Hawaiian Airlines flights in the air for longer than Alaska Airlines flights? Furthermore, let's assume that 2013 flights are a representative sample of all such flights. Then we can use the `flights` data frame in the `nycflights13` package we introduced in Section 1.4 to answer our question. Let's filter this data frame to only include Hawaiian and Alaska Airlines using their `carrier` codes `HA` and `AS`:

```
flights_sample <- flights %>%
  filter(carrier %in% c("HA", "AS"))
```

There are two possible statistical inference methods we could use to answer such questions. First, we could construct a 95% confidence interval for the difference in population means $\mu_{HA} - \mu_{AS}$, where μ_{HA} is the mean air time of all Hawaiian Airlines flights and μ_{AS} is the mean air time of all Alaska Airlines flights. We could then check if the entirety of the interval is greater than 0, suggesting that $\mu_{HA} - \mu_{AS} > 0$, or, in other words suggesting that $\mu_{HA} > \mu_{AS}$. Second, we could perform a hypothesis test of the null hypothesis $H_0 : \mu_{HA} - \mu_{AS} = 0$ versus the alternative hypothesis $H_A : \mu_{HA} - \mu_{AS} > 0$.

However, let's first construct an exploratory visualization as we suggested earlier. Since `air_time` is numerical and `carrier` is categorical, a boxplot can display the relationship between these two variables, which we display in Figure 9.24.

```
ggplot(data = flights_sample, mapping = aes(x = carrier, y = air_time)) +
  geom_boxplot() +
  labs(x = "Carrier", y = "Air Time")
```

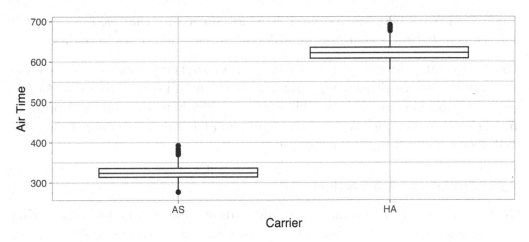

FIGURE 9.24: Air time for Hawaiian and Alaska Airlines flights departing NYC in 2013.

This is what we like to call "no PhD in Statistics needed" moments. You don't have to be an expert in statistics to know that Alaska Airlines and Hawaiian Airlines have *significantly* different air times. The two boxplots don't even overlap! Constructing a confidence interval or conducting a hypothesis test would frankly not provide much more insight than Figure 9.24.

Let's investigate why we observe such a clear cut difference between these two airlines using data wrangling. Let's first group by the rows of flights_sample not only by carrier but also by destination dest. Subsequently, we'll compute two summary statistics: the number of observations using n() and the mean airtime:

```
flights_sample %>%
  group_by(carrier, dest) %>%
  summarize(n = n(), mean_time = mean(air_time, na.rm = TRUE))
```

```
# A tibble: 2 x 4
# Groups:   carrier [2]
  carrier dest       n mean_time
  <chr>   <chr> <int>     <dbl>
1 AS      SEA     714   325.618
2 HA      HNL     342   623.088
```

It turns out that from New York City in 2013, Alaska only flew to SEA (Seattle) from New York City (NYC) while Hawaiian only flew to HNL (Honolulu) from NYC. Given the clear difference in distance from New York City to Seattle versus New York City to Honolulu, it is not surprising that we observe such different (*statistically significantly different*, in fact) air times in flights.

This is a clear example of not needing to do anything more than a simple exploratory data analysis using data visualization and descriptive statistics to get an appropriate conclusion. This is why we highly recommend you perform an EDA of any sample data before running statistical inference methods like confidence intervals and hypothesis tests.

9.6.3   Problems with p-values

On top of the many common misunderstandings about hypothesis testing and p-values we listed in Section 9.4, another unfortunate consequence of the expanded use of p-values and hypothesis testing is a phenomenon known as "p-hacking." p-hacking is the act of "cherry-picking" only results that are "statistically significant" while dismissing those that aren't, even if at the expense of the scientific ideas. There are lots of articles written recently about misunderstandings and the problems with p-values. We encourage you to check some of them out:

1.   Misunderstandings of p-values[7]
2.   What a nerdy debate about p-values shows about science - and how to fix it[8]
3.   Statisticians issue warning over misuse of P values[9]
4.   You Can't Trust What You Read About Nutrition[10]
5.   A Litany of Problems with p-values[11]

Such issues were getting so problematic that the American Statistical Association (ASA) put out a statement in 2016 titled, "The ASA Statement on Statistical Significance and P-Values,"[12] with six principles underlying the proper use and interpretation of p-values. The ASA released this guidance on p-values to improve the conduct and interpretation of quantitative science and to inform the growing emphasis on reproducibility of science research.

We as authors much prefer the use of confidence intervals for statistical inference, since in our opinion they are much less prone to large misinterpretation. However, many fields still exclusively use p-values for statistical inference and

---

[7]https://en.wikipedia.org/wiki/Misunderstandings_of_p-values

[8]https://www.vox.com/science-and-health/2017/7/31/16021654/p-values-statistical-significance-redefine-0005

[9]https://www.nature.com/news/statisticians-issue-warning-over-misuse-of-p-values-1.19503

[10]https://fivethirtyeight.com/features/you-cant-trust-what-you-read-about-nutrition/

[11]http://www.fharrell.com/post/pval-litany/

[12]https://www.amstat.org/asa/files/pdfs/P-ValueStatement.pdf

this is one reason for including them in this text. We encourage you to learn more about "p-hacking" as well and its implication for science.

9.6.4 Additional resources

Solutions to all *Learning checks* can be found online in Appendix D[13].

An R script file of all R code used in this chapter is available at `https://www.moderndive.com/scripts/09-hypothesis-testing.R`.

If you want more examples of the `infer` workflow for conducting hypothesis tests, we suggest you check out the `infer` package homepage, in particular, a series of example analyses available at `https://infer.netlify.com/articles/`.

9.6.5 What's to come

We conclude with the `infer` pipeline for hypothesis testing in Figure 9.25.

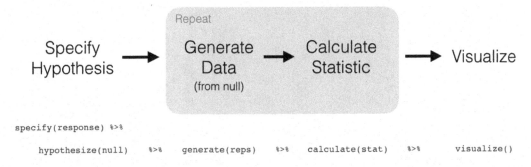

FIGURE 9.25: infer package workflow for hypothesis testing.

Now that we've armed ourselves with an understanding of confidence intervals from Chapter 8 and hypothesis tests from this chapter, we'll now study inference for regression in the upcoming Chapter 10.

We'll revisit the regression models we studied in Chapter 5 on basic regression and Chapter 6 on multiple regression. For example, recall Table 5.2 (shown again here in Table 9.4), corresponding to our regression model for an instructor's teaching score as a function of their "beauty" score.

```
# Fit regression model:
score_model <- lm(score ~ bty_avg, data = evals)
```

---

[13]`https://moderndive.com/D-appendixD.html`

```
# Get regression table:
get_regression_table(score_model)
```

TABLE 9.4: Linear regression table

term	estimate	std_error	statistic	p_value	lower_ci	upper_ci
intercept	3.880	0.076	50.96	0	3.731	4.030
bty_avg	0.067	0.016	4.09	0	0.035	0.099

We previously saw in Subsection 5.1.2 that the values in the estimate column are the fitted intercept b_0 and fitted slope for beauty score b_1. In Chapter 10, we'll unpack the remaining columns: std_error which is the standard error, statistic which is the observed *standardized* test statistic to compute the p_value, and the 95% confidence intervals as given by lower_ci and upper_ci.

10

Inference for Regression

In our penultimate chapter, we'll revisit the regression models we first studied in Chapters 5 and 6. Armed with our knowledge of confidence intervals and hypothesis tests from Chapters 8 and 9, we'll be able to apply statistical inference to further our understanding of relationships between outcome and explanatory variables.

Needed packages

Let's load all the packages needed for this chapter (this assumes you've already installed them). Recall from our discussion in Section 4.4 that loading the `tidyverse` package by running `library(tidyverse)` loads the following commonly used data science packages all at once:

- `ggplot2` for data visualization
- `dplyr` for data wrangling
- `tidyr` for converting data to "tidy" format
- `readr` for importing spreadsheet data into R
- As well as the more advanced `purrr`, `tibble`, `stringr`, and `forcats` packages

If needed, read Section 1.3 for information on how to install and load R packages.

```
library(tidyverse)
library(moderndive)
library(infer)
```

10.1 Regression refresher

Before jumping into inference for regression, let's remind ourselves of the University of Texas Austin teaching evaluations analysis in Section 5.1.

10.1.1 Teaching evaluations analysis

Recall using simple linear regression we modeled the relationship between

1. A numerical outcome variable y (the instructor's teaching score) and
2. A single numerical explanatory variable x (the instructor's "beauty" score).

We first created an `evals_ch5` data frame that selected a subset of variables from the `evals` data frame included in the `moderndive` package. This `evals_ch5` data frame contains only the variables of interest for our analysis, in particular the instructor's teaching `score` and the "beauty" rating `bty_avg`:

```
evals_ch5 <- evals %>%
  select(ID, score, bty_avg, age)
glimpse(evals_ch5)
```

```
Observations: 463
Variables: 4
$ ID      <int> 1, 2, 3, 4, 5, 6, 7, 8, 9, 10, 11, 12, 13, 14, 15, 16,...
$ score   <dbl> 4.7, 4.1, 3.9, 4.8, 4.6, 4.3, 2.8, 4.1, 3.4, 4.5, 3.8,...
$ bty_avg <dbl> 5.00, 5.00, 5.00, 5.00, 3.00, 3.00, 3.00, 3.33, 3.33, ...
$ age     <int> 36, 36, 36, 36, 59, 59, 59, 51, 51, 40, 40, 40, 40, 40...
```

In Subsection 5.1.1, we performed an exploratory data analysis of the relationship between these two variables of `score` and `bty_avg`. We saw there that a weakly positive correlation of 0.187 existed between the two variables.

This was evidenced in Figure 10.1 of the scatterplot along with the "best-fitting" regression line that summarizes the linear relationship between the two variables of `score` and `bty_avg`. Recall in Subsection 5.3.2 that we defined a "best-fitting" line as the line that minimizes the *sum of squared residuals*.

```
ggplot(evals_ch5,
       aes(x = bty_avg, y = score)) +
  geom_point() +
  labs(x = "Beauty Score",
       y = "Teaching Score",
       title = "Relationship between teaching and beauty scores") +
  geom_smooth(method = "lm", se = FALSE)
```

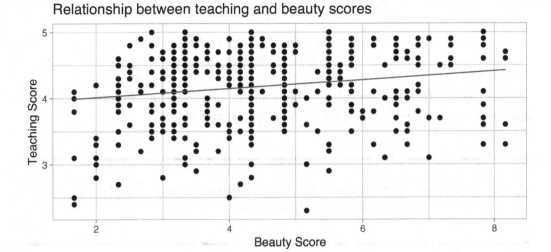

FIGURE 10.1: Relationship with regression line.

Looking at this plot again, you might be asking, "Does that line really have all that positive of a slope?". It does increase from left to right as the bty_avg variable increases, but by how much? To get to this information, recall that we followed a two-step procedure:

1. We first "fit" the linear regression model using the lm() function with the formula score ~ bty_avg. We saved this model in score_model.
2. We get the regression table by applying the get_regression_table() function from the moderndive package to score_model.

```
# Fit regression model:
score_model <- lm(score ~ bty_avg, data = evals_ch5)
# Get regression table:
get_regression_table(score_model)
```

TABLE 10.1: Previously seen linear regression table

term	estimate	std_error	statistic	p_value	lower_ci	upper_ci
intercept	3.880	0.076	50.96	0	3.731	4.030
bty_avg	0.067	0.016	4.09	0	0.035	0.099

Using the values in the estimate column of the resulting regression table in Table 10.1, we could then obtain the equation of the "best-fitting" regression line in Figure 10.1:

$$\hat{y} = b_0 + b_1 \cdot x$$
$$\widehat{\text{score}} = b_0 + b_{\text{bty\_avg}} \cdot \text{bty\_avg}$$
$$= 3.880 + 0.067 \cdot \text{bty\_avg}$$

where b_0 is the fitted intercept and b_1 is the fitted slope for `bty_avg`. Recall the interpretation of the $b_1 = 0.067$ value of the fitted slope:

For every increase of one unit in "beauty" rating, there is an associated increase, on average, of 0.067 units of evaluation score.

Thus, the slope value quantifies the relationship between the y variable `score` and the x variable `bty_avg`. We also discussed the intercept value of $b_0 = 3.88$ and its lack of practical interpretation, since the range of possible "beauty" scores does not include 0.

10.1.2 Sampling scenario

Let's now revisit this study in terms of the terminology and notation related to sampling we studied in Subsection 7.3.1.

First, let's view the instructors for these 463 courses as a *representative sample* from a greater *study population*. In our case, let's assume that the study population is *all* instructors at UT Austin and that the sample of instructors who taught these 463 courses is a representative sample. Unfortunately, we can only *assume* these two facts without more knowledge of the *sampling methodology* used by the researchers.

Since we are viewing these $n = 463$ courses as a sample, we can view our fitted slope $b_1 = 0.067$ as a *point estimate* of the *population slope* β_1. In other words, β_1 quantifies the relationship between teaching `score` and "beauty" average `bty_avg` for *all* instructors at UT Austin. Similarly, we can view our fitted intercept $b_0 = 3.88$ as a *point estimate* of the *population intercept* β_0 for *all* instructors at UT Austin.

Putting these two ideas together, we can view the equation of the fitted line $\hat{y} = b_0 + b_1 \cdot x = 3.880 + 0.067 \cdot \text{bty\_avg}$ as an estimate of some true and unknown *population line* $y = \beta_0 + \beta_1 \cdot x$. Thus we can draw parallels between our teaching evaluations analysis and all the sampling scenarios we've seen previously. In this chapter, we'll focus on the final scenario of regression slopes as shown in Table 10.2.

TABLE 10.2: Scenarios of sampling for inference

Scenario	Population parameter	Notation	Point estimate	Symbol(s)
1	Population proportion	p	Sample proportion	$\hat{p}$
2	Population mean	μ	Sample mean	$\overline{x}$ or $\widehat{\mu}$
3	Difference in population proportions	$p_1 - p_2$	Difference in sample proportions	$\hat{p}_1 - \hat{p}_2$
4	Difference in population means	$\mu_1 - \mu_2$	Difference in sample means	$\overline{x}_1 - \overline{x}_2$
5	Population regression slope	β_1	Fitted regression slope	b_1 or $\widehat{\beta}_1$

Since we are now viewing our fitted slope b_1 and fitted intercept b_0 as *point estimates* based on a *sample*, these estimates will again be subject to *sampling variability*. In other words, if we collected a new sample of data on a different set of $n = 463$ courses and their instructors, the new fitted slope b_1 will likely differ from 0.067. The same goes for the new fitted intercept b_0. But by how much will these estimates *vary*? This information is in the remaining columns of the regression table in Table 10.1. Our knowledge of sampling from Chapter 7, confidence intervals from Chapter 8, and hypothesis tests from Chapter 9 will help us interpret these remaining columns.

10.2 Interpreting regression tables

We've so far focused only on the two leftmost columns of the regression table in Table 10.1: `term` and `estimate`. Let's now shift our attention to the remaining columns: `std_error`, `statistic`, `p_value`, `lower_ci` and `upper_ci` in Table 10.3.

TABLE 10.3: Previously seen regression table

term	estimate	std_error	statistic	p_value	lower_ci	upper_ci
intercept	3.880	0.076	50.96	0	3.731	4.030
bty_avg	0.067	0.016	4.09	0	0.035	0.099

Given the lack of practical interpretation for the fitted intercept b_0, in this section we'll focus only on the second row of the table corresponding to the fitted slope b_1. We'll first interpret the std_error, statistic, p_value, lower_ci and upper_ci columns. Afterwards in the upcoming Subsection 10.2.5, we'll discuss how R computes these values.

10.2.1   Standard error

The third column of the regression table in Table 10.1 std_error corresponds to the *standard error* of our estimates. Recall the definition of **standard error** we saw in Subsection 7.3.2:

The *standard error* is the standard deviation of any point estimate computed from a sample.

So what does this mean in terms of the fitted slope $b_1 = 0.067$? This value is just one possible value of the fitted slope resulting from *this particular sample* of $n = 463$ pairs of teaching and beauty scores. However, if we collected a different sample of $n = 463$ pairs of teaching and beauty scores, we will almost certainly obtain a different fitted slope b_1. This is due to *sampling variability*.

Say we hypothetically collected 1000 such samples of pairs of teaching and beauty scores, computed the 1000 resulting values of the fitted slope b_1, and visualized them in a histogram. This would be a visualization of the *sampling distribution* of b_1, which we defined in Subsection 7.3.2. Further recall that the standard deviation of the *sampling distribution* of b_1 has a special name: the *standard error*.

Recall that we constructed three sampling distributions for the sample proportion $\hat{p}$ using shovels of size 25, 50, and 100 in Figure 7.12. We observed that as the sample size increased, the standard error decreased as evidenced by the narrowing sampling distribution.

The *standard error* of b_1 similarly quantifies how much variation in the fitted slope b_1 one would expect between different samples. So in our case, we can expect about 0.016 units of variation in the bty_avg slope variable. Recall that the estimate and std_error values play a key role in *inferring* the value of the unknown population slope β_1 relating to *all* instructors.

In Section 10.4, we'll perform a simulation using the infer package to construct the bootstrap distribution for b_1 in this case. Recall from Subsection 8.7.1 that the bootstrap distribution is an *approximation* to the sampling distribution in that they have a similar shape. Since they have a similar shape, they have similar *standard errors*. However, unlike the sampling distribution, the bootstrap distribution is constructed from a *single* sample, which is a practice more aligned with what's done in real life.

10.2.2   Test statistic

The fourth column of the regression table in Table 10.1 `statistic` corresponds to a *test statistic* relating to the following *hypothesis test*:

$$H_0 : \beta_1 = 0$$
$$\text{vs } H_A : \beta_1 \neq 0.$$

Recall our terminology, notation, and definitions related to hypothesis tests we introduced in Section 9.2.

A *hypothesis test* consists of a test between two competing hypotheses: (1) a *null hypothesis* H_0 versus (2) an *alternative hypothesis* H_A.

A *test statistic* is a point estimate/sample statistic formula used for hypothesis testing.

Here, our *null hypothesis* H_0 assumes that the population slope β_1 is 0. If the population slope β_1 is truly 0, then this is saying that there is *no true relationship* between teaching and "beauty" scores for *all* the instructors in our population. In other words, $x = $ "beauty" score would have no associated effect on $y = $ teaching score. The *alternative hypothesis* H_A, on the other hand, assumes that the population slope β_1 is not 0, meaning it could be either positive or negative. This suggests either a positive or negative relationship between teaching and "beauty" scores. Recall we called such alternative hypotheses *two-sided*. By convention, all hypothesis testing for regression assumes two-sided alternatives.

Recall our "hypothesized universe" of no gender discrimination we *assumed* in our `promotions` activity in Section 9.1. Similarly here when conducting this hypothesis test, we'll assume a "hypothesized universe" where there is no relationship between teaching and "beauty" scores. In other words, we'll assume the null hypothesis $H_0 : \beta_1 = 0$ is true.

The statistic column in the regression table is a tricky one, however. It corresponds to a standardized *t-test statistic*, much like the *two-sample t statistic* we saw in Subsection 9.6.1 where we used a theory-based method for conducting hypothesis tests. In both these cases, the *null distribution* can be mathematically proven to be a *t-distribution*. Since such test statistics are tricky for individuals new to statistical inference to study, we'll skip this and jump into interpreting the *p*-value. If you're curious, we have included a discussion of this standardized *t-test statistic* in Subsection 10.5.1.

10.2.3   p-value

The fifth column of the regression table in Table 10.1 p_value corresponds to the *p-value* of the hypothesis test $H_0 : \beta_1 = 0$ versus $H_A : \beta_1 \neq 0$.

Again recalling our terminology, notation, and definitions related to hypothesis tests we introduced in Section 9.2, let's focus on the definition of the *p*-value:

A *p-value* is the probability of obtaining a test statistic just as extreme or more extreme than the observed test statistic *assuming the null hypothesis* H_0 *is true.*

Recall that you can intuitively think of the *p*-value as quantifying how "extreme" the observed fitted slope of $b_1 = 0.067$ is in a "hypothesized universe" where there is no relationship between teaching and "beauty" scores.

Following the hypothesis testing procedure we outlined in Section 9.4, since the *p*-value in this case is 0, for any choice of significance level α we would reject H_0 in favor of H_A. Using non-statistical language, this is saying: we reject the hypothesis that there is no relationship between teaching and "beauty" scores in favor of the hypothesis that there is. That is to say, the evidence suggests there is a significant relationship, one that is positive.

More precisely, however, the *p*-value corresponds to how extreme the observed test statistic of 4.09 is when compared to the appropriate *null distribution*. In Section 10.4, we'll perform a simulation using the infer package to construct the null distribution in this case.

An extra caveat here is that the results of this hypothesis test are only valid if certain "conditions for inference for regression" are met, which we'll introduce shortly in Section 10.3.

10.2.4 Confidence interval

The two rightmost columns of the regression table in Table 10.1 (lower_ci and upper_ci) correspond to the endpoints of the 95% *confidence interval* for the population slope β_1. Recall our analogy of "nets are to fish" what "confidence intervals are to population parameters" from Section 8.3. The resulting 95% confidence interval for β_1 of (0.035, 0.099) can be thought of as a range of plausible values for the population slope β_1 of the linear relationship between teaching and "beauty" scores.

As we introduced in Subsection 8.5.2 on the precise and shorthand interpretation of confidence intervals, the statistically precise interpretation of this confidence interval is: "if we repeated this sampling procedure a large number of times, we expect about 95% of the resulting confidence intervals to capture the value of the population slope β_1." However, we'll summarize this using our shorthand interpretation that "we're 95% 'confident' that the true population slope β_1 lies between 0.035 and 0.099."

Notice in this case that the resulting 95% confidence interval for β_1 of (0.035, 0.099) does not contain a very particular value: β_1 equals 0. Recall we mentioned that if the population regression slope β_1 is 0, this is equivalent to saying there is *no* relationship between teaching and "beauty" scores. Since $\beta_1 = 0$ is not in our plausible range of values for β_1, we are inclined to believe that there, in fact, *is* a relationship between teaching and "beauty" scores and a positive one at that. So in this case, the conclusion about the population slope β_1 from the 95% confidence interval matches the conclusion from the hypothesis test: evidence suggests that there is a meaningful relationship between teaching and "beauty" scores.

Recall from Subsection 8.5.3, however, that the *confidence level* is one of many factors that determine confidence interval widths. So for example, say we used a higher confidence level of 99% instead of 95%. The resulting confidence interval for β_1 would be wider and thus might now include 0. The lesson to remember here is that any confidence-interval-based conclusion depends highly on the confidence level used.

What are the calculations that went into computing the two endpoints of the 95% confidence interval for β_1?

Recall our sampling bowl example from Subsection 8.7.2 discussing lower_ci and upper_ci. Since the sampling and bootstrap distributions of the sample proportion $\hat{p}$ were roughly normal, we could use the rule of thumb for bell-shaped distributions from Appendix A.2 to create a 95% confidence interval for p with the following equation:

$$\hat{p} \pm \text{MoE}_{\hat{p}} = \hat{p} \pm 1.96 \cdot \text{SE}_{\hat{p}} = \hat{p} \pm 1.96 \cdot \sqrt{\frac{\hat{p}(1-\hat{p})}{n}}$$

We can generalize this to other point estimates that have roughly normally shaped sampling and/or bootstrap distributions:

$$\text{point estimate} \pm \text{MoE} = \text{point estimate} \pm 1.96 \cdot \text{SE}.$$

We'll show in Section 10.4 that the sampling/bootstrap distribution for the fitted slope b_1 is in fact bell-shaped as well. Thus we can construct a 95% confidence interval for β_1 with the following equation:

$$b_1 \pm \text{MoE}_{b_1} = b_1 \pm 1.96 \cdot \text{SE}_{b_1}.$$

What is the value of the standard error SE_{b_1}? It is in fact in the third column of the regression table in Table 10.1: 0.016. Thus

$$b_1 \pm 1.96 \cdot \text{SE}_{b_1} = 0.067 \pm 1.96 \cdot 0.016 = 0.067 \pm 0.031$$
$$= (0.036, 0.098)$$

This closely matches the $(0.035, 0.099)$ confidence interval in the last two columns of Table 10.1.

Much like hypothesis tests, however, the results of this confidence interval also are only valid if the "conditions for inference for regression" to be discussed in Section 10.3 are met.

10.2.5 How does R compute the table?

Since we didn't perform the simulation to get the values of the standard error, test statistic, p-value, and endpoints of the 95% confidence interval in Table 10.1, you might be wondering how were these values computed. What did R do behind the scenes? Does R run simulations like we did using the `infer` package in Chapters 8 and 9 on confidence intervals and hypothesis testing?

The answer is no! Much like the theory-based method for constructing confidence intervals you saw in Subsection 8.7.2 and the theory-based hypothesis test you saw in Subsection 9.6.1, there exist mathematical formulas that allow you to construct confidence intervals and conduct hypothesis tests for inference for regression. These formulas were derived in a time when computers didn't exist, so it would've been impossible to run the extensive computer simulations we have in this book. We present these formulas in Subsection 10.5.1 on "theory-based inference for regression."

In Section 10.4, we'll go over a simulation-based approach to constructing confidence intervals and conducting hypothesis tests using the infer package. In particular, we'll convince you that the bootstrap distribution of the fitted slope b_1 is indeed bell-shaped.

10.3 Conditions for inference for regression

Recall in Subsection 8.3.2 we stated that we could only use the standard-error-based method for constructing confidence intervals if the bootstrap distribution was bell shaped. Similarly, there are certain conditions that need to be met in order for the results of our hypothesis tests and confidence intervals we described in Section 10.2 to have valid meaning. These conditions must be met for the assumed underlying mathematical and probability theory to hold true.

For inference for regression, there are four conditions that need to be met. Note the first four letters of these conditions are highlighted in bold in what follows: **LINE**. This can serve as a nice reminder of what to check for whenever you perform linear regression.

1. **L**inearity of relationship between variables
2. **I**ndependence of the residuals
3. **N**ormality of the residuals
4. **E**quality of variance of the residuals

Conditions **L**, **N**, and **E** can be verified through what is known as a *residual analysis*. Condition **I** can only be verified through an understanding of how the data was collected.

In this section, we'll go over a refresher on residuals, verify whether each of the four **LINE** conditions hold true, and then discuss the implications.

10.3.1 Residuals refresher

Recall our definition of a residual from Subsection 5.1.3: it is the *observed value* minus the *fitted value* denoted by $y - \hat{y}$. Recall that residuals can be thought of as the error or the "lack-of-fit" between the observed value y and the fitted value $\hat{y}$ on the regression line in Figure 10.1. In Figure 10.2, we illustrate one particular residual out of 463 using an arrow, as well as its corresponding observed and fitted values using a circle and a square, respectively.

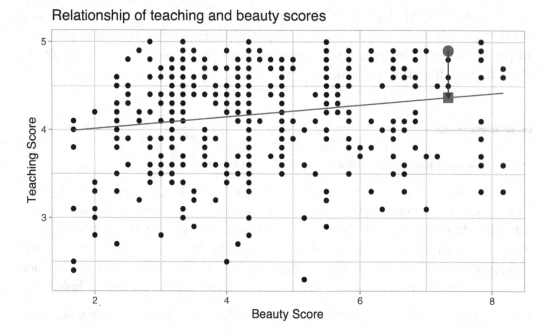

FIGURE 10.2: Example of observed value, fitted value, and residual.

Furthermore, we can automate the calculation of all $n = 463$ residuals by applying the get_regression_points() function to our saved regression model in score_model. Observe how the resulting values of residual are roughly equal to score - score_hat (there is potentially a slight difference due to rounding error).

```
# Fit regression model:
score_model <- lm(score ~ bty_avg, data = evals_ch5)
# Get regression points:
regression_points <- get_regression_points(score_model)
regression_points
```

```
# A tibble: 463 x 5
     ID score bty_avg score_hat residual
  <int> <dbl>   <dbl>     <dbl>    <dbl>
1     1   4.7       5     4.214    0.486
2     2 4.100       5     4.214   -0.114
3     3   3.9       5     4.214   -0.314
4     4   4.8       5     4.214    0.586
5     5 4.600       3     4.08      0.52
6     6   4.3       3     4.08      0.22
7     7   2.8       3     4.08     -1.28
```

```
8        8 4.100 3.333       4.102 -0.002
9        9 3.4   3.333       4.102 -0.702
10      10 4.5   3.16700     4.091  0.40900
# ... with 453 more rows
```

A *residual analysis* is used to verify conditions **L**, **N**, and **E** and can be performed using appropriate data visualizations. While there are more sophisticated statistical approaches that can also be done, we'll focus on the much simpler approach of looking at plots.

10.3.2 Linearity of relationship

The first condition is that the relationship between the outcome variable y and the explanatory variable x must be **Linear**. Recall the scatterplot in Figure 10.1 where we had the explanatory variable x as "beauty" score and the outcome variable y as teaching score. Would you say that the relationship between x and y is linear? It's hard to say because of the scatter of the points about the line. In the authors' opinions, we feel this relationship is "linear enough."

Let's present an example where the relationship between x and y is clearly not linear in Figure 10.3. In this case, the points clearly do not form a line, but rather a U-shaped polynomial curve. In this case, any results from an inference for regression would not be valid.

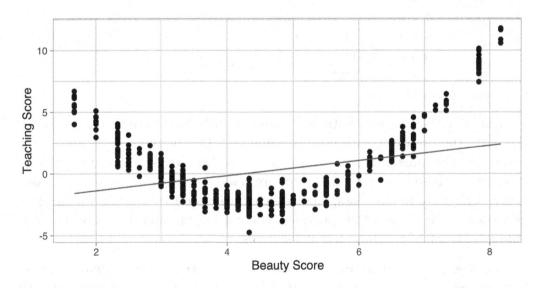

FIGURE 10.3: Example of a clearly non-linear relationship.

10.3.3   Independence of residuals

The second condition is that the residuals must be **I**ndependent. In other words, the different observations in our data must be independent of one another.

For our UT Austin data, while there is data on 463 courses, these 463 courses were actually taught by 94 unique instructors. In other words, the same professor is often included more than once in our data. The original `evals` data frame that we used to construct the `evals_ch5` data frame has a variable `prof_ID`, which is an anonymized identification variable for the professor:

```
evals %>%
   select(ID, prof_ID, score, bty_avg)
```

```
# A tibble: 463 x 4
      ID prof_ID score bty_avg
   <int>   <int> <dbl>   <dbl>
1      1       1 1 4.7        5
2      2       2 1 4.100      5
3      3       3 1 3.9        5
4      4       4 1 4.8        5
5      5       5 2 4.600      3
6      6       6 2 4.3        3
7      7       7 2 2.8        3
8      8       8 3 4.100  3.333
9      9       9 3 3.4    3.333
10    10      10 4 4.5  3.16700
# ... with 453 more rows
```

For example, the professor with `prof_ID` equal to 1 taught the first 4 courses in the data, the professor with `prof_ID` equal to 2 taught the next 3, and so on. Given that the same professor taught these first four courses, it is reasonable to expect that these four teaching scores are related to each other. If a professor gets a high score in one class, chances are fairly good they'll get a high score in another. This dataset thus provides different information than if we had 463 unique instructors teaching the 463 courses.

In this case, we say there exists *dependence* between observations. The first four courses taught by professor 1 are dependent, the next 3 courses taught by professor 2 are related, and so on. Any proper analysis of this data needs to take into account that we have *repeated measures* for the same profs.

So in this case, the independence condition is not met. What does this mean for our analysis? We'll address this in Subsection 10.3.6 coming up, after we check the remaining two conditions.

10.3.4   Normality of residuals

The third condition is that the residuals should follow a **Normal** distribution. Furthermore, the center of this distribution should be 0. In other words, sometimes the regression model will make positive errors: $y - \hat{y} > 0$. Other times, the regression model will make equally negative errors: $y - \hat{y} < 0$. However, *on average* the errors should equal 0 and their shape should be similar to that of a bell.

The simplest way to check the normality of the residuals is to look at a histogram, which we visualize in Figure 10.4.

```
ggplot(regression_points, aes(x = residual)) +
  geom_histogram(binwidth = 0.25, color = "white") +
  labs(x = "Residual")
```

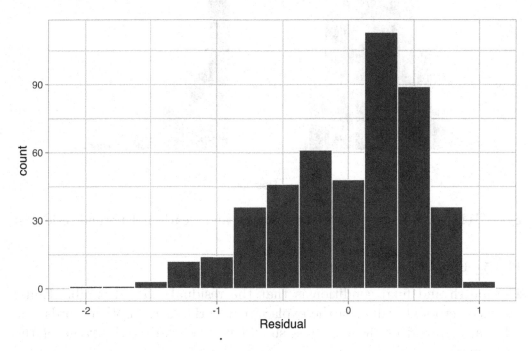

FIGURE 10.4: Histogram of residuals.

This histogram shows that we have more positive residuals than negative. Since the residual $y - \hat{y}$ is positive when $y > \hat{y}$, it seems our regression model's fitted teaching scores $\hat{y}$ tend to *underestimate* the true teaching scores y. Furthermore, this histogram has a slight *left-skew* in that there is a tail on the left. This is another way to say the residuals exhibit a *negative skew.*

Is this a problem? Again, there is a certain amount of subjectivity in the response. In the authors' opinion, while there is a slight skew to the residuals, we feel it isn't drastic. On the other hand, others might disagree with our assessment.

Let's present examples where the residuals clearly do and don't follow a normal distribution in Figure 10.5. In this case of the model yielding the clearly non-normal residuals on the right, any results from an inference for regression would not be valid.

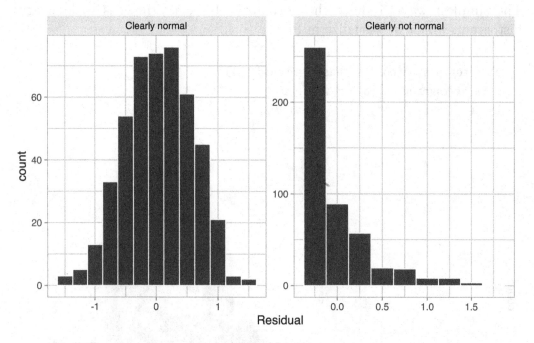

FIGURE 10.5: Example of clearly normal and clearly not normal residuals.

10.3.5 Equality of variance

The fourth and final condition is that the residuals should exhibit Equal variance across all values of the explanatory variable x. In other words, the value and spread of the residuals should not depend on the value of the explanatory variable x.

Recall the scatterplot in Figure 10.1: we had the explanatory variable x of "beauty" score on the x-axis and the outcome variable y of teaching score on the y-axis. Instead, let's create a scatterplot that has the same values on the x-axis, but now with the residual $y - \hat{y}$ on the y-axis as seen in Figure 10.6.

```
ggplot(regression_points, aes(x = bty_avg, y = residual)) +
  geom_point() +
  labs(x = "Beauty Score", y = "Residual") +
  geom_hline(yintercept = 0, col = "blue", size = 1)
```

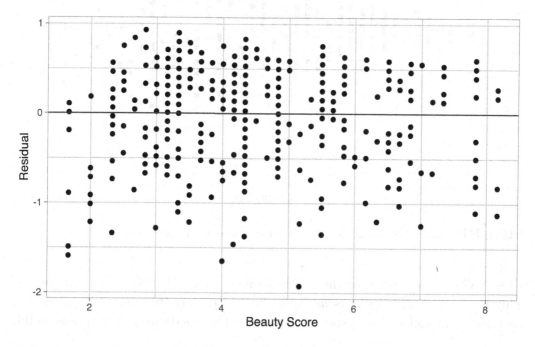

FIGURE 10.6: Plot of residuals over beauty score.

You can think of Figure 10.6 as a modified version of the plot with the regression line in Figure 10.1, but with the regression line flattened out to $y = 0$. Looking at this plot, would you say that the spread of the residuals around the line at $y = 0$ is constant across all values of the explanatory variable x of "beauty" score? This question is rather qualitative and subjective in nature, thus different people may respond with different answers. For example, some people might say that there is slightly more variation in the residuals for smaller values of x than for higher ones. However, it can be argued that there isn't a *drastic* non-constancy.

In Figure 10.7 let's present an example where the residuals clearly do not have equal variance across all values of the explanatory variable x.

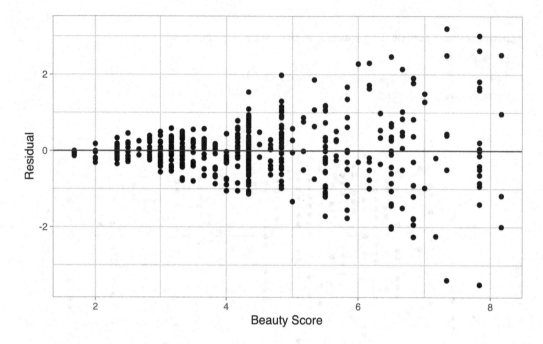

FIGURE 10.7: Example of clearly non-equal variance.

Observe how the spread of the residuals increases as the value of x increases. This is a situation known as *heteroskedasticity*. Any inference for regression based on a model yielding such a pattern in the residuals would not be valid.

10.3.6   What's the conclusion?

Let's list our four conditions for inference for regression again and indicate whether or not they were satisfied in our analysis:

1.   Linearity of relationship between variables: Yes
2.   Independence of residuals: No
3.   Normality of residuals: Somewhat
4.   Equality of variance: Yes

So what does this mean for the results of our confidence intervals and hypothesis tests in Section 10.2?

First, the Independence condition. The fact that there exist dependencies between different rows in `evals_ch5` must be addressed. In more advanced statistics courses, you'll learn how to incorporate such dependencies into your regression models. One such technique is called *hierarchical/multilevel modeling*.

Second, when conditions **L**, **N**, **E** are not met, it often means there is a shortcoming in our model. For example, it may be the case that using only a single explanatory variable is insufficient, as we did with "beauty" score. We may need to incorporate more explanatory variables in a multiple regression model as we did in Chapter 6.

In our case, the best we can do is view the results suggested by our confidence intervals and hypothesis tests as preliminary. While a preliminary analysis suggests that there is a significant relationship between teaching and "beauty" scores, further investigation is warranted; in particular, by improving the preliminary score ~ bty_avg model so that the four conditions are met. When the four conditions are roughly met, then we can put more faith into our confidence intervals and *p*-values.

The conditions for inference in regression problems are a key part of regression analysis that are of vital importance to the processes of constructing confidence intervals and conducting hypothesis tests. However, it is often the case with regression analysis in the real world that not all the conditions are completely met. Furthermore, as you saw, there is a level of subjectivity in the residual analyses to verify the **L, N,** and **E** conditions. So what can you do? We as authors advocate for transparency in communicating all results. This lets the stakeholders of any analysis know about a model's shortcomings or whether the model is "good enough." So while this checking of assumptions has lead to some fuzzy "it depends" results, we decided as authors to show you these scenarios to help prepare you for difficult statistical decisions you may need to make down the road.

Learning check

(**LC10.1**) Continuing with our regression using age as the explanatory variable and teaching score as the outcome variable.

- Use the get_regression_points() function to get the observed values, fitted values, and residuals for all 463 instructors.
- Perform a residual analysis and look for any systematic patterns in the residuals. Ideally, there should be little to no pattern but comment on what you find here.

10.4   Simulation-based inference for regression

Recall in Subsection 10.2.5 when we interpreted the third through seventh columns of a regression table, we stated that R doesn't do simulations to compute these values. Rather R uses theory-based methods that involve mathematical formulas.

In this section, we'll use the simulation-based methods you previously learned in Chapters 8 and 9 to recreate the values in the regression table in Table 10.1. In particular, we'll use the `infer` package workflow to

- Construct a 95% confidence interval for the population slope β_1 using bootstrap resampling with replacement. We did this previously in Sections 8.4 with the `pennies` data and 8.6 with the `mythbusters_yawn` data.
- Conduct a hypothesis test of $H_0 : \beta_1 = 0$ versus $H_A : \beta_1 \neq 0$ using a permutation test. We did this previously in Sections 9.3 with the `promotions` data and 9.5 with the `movies_sample` IMDb data.

10.4.1 Confidence interval for slope

We'll construct a 95% confidence interval for β_1 using the `infer` workflow outlined in Subsection 8.4.2. Specifically, we'll first construct the bootstrap distribution for the fitted slope b_1 using our single sample of 463 courses:

1. `specify()` the variables of interest in `evals_ch5` with the formula: `score ~ bty_avg`.
2. `generate()` replicates by using `bootstrap` resampling with replacement from the original sample of 463 courses. We generate `reps = 1000` replicates using `type = "bootstrap"`.
3. `calculate()` the summary statistic of interest: the fitted `slope` b_1.

Using this bootstrap distribution, we'll construct the 95% confidence interval using the percentile method and (if appropriate) the standard error method as well. It is important to note in this case that the bootstrapping with replacement is done *row-by-row*. Thus, the original pairs of `score` and `bty_avg` values are always kept together, but different pairs of `score` and `bty_avg` values may be resampled multiple times. The resulting confidence interval will denote a range of plausible values for the unknown population slope β_1 quantifying the relationship between teaching and "beauty" scores for *all* professors at UT Austin.

Let's first construct the bootstrap distribution for the fitted slope b_1:

```
bootstrap_distn_slope <- evals_ch5 %>%
  specify(formula = score ~ bty_avg) %>%
  generate(reps = 1000, type = "bootstrap") %>%
```

```
calculate(stat = "slope")
bootstrap_distn_slope
```

```
# A tibble: 1,000 x 2
   replicate       stat
       <int>      <dbl>
1          1  0.0651055
2          2  0.0382313
3          3  0.108056
4          4  0.0666601
5          5  0.0715932
6          6  0.0854565
7          7  0.0624868
8          8  0.0412859
9          9  0.0796269
10        10  0.0761299
# ... with 990 more rows
```

Observe how we have 1000 values of the bootstrapped slope b_1 in the stat column. Let's visualize the 1000 bootstrapped values in Figure 10.8.

```
visualize(bootstrap_distn_slope)
```

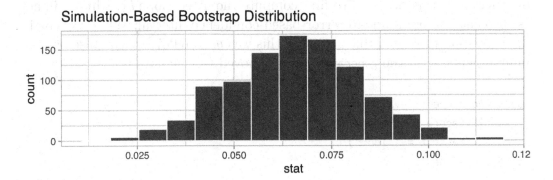

FIGURE 10.8: Bootstrap distribution of slope.

Observe how the bootstrap distribution is roughly bell-shaped. Recall from Subsection 8.7.1 that the shape of the bootstrap distribution of b_1 closely approximates the shape of the sampling distribution of b_1.

Percentile-method

First, let's compute the 95% confidence interval for β_1 using the percentile method. We'll do so by identifying the 2.5th and 97.5th percentiles which include the middle 95% of values. Recall that this method does not require the bootstrap distribution to be normally shaped.

```
percentile_ci <- bootstrap_distn_slope %>%
  get_confidence_interval(type = "percentile", level = 0.95)
percentile_ci
```

```
# A tibble: 1 x 2
    `2.5%`    `97.5%`
     <dbl>      <dbl>
1 0.0323411 0.0990027
```

The resulting percentile-based 95% confidence interval for β_1 of (0.032, 0.099) is similar to the confidence interval in the regression Table 10.1 of (0.035, 0.099).

Standard error method

Since the bootstrap distribution in Figure 10.8 appears to be roughly bell-shaped, we can also construct a 95% confidence interval for β_1 using the standard error method.

In order to do this, we need to first compute the fitted slope b_1, which will act as the center of our standard error-based confidence interval. While we saw in the regression table in Table 10.1 that this was $b_1 = 0.067$, we can also use the infer pipeline with the generate() step removed to calculate it:

```
observed_slope <- evals %>%
  specify(score ~ bty_avg) %>%
  calculate(stat = "slope")
observed_slope
```

```
# A tibble: 1 x 1
      stat
     <dbl>
1 0.0666370
```

We then use the get_ci() function with level = 0.95 to compute the 95% confidence interval for β_1. Note that setting the point_estimate argument to the observed_slope of 0.067 sets the center of the confidence interval.

```
se_ci <- bootstrap_distn_slope %>%
  get_ci(level = 0.95, type = "se", point_estimate = observed_slope)
se_ci
```

```
# A tibble: 1 x 2
      lower      upper
      <dbl>      <dbl>
1 0.0333767  0.0998974
```

The resulting standard error-based 95% confidence interval for β_1 of $(0.033, 0.1)$ is slightly different than the confidence interval in the regression Table 10.1 of $(0.035, 0.099)$.

Comparing all three

Let's compare all three confidence intervals in Figure 10.9, where the percentile-based confidence interval is marked with solid lines, the standard error based confidence interval is marked with dashed lines, and the theory-based confidence interval $(0.035, 0.099)$ from the regression table in Table 10.1 is marked with dotted lines.

```
visualize(bootstrap_distn_slope) +
  shade_confidence_interval(endpoints = percentile_ci, fill = NULL,
                            linetype = "solid", color = "grey90") +
  shade_confidence_interval(endpoints = se_ci, fill = NULL,
                            linetype = "dashed", color = "grey60") +
  shade_confidence_interval(endpoints = c(0.035, 0.099), fill = NULL,
                            linetype = "dotted", color = "black")
```

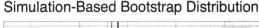

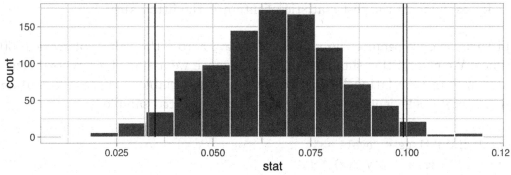

FIGURE 10.9: Comparing three confidence intervals for the slope.

Observe that all three are quite similar! Furthermore, none of the three confidence intervals for β_1 contain 0 and are entirely located above 0. This is suggesting that there is in fact a meaningful positive relationship between teaching and "beauty" scores.

10.4.2 Hypothesis test for slope

Let's now conduct a hypothesis test of $H_0 : \beta_1 = 0$ vs. $H_A : \beta_1 \neq 0$. We will use the infer package, which follows the hypothesis testing paradigm in the "There is only one test" diagram in Figure 9.14.

Let's first think about what it means for β_1 to be zero as assumed in the null hypothesis H_0. Recall we said if $\beta_1 = 0$, then this is saying there is no relationship between the teaching and "beauty" scores. Thus assuming this particular null hypothesis H_0 means that in our "hypothesized universe" there is no relationship between score and bty_avg. We can therefore shuffle/permute the bty_avg variable to no consequence.

We construct the null distribution of the fitted slope b_1 by performing the steps that follow. Recall from Section 9.2 on terminology, notation, and definitions related to hypothesis testing where we defined the *null distribution*: the sampling distribution of our test statistic b_1 assuming the null hypothesis H_0 is true.

1. specify() the variables of interest in evals_ch5 with the formula: score ~ bty_avg.
2. hypothesize() the null hypothesis of independence. Recall from Section 9.3 that this is an additional step that needs to be added for hypothesis testing.
3. generate() replicates by permuting/shuffling values from the original sample of 463 courses. We generate reps = 1000 replicates using type = "permute" here.
4. calculate() the test statistic of interest: the fitted slope b_1.

In this case, we permute the values of bty_avg across the values of score 1000 times. We can do this shuffling/permuting since we assumed a "hypothesized universe" of no relationship between these two variables. Then we calculate the "slope" coefficient for each of these 1000 generated samples.

```
null_distn_slope <- evals %>%
  specify(score ~ bty_avg) %>%
  hypothesize(null = "independence") %>%
```

```
generate(reps = 1000, type = "permute") %>%
calculate(stat = "slope")
```

Observe the resulting null distribution for the fitted slope b_1 in Figure 10.10.

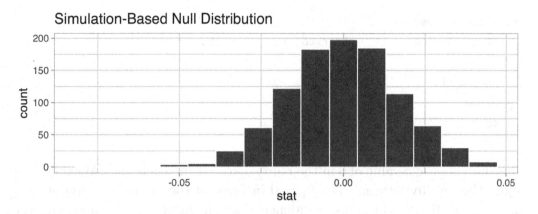

FIGURE 10.10: Null distribution of slopes.

Notice how it is centered at $b_1 = 0$. This is because in our hypothesized universe, there is no relationship between score and bty_avg and so $\beta_1 = 0$. Thus, the most typical fitted slope b_1 we observe across our simulations is 0. Observe, furthermore, how there is variation around this central value of 0.

Let's visualize the p-value in the null distribution by comparing it to the observed test statistic of $b_1 = 0.067$ in Figure 10.11. We'll do this by adding a shade_p_value() layer to the previous visualize() code.

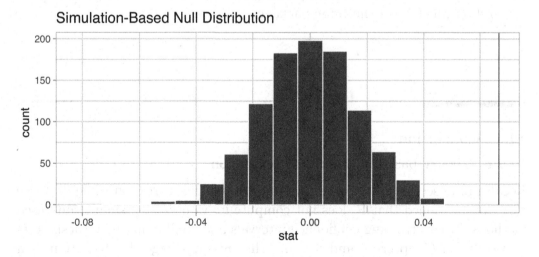

FIGURE 10.11: Null distribution and p-value.

Since the observed fitted slope 0.067 falls far to the right of this null distribution and thus the shaded region doesn't overlap it, we'll have a p-value of 0. For completeness, however, let's compute the numerical value of the p-value anyways using the `get_p_value()` function. Recall that it takes the same inputs as the `shade_p_value()` function:

```
null_distn_slope %>%
  get_p_value(obs_stat = observed_slope, direction = "both")
```

```
# A tibble: 1 x 1
  p_value
    <dbl>
1       0
```

This matches the p-value of 0 in the regression table in Table 10.1. We therefore reject the null hypothesis $H_0 : \beta_1 = 0$ in favor of the alternative hypothesis $H_A : \beta_1 \neq 0$. We thus have evidence that suggests there is a significant relationship between teaching and "beauty" scores for *all* instructors at UT Austin.

When the conditions for inference for regression are met and the null distribution has a bell shape, we are likely to see similar results between the simulation-based results we just demonstrated and the theory-based results shown in the regression table in Table 10.1.

Learning check

(LC10.2) Repeat the inference but this time for the correlation coefficient instead of the slope. Note the implementation of `stat = "correlation"` in the `calculate()` function of the `infer` package.

10.5 Conclusion

10.5.1 Theory-based inference for regression

Recall in Subsection 10.2.5 when we interpreted the regression table in Table 10.1, we mentioned that R does not compute its values using simulation-based methods for constructing confidence intervals and conducting hypothesis tests as we did in Chapters 8 and 9 using the `infer` package. Rather, R uses a theory-based approach using mathematical formulas, much like the theory-based confidence intervals you saw in Subsection 8.7.2 and the theory-based hypothesis tests you saw in Subsection 9.6.1. These formulas were derived in a time when computers didn't exist, so it would've been incredibly labor intensive to run extensive simulations.

In particular, there is a formula for the *standard error* of the fitted slope b_1:

$$\text{SE}_{b_1} = \frac{\frac{s_y}{s_x} \cdot \sqrt{1 - r^2}}{\sqrt{n - 2}}$$

As with many formulas in statistics, there's a lot going on here, so let's first break down what each symbol represents. First s_x and s_y are the *sample standard deviations* of the explanatory variable bty_avg and the response variable score, respectively. Second, r is the sample *correlation coefficient* between score and bty_avg. This was computed as 0.187 in Chapter 5. Lastly, n is the number of pairs of points in the evals_ch5 data frame, here 463.

To put this formula into words, the standard error of b_1 depends on the relationship between the variability of the response variable and the variability of the explanatory variable as measured in the s_y/s_x term. Next, it looks into how the two variables relate to each other in the $\sqrt{1 - r^2}$ term.

However, the most important observation to make in the previous formula is that there is an $n - 2$ in the denominator. In other words, as the sample size n increases, the standard error SE_{b_1} decreases. Just as we demonstrated in Subsection 7.3.3 when we used shovels with $n = 25$, 50, and 100 slots, the amount of sampling variation of the fitted slope b_1 will depend on the sample size n. In particular, as the sample size increases, both the sampling and bootstrap distributions narrow and the standard error SE_{b_1} decreases. Hence, our estimates of b_1 for the true population slope β_1 get more and more *precise*.

R then uses this formula for the standard error of b_1 in the third column of the regression table and subsequently to construct 95% confidence intervals. But what about the hypothesis test? Much like with our theory-based hypothesis test in Subsection 9.6.1, R uses the following *t-statistic* as the test statistic for hypothesis testing:

$$t = \frac{b_1 - \beta_1}{\text{SE}_{b_1}}$$

And since the null hypothesis $H_0 : \beta_1 = 0$ is assumed during the hypothesis test, the *t*-statistic becomes

$$t = \frac{b_1 - 0}{SE_{b_1}} = \frac{b_1}{SE_{b_1}}$$

What are the values of b_1 and SE_{b_1}? They are in the `estimate` and `std_error` column of the regression table in Table 10.1. Thus the value of 4.09 in the table is computed as $0.067/0.016 = 4.188$. Note there is a difference due to some rounding error here.

Lastly, to compute the p-value, we need to compare the observed test statistic of 4.09 to the appropriate null distribution. Recall from Section 9.2, that a null distribution is the sampling distribution of the test statistic *assuming the null hypothesis H_0 is true*. Much like in our theory-based hypothesis test in Subsection 9.6.1, it can be mathematically proven that this distribution is a t-distribution with degrees of freedom equal to $df = n - 2 = 463 - 2 = 461$.

Don't worry if you're feeling a little overwhelmed at this point. There is a lot of background theory to understand before you can fully make sense of the equations for theory-based methods. That being said, theory-based methods and simulation-based methods for constructing confidence intervals and conducting hypothesis tests often yield consistent results. As mentioned before, in our opinion, two large benefits of simulation-based methods over theory-based are that (1) they are easier for people new to statistical inference to understand, and (2) they also work in situations where theory-based methods and mathematical formulas don't exist.

10.5.2 Summary of statistical inference

We've finished the last two scenarios from the "Scenarios of sampling for inference" table in Subsection 7.5.1, which we re-display in Table 10.4.

TABLE 10.4: Scenarios of sampling for inference

Scenario	Population parameter	Notation	Point estimate	Symbol(s)
1	Population proportion	p	Sample proportion	$\hat{p}$
2	Population mean	μ	Sample mean	$\bar{x}$ or $\hat{\mu}$
3	Difference in population proportions	$p_1 - p_2$	Difference in sample proportions	$\hat{p}_1 - \hat{p}_2$
4	Difference in population means	$\mu_1 - \mu_2$	Difference in sample means	$\bar{x}_1 - \bar{x}_2$
5	Population regression slope	β_1	Fitted regression slope	b_1 or $\hat{\beta}_1$

Armed with the regression modeling techniques you learned in Chapters 5 and 6, your understanding of sampling for inference in Chapter 7, and the tools for statistical inference like confidence intervals and hypothesis tests in Chapters 8 and 9, you're now equipped to study the significance of relationships between variables in a wide array of data! Many of the ideas presented here can be extended into multiple regression and other more advanced modeling techniques.

10.5.3 Additional resources

Solutions to all *Learning checks* can be found online in Appendix D[1].

An R script file of all R code used in this chapter is available at `https://www.moderndive.com/scripts/10-inference-for-regression.R`.

10.5.4 What's to come

You've now concluded the last major part of the book on "Statistical Inference with `infer`." The closing Chapter 11 concludes this book with various short case studies involving real data, such as house prices in the city of Seattle, Washington in the US. You'll see how the principles in this book can help you become a great storyteller with data!

---

[1] `https://moderndive.com/D-appendixD.html`

Part IV

Conclusion

11

Tell Your Story with Data

Recall in the Preface and at the end of chapters throughout this book, we displayed the *"ModernDive flowchart"* mapping your journey through this book.

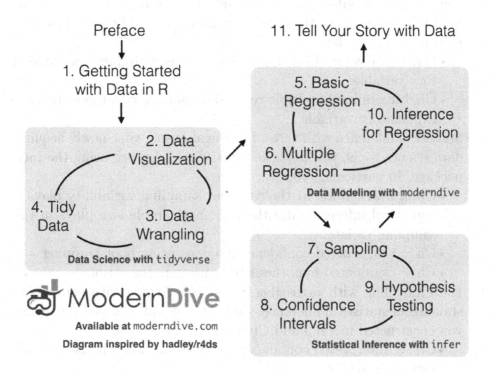

FIGURE 11.1: *ModernDive* flowchart.

11.1 Review

Let's go over a refresher of what you've covered so far. You first got started with data in Chapter 1 where you learned about the difference between R and RStudio, started coding in R, installed and loaded your first R packages, and explored your first dataset: all domestic departure `flights` from a major New York City airport in 2013. Then you covered the following three parts of this book (Parts 2 and 4 are combined into a single portion):

1. Data science with `tidyverse`. You assembled your data science toolbox using `tidyverse` packages. In particular, you
    - Ch.2: Visualized data using the `ggplot2` package.
    - Ch.3: Wrangled data using the `dplyr` package.
    - Ch.4: Learned about the concept of "tidy" data as a standardized data frame input and output format for all packages in the `tidyverse`. Furthermore, you learned how to import spreadsheet files into R using the `readr` package.
2. Data modeling with `moderndive`. Using these data science tools and helper functions from the `moderndive` package, you fit your first data models. In particular, you
    - Ch.5: Discovered basic regression models with only one explanatory variable.
    - Ch.6: Examined multiple regression models with more than one explanatory variable.
3. Statistical inference with `infer`. Once again using your newly acquired data science tools, you unpacked statistical inference using the `infer` package. In particular, you
    - Ch.7: Learned about the role that sampling variability plays in statistical inference and the role that sample size plays in this sampling variability.
    - Ch.8: Constructed confidence intervals using bootstrapping.
    - Ch.9: Conducted hypothesis tests using permutation.
4. Data modeling with `moderndive` (revisited): Armed with your understanding of statistical inference, you revisited and reviewed the models you constructed in Ch.5 and Ch.6. In particular, you
    - Ch.10: Interpreted confidence intervals and hypothesis tests in a regression setting.

We've guided you through your first experiences of "thinking with data,"[1] an expression originally coined by Dr. Diane Lambert. The philosophy underlying this expression guided your path in the flowchart in Figure 11.1.

This philosophy is also well-summarized in "Practical Data Science for Stats"[2]: a collection of pre-prints focusing on the practical side of data science work-

---

[1] https://arxiv.org/pdf/1410.3127.pdf
[2] https://peerj.com/collections/50-practicaldatascistats/

flows and statistical analysis curated by Dr. Jennifer Bryan[3] and Dr. Hadley Wickham[4]. They quote:

There are many aspects of day-to-day analytical work that are almost absent from the conventional statistics literature and curriculum. And yet these activities account for a considerable share of the time and effort of data analysts and applied statisticians. The goal of this collection is to increase the visibility and adoption of modern data analytical workflows. We aim to facilitate the transfer of tools and frameworks between industry and academia, between software engineering and statistics and computer science, and across different domains.

In other words, to be equipped to "think with data" in the 21st century, analysts need practice going through the "data/science pipeline"[5] we saw in the Preface (re-displayed in Figure 11.2). It is our opinion that, for too long, statistics education has only focused on parts of this pipeline, instead of going through it in its *entirety*.

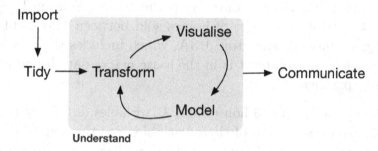

FIGURE 11.2: Data/science pipeline.

To conclude this book, we'll present you with some additional case studies of working with data. In Section 11.2 we'll take you through a full-pass of the "Data/Science Pipeline" in order to analyze the sale price of houses in Seattle, WA, USA. In Section 11.3, we'll present you with some examples of effective data storytelling drawn from the data journalism website, FiveThirtyEight.com[6]. We present these case studies to you because we believe that you should not

---

[3]https://twitter.com/jennybryan
[4]https://twitter.com/hadleywickham
[5]http://r4ds.had.co.nz/explore-intro.html
[6]https://fivethirtyeight.com/

only be able to "think with data," but also be able to "tell your story with data." Let's explore how to do this!

Needed packages

Let's load all the packages needed for this chapter (this assumes you've already installed them). Read Section 1.3 for information on how to install and load R packages.

```
library(tidyverse)
library(moderndive)
library(skimr)
library(fivethirtyeight)
```

11.2   Case study: Seattle house prices

Kaggle.com[7] is a machine learning and predictive modeling competition website that hosts datasets uploaded by companies, governmental organizations, and other individuals. One of their datasets is the "House Sales in King County, USA"[8]. It consists of sale prices of homes sold between May 2014 and May 2015 in King County, Washington, USA, which includes the greater Seattle metropolitan area. This dataset is in the house_prices data frame included in the moderndive package.

The dataset consists of 21,613 houses and 21 variables describing these houses (for a full list and description of these variables, see the help file by running ?house_prices in the console). In this case study, we'll create a multiple regression model where:

- The outcome variable y is the sale price of houses.
- Two explanatory variables:

    1. A numerical explanatory variable x_1: house size sqft_living as measured in square feet of living space. Note that 1 square foot is about 0.09 square meters.
    2. A categorical explanatory variable x_2: house condition, a categorical variable with five levels where 1 indicates "poor" and 5 indicates "excellent."

---

[7]https://www.kaggle.com/
[8]https://www.kaggle.com/harlfoxem/housesalesprediction

11.2.1 Exploratory data analysis: Part I

As we've said numerous times throughout this book, a crucial first step when presented with data is to perform an exploratory data analysis (EDA). Exploratory data analysis can give you a sense of your data, help identify issues with your data, bring to light any outliers, and help inform model construction.

Recall the three common steps in an exploratory data analysis we introduced in Subsection 5.1.1:

1. Looking at the raw data values.
2. Computing summary statistics.
3. Creating data visualizations.

First, let's look at the raw data using `View()` to bring up RStudio's spreadsheet viewer and the `glimpse()` function from the `dplyr` package:

```
View(house_prices)
glimpse(house_prices)
```

```
Observations: 21,613
Variables: 21
$ id             <chr> "7129300520", "6414100192", "5631500400", "24872...
$ date           <date> 2014-10-13, 2014-12-09, 2015-02-25, 2014-12-09,...
$ price          <dbl> 221900, 538000, 180000, 604000, 510000, 1225000,...
$ bedrooms       <int> 3, 3, 2, 4, 3, 4, 3, 3, 3, 3, 3, 2, 3, 3, 5, 4, ...
$ bathrooms      <dbl> 1.00, 2.25, 1.00, 3.00, 2.00, 4.50, 2.25, 1.50, ...
$ sqft_living    <int> 1180, 2570, 770, 1960, 1680, 5420, 1715, 1060, 1...
$ sqft_lot       <int> 5650, 7242, 10000, 5000, 8080, 101930, 6819, 971...
$ floors         <dbl> 1.0, 2.0, 1.0, 1.0, 1.0, 1.0, 2.0, 1.0, 1.0, 2.0...
$ waterfront     <lgl> FALSE, FALSE, FALSE, FALSE, FALSE, FALSE, FALSE,...
$ view           <int> 0, 0, 0, 0, 0, 0, 0, 0, 0, 0, 0, 0, 0, 0, 3, ...
$ condition      <fct> 3, 3, 3, 5, 3, 3, 3, 3, 3, 3, 3, 4, 4, 4, 3, 3, ...
$ grade          <fct> 7, 7, 6, 7, 8, 11, 7, 7, 7, 7, 8, 7, 7, 7, 7, 9,...
$ sqft_above     <int> 1180, 2170, 770, 1050, 1680, 3890, 1715, 1060, 1...
$ sqft_basement  <int> 0, 400, 0, 910, 0, 1530, 0, 0, 730, 0, 1700, 300...
$ yr_built       <int> 1955, 1951, 1933, 1965, 1987, 2001, 1995, 1963, ...
$ yr_renovated   <int> 0, 1991, 0, 0, 0, 0, 0, 0, 0, 0, 0, 0, 0, 0, 0, ...
$ zipcode        <fct> 98178, 98125, 98028, 98136, 98074, 98053, 98003,...
$ lat            <dbl> 47.5, 47.7, 47.7, 47.5, 47.6, 47.7, 47.3, 47.4, ...
$ long           <dbl> -122, -122, -122, -122, -122, -122, -122, -122, ...
$ sqft_living15  <int> 1340, 1690, 2720, 1360, 1800, 4760, 2238, 1650, ...
```

```
$ sqft_lot15    <int> 5650, 7639, 8062, 5000, 7503, 101930, 6819, 9711...
```

Here are some questions you can ask yourself at this stage of an EDA: Which variables are numerical? Which are categorical? For the categorical variables, what are their levels? Besides the variables we'll be using in our regression model, what other variables do you think would be useful to use in a model for house price?

Observe, for example, that while the condition variable has values 1 through 5, these are saved in R as fct standing for "factors." This is one of R's ways of saving categorical variables. So you should think of these as the "labels" 1 through 5 and not the numerical values 1 through 5.

Let's now perform the second step in an EDA: computing summary statistics. Recall from Section 3.3 that *summary statistics* are single numerical values that summarize a large number of values. Examples of summary statistics include the mean, the median, the standard deviation, and various percentiles.

We could do this using the summarize() function in the dplyr package along with R's built-in *summary functions*, like mean() and median(). However, recall in Section 3.5, we saw the following code that computes a variety of summary statistics of the variable gain, which is the amount of time that a flight makes up mid-air:

```
gain_summary <- flights %>%
  summarize(
    min = min(gain, na.rm = TRUE),
    q1 = quantile(gain, 0.25, na.rm = TRUE),
    median = quantile(gain, 0.5, na.rm = TRUE),
    q3 = quantile(gain, 0.75, na.rm = TRUE),
    max = max(gain, na.rm = TRUE),
    mean = mean(gain, na.rm = TRUE),
    sd = sd(gain, na.rm = TRUE),
    missing = sum(is.na(gain))
  )
```

To repeat this for all three price, sqft_living, and condition variables would be tedious to code up. So instead, let's use the convenient skim() function from the skimr package we first used in Subsection 6.1.1, being sure to only select() the variables of interest for our model:

```
house_prices %>%
  select(price, sqft_living, condition) %>%
  skim()
```

```
Skim summary statistics
 n obs: 21613
 n variables: 3
```

```
── Variable type:factor
  variable missing complete     n n_unique                      top_counts ordered
 condition     0    21613 21613     5 3: 14031, 4: 5679, 5: 1701, 2: 172   FALSE
```

```
── Variable type:integer
     variable missing complete     n    mean     sd  p0  p25  p50  p75  p100
  sqft_living       0    21613 21613 2079.9 918.44 290 1427 1910 2550 13540
```

```
── Variable type:numeric
 variable missing complete  n    mean       sd    p0    p25    p50    p75    p100
    price    0   21613 21613 540088.14 367127.2 75000 321950 450000 645000 7700000
```

Observe that the mean `price` of $540,088 is larger than the median of $450,000. This is because a small number of very expensive houses are inflating the average. In other words, there are "outlier" house prices in our dataset. (This fact will become even more apparent when we create our visualizations next.)

However, the median is not as sensitive to such outlier house prices. This is why news about the real estate market generally report median house prices and not mean/average house prices. We say here that the median is more *robust to outliers* than the mean. Similarly, while both the standard deviation and interquartile-range (IQR) are both measures of spread and variability, the IQR is more *robust to outliers*.

Let's now perform the last of the three common steps in an exploratory data analysis: creating data visualizations. Let's first create *univariate* visualizations. These are plots focusing on a single variable at a time. Since `price` and `sqft_living` are numerical variables, we can visualize their distributions using a `geom_histogram()` as seen in Section 2.5 on histograms. On the other hand, since `condition` is categorical, we can visualize its distribution using a `geom_bar()`. Recall from Section 2.8 on barplots that since `condition` is not "pre-counted", we use a `geom_bar()` and not a `geom_col()`.

```
# Histogram of house price:
ggplot(house_prices, aes(x = price)) +
  geom_histogram(color = "white") +
  labs(x = "price (USD)", title = "House price")

# Histogram of sqft_living:
ggplot(house_prices, aes(x = sqft_living)) +
  geom_histogram(color = "white") +
  labs(x = "living space (square feet)", title = "House size")

# Barplot of condition:
ggplot(house_prices, aes(x = condition)) +
  geom_bar() +
  labs(x = "condition", title = "House condition")
```

In Figure 11.3, we display all three of these visualizations at once.

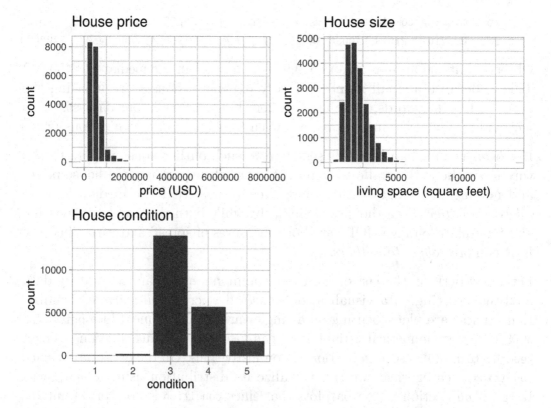

FIGURE 11.3: Exploratory visualizations of Seattle house prices data.

First, observe in the bottom plot that most houses are of condition "3", with a few more of conditions "4" and "5", and almost none that are "1" or "2".

Next, observe in the histogram for `price` in the top-left plot that a majority of houses are less than two million dollars. Observe also that the x-axis stretches out to 8 million dollars, even though there does not appear to be any houses close to that price. This is because there are a *very small number* of houses with prices closer to 8 million. These are the outlier house prices we mentioned earlier. We say that the variable `price` is *right-skewed* as exhibited by the long right tail.

Further, observe in the histogram of `sqft_living` in the middle plot as well that most houses appear to have less than 5000 square feet of living space. For comparison, a football field in the US is about 57,600 square feet, whereas a standard soccer/association football field is about 64,000 square feet. Observe also that this variable is also right-skewed, although not as drastically as the `price` variable.

For both the `price` and `sqft_living` variables, the right-skew makes distinguishing houses at the lower end of the x-axis hard. This is because the scale of the x-axis is compressed by the small number of quite expensive and immensely-sized houses.

So what can we do about this skew? Let's apply a *log10 transformation* to these variables. If you are unfamiliar with such transformations, we highly recommend you read Appendix A.3 on logarithmic (log) transformations. In summary, log transformations allow us to alter the scale of a variable to focus on *multiplicative* changes instead of *additive* changes. In other words, they shift the view to be on *relative* changes instead of *absolute* changes. Such multiplicative/relative changes are also called changes in *orders of magnitude*.

Let's create new log10 transformed versions of the right-skewed variable `price` and `sqft_living` using the `mutate()` function from Section 3.5, but we'll give the latter the name `log10_size`, which is shorter and easier to understand than the name `log10_sqft_living`.

```
house_prices <- house_prices %>%
  mutate(
    log10_price = log10(price),
    log10_size = log10(sqft_living)
    )
```

Let's display the before and after effects of this transformation on these variables for only the first 10 rows of `house_prices`:

```
house_prices %>%
  select(price, log10_price, sqft_living, log10_size)
```

```
# A tibble: 21,613 x 4
     price log10_price sqft_living log10_size
     <dbl>       <dbl>       <int>      <dbl>
1   221900     5.34616        1180    3.07188
2   538000     5.73078        2570    3.40993
3   180000     5.25527         770    2.88649
4   604000     5.78104        1960    3.29226
5   510000     5.70757        1680    3.22531
6  1225000     6.08814        5420    3.73400
7   257500     5.41078        1715    3.23426
8   291850     5.46516        1060    3.02531
9   229500     5.36078        1780    3.25042
10  323000     5.50920        1890    3.27646
# ... with 21,603 more rows
```

Observe in particular the houses in the sixth and third rows. The house in the sixth row has price $1,225,000, which is just above one million dollars. Since 10^6 is one million, its log10_price is around 6.09.

Contrast this with all other houses with log10_price less than six, since they all have price less than $1,000,000. The house in the third row is the only house with sqft_living less than 1000. Since $1000 = 10^3$, it's the lone house with log10_size less than 3.

Let's now visualize the before and after effects of this transformation for price in Figure 11.4.

```
# Before log10 transformation:
ggplot(house_prices, aes(x = price)) +
  geom_histogram(color = "white") +
  labs(x = "price (USD)", title = "House price: Before")


# After log10 transformation:
ggplot(house_prices, aes(x = log10_price)) +
  geom_histogram(color = "white") +
  labs(x = "log10 price (USD)", title = "House price: After")
```

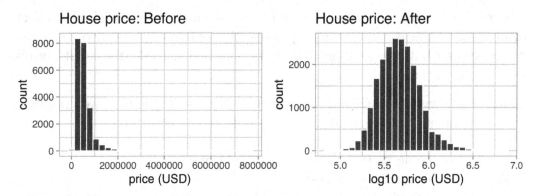

FIGURE 11.4: House price before and after log10 transformation.

Observe that after the transformation, the distribution is much less skewed, and in this case, more symmetric and more bell-shaped. Now you can more easily distinguish the lower priced houses.

Let's do the same for house size, where the variable `sqft_living` was log10 transformed to `log10_size`.

```
# Before log10 transformation:
ggplot(house_prices, aes(x = sqft_living)) +
  geom_histogram(color = "white") +
  labs(x = "living space (square feet)", title = "House size: Before")

# After log10 transformation:
ggplot(house_prices, aes(x = log10_size)) +
  geom_histogram(color = "white") +
  labs(x = "log10 living space (square feet)", title = "House size: After")
```

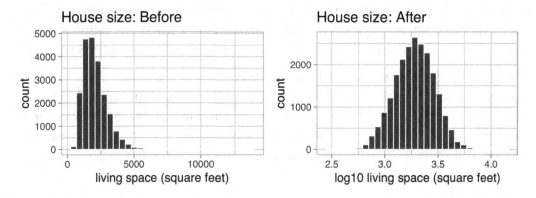

FIGURE 11.5: House size before and after log10 transformation.

Observe in Figure 11.5 that the log10 transformation has a similar effect of unskewing the variable. We emphasize that while in these two cases the resulting distributions are more symmetric and bell-shaped, this is not always necessarily the case.

Given the now symmetric nature of `log10_price` and `log10_size`, we are going to revise our multiple regression model to use our new variables:

1. The outcome variable y is the sale `log10_price` of houses.
2. Two explanatory variables:
    1. A numerical explanatory variable x_1: house size `log10_size` as measured in log base 10 square feet of living space.
    2. A categorical explanatory variable x_2: house `condition`, a categorical variable with five levels where 1 indicates "poor" and 5 indicates "excellent."

11.2.2 Exploratory data analysis: Part II

Let's now continue our EDA by creating *multivariate* visualizations. Unlike the *univariate* histograms and barplot in the earlier Figures 11.3, 11.4, and 11.5, *multivariate* visualizations show relationships between more than one variable. This is an important step of an EDA to perform since the goal of modeling is to explore relationships between variables.

Since our model involves a numerical outcome variable, a numerical explanatory variable, and a categorical explanatory variable, we are in a similar regression modeling situation as in Section 6.1 where we studied the UT Austin teaching scores dataset. Recall in that case the numerical outcome variable was teaching `score`, the numerical explanatory variable was instructor `age`, and the categorical explanatory variable was (binary) `gender`.

We thus have two choices of models we can fit: either (1) an *interaction model* where the regression line for each `condition` level will have both a different slope and a different intercept or (2) a *parallel slopes model* where the regression line for each `condition` level will have the same slope but different intercepts.

Recall from Subsection 6.1.3 that the `geom_parallel_slopes()` function is a special purpose function that Evgeni Chasnovski created and included in the `moderndive` package, since the `geom_smooth()` method in the `ggplot2` package does not have a convenient way to plot parallel slopes models. We plot both resulting models in Figure 11.6, with the interaction model on the left.

```
# Plot interaction model
ggplot(house_prices,
```

```
        aes(x = log10_size, y = log10_price, col = condition)) +
  geom_point(alpha = 0.05) +
  geom_smooth(method = "lm", se = FALSE) +
  labs(y = "log10 price",
       x = "log10 size",
       title = "House prices in Seattle")
# Plot parallel slopes model
ggplot(house_prices,
        aes(x = log10_size, y = log10_price, col = condition)) +
  geom_point(alpha = 0.05) +
  geom_parallel_slopes(se = FALSE) +
  labs(y = "log10 price",
       x = "log10 size",
       title = "House prices in Seattle")
```

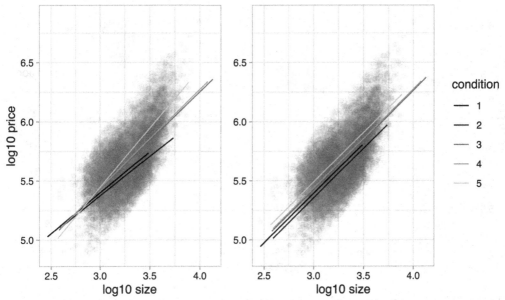

FIGURE 11.6: Interaction and parallel slopes models.

In both cases, we see there is a positive relationship between house price and size, meaning as houses are larger, they tend to be more expensive. Furthermore, in both plots it seems that houses of condition 5 tend to be the most expensive for most house sizes as evidenced by the fact that the line for condition 5 is highest, followed by conditions 4 and 3. As for conditions 1 and 2, this pattern isn't as clear. Recall from the univariate barplot of condition in Figure 11.3, there are only a few houses of condition 1 or 2.

Let's also show a faceted version of just the interaction model in Figure 11.7. It is now much more apparent just how few houses are of condition 1 or 2.

```
ggplot(house_prices,
       aes(x = log10_size, y = log10_price, col = condition)) +
  geom_point(alpha = 0.4) +
  geom_smooth(method = "lm", se = FALSE) +
  labs(y = "log10 price",
       x = "log10 size",
       title = "House prices in Seattle") +
  facet_wrap(~ condition)
```

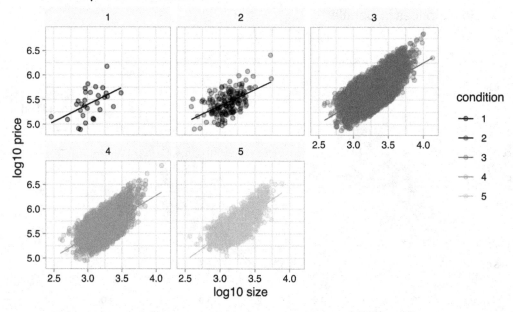

FIGURE 11.7: Faceted plot of interaction model.

Which exploratory visualization of the interaction model is better, the one in the left-hand plot of Figure 11.6 or the faceted version in Figure 11.7? There is no universal right answer. You need to make a choice depending on what you want to convey, and own that choice, with including and discussing both also as an option as needed.

11.2.3   Regression modeling

Which of the two models in Figure 11.6 is "better"? The interaction model in the left-hand plot or the parallel slopes model in the right-hand plot?

We had a similar discussion in Subsection 6.3.1 on *model selection*. Recall that we stated that we should only favor more complex models if the additional complexity is *warranted*. In this case, the more complex model is the interaction model since it considers five intercepts and five slopes total. This is in contrast to the parallel slopes model which considers five intercepts but only one common slope.

Is the additional complexity of the interaction model warranted? Looking at the left-hand plot in Figure 11.6, we're of the opinion that it is, as evidenced by the slight x-like pattern to some of the lines. Therefore, we'll focus the rest of this analysis only on the interaction model. This visual approach is somewhat subjective, however, so feel free to disagree! What are the five different slopes and five different intercepts for the interaction model? We can obtain these values from the regression table. Recall our two-step process for getting the regression table:

```
# Fit regression model:
price_interaction <- lm(log10_price ~ log10_size * condition,
                        data = house_prices)

# Get regression table:
get_regression_table(price_interaction)
```

TABLE 11.1: Regression table for interaction model

term	estimate	std_error	statistic	p_value	lower_ci	upper_ci
intercept	3.330	0.451	7.380	0.000	2.446	4.215
log10_size	0.690	0.148	4.652	0.000	0.399	0.980
condition2	0.047	0.498	0.094	0.925	-0.930	1.024
condition3	-0.367	0.452	-0.812	0.417	-1.253	0.519
condition4	-0.398	0.453	-0.879	0.380	-1.286	0.490
condition5	-0.883	0.457	-1.931	0.053	-1.779	0.013
log10_size:condition2	-0.024	0.163	-0.148	0.882	-0.344	0.295
log10_size:condition3	0.133	0.148	0.893	0.372	-0.158	0.424
log10_size:condition4	0.146	0.149	0.979	0.328	-0.146	0.437
log10_size:condition5	0.310	0.150	2.067	0.039	0.016	0.604

Recall we saw in Subsection 6.1.2 how to interpret a regression table when there are both numerical and categorical explanatory variables. Let's now do the same for all 10 values in the `estimate` column of Table 11.1.

In this case, the "baseline for comparison" group for the categorical variable `condition` are the condition 1 houses, since "1" comes first alphanumerically. Thus, the `intercept` and `log10_size` values are the intercept and slope for `log10_size` for this baseline group. Next, the `condition2` through `condition5` terms are the *offsets* in intercepts relative to the condition 1 intercept. Finally, the `log10_size:condition2` through `log10_size:condition5` are the *offsets* in slopes for `log10_size` relative to the condition 1 slope for `log10_size`.

Let's simplify this by writing out the equation of each of the five regression lines using these 10 `estimate` values. We'll write out each line in the following format:

$$\log \widehat{10(\text{price})} = \hat{\beta}_0 + \hat{\beta}_{\text{size}} \cdot \log 10(\text{size})$$

1. Condition 1:

$$\log \widehat{10(\text{price})} = 3.33 + 0.69 \cdot \log 10(\text{size})$$

2. Condition 2:

$$\log \widehat{10(\text{price})} = (3.33 + 0.047) + (0.69 - 0.024) \cdot \log 10(\text{size})$$
$$= 3.377 + 0.666 \cdot \log 10(\text{size})$$

3. Condition 3:

$$\log \widehat{10(\text{price})} = (3.33 - 0.367) + (0.69 + 0.133) \cdot \log 10(\text{size})$$
$$= 2.963 + 0.823 \cdot \log 10(\text{size})$$

4. Condition 4:

$$\log \widehat{10(\text{price})} = (3.33 - 0.398) + (0.69 + 0.146) \cdot \log 10(\text{size})$$
$$= 2.932 + 0.836 \cdot \log 10(\text{size})$$

5. Condition 5:

$$\log \widehat{10(\text{price})} = (3.33 - 0.883) + (0.69 + 0.31) \cdot \log 10(\text{size})$$
$$= 2.447 + 1 \cdot \log 10(\text{size})$$

These correspond to the regression lines in the left-hand plot of Figure 11.6 and the faceted plot in Figure 11.7. For homes of all five condition types, as the size of the house increases, the price increases. This is what most would expect. However, the rate of increase of price with size is fastest for the homes with conditions 3, 4, and 5 of 0.823, 0.836, and 1, respectively. These are the three largest slopes out of the five.

11.2.4 Making predictions

Say you're a realtor and someone calls you asking you how much their home will sell for. They tell you that it's in condition = 5 and is sized 1900 square feet. What do you tell them? Let's use the interaction model we fit to make predictions!

We first make this prediction visually in Figure 11.8. The predicted log10_price of this house is marked with a black dot. This is where the following two lines intersect:

- The regression line for the condition = 5 homes and
- The vertical dashed black line at log10_size equals 3.28, since our predictor variable is the log10 transformed square feet of living space of $\log 10(1900) = 3.28$.

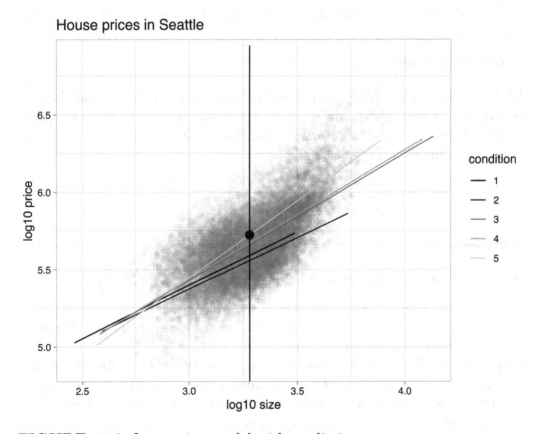

FIGURE 11.8: Interaction model with prediction.

Eyeballing it, it seems the predicted log10_price seems to be around 5.75. Let's now obtain the exact numerical value for the prediction using the equation of the regression line for the condition = 5 houses, being sure to log10() the square footage first.

```
2.45 + 1 * log10(1900)
```

```
[1] 5.73
```

This value is very close to our earlier visually made prediction of 5.75. But wait! Is our prediction for the price of this house $5.75? No! Remember that we are using log10_price as our outcome variable! So, if we want a prediction in dollar units of price, we need to unlog this by taking a power of 10 as described in Appendix A.3.

```
10^(2.45 + 1 * log10(1900))
```

```
[1] 535493
```

So our predicted price for this home of condition 5 and of size 1900 square feet is $535,493.

Learning check

(LC11.1) Repeat the regression modeling in Subsection 11.2.3 and the prediction making you just did on the house of condition 5 and size 1900 square feet in Subsection 11.2.4, but using the parallel slopes model you visualized in Figure 11.6. Show that it's $524,807!

11.3   Case study: Effective data storytelling

As we've progressed throughout this book, you've seen how to work with data in a variety of ways. You've learned effective strategies for plotting data by understanding which types of plots work best for which combinations of variable types. You've summarized data in spreadsheet form and calculated summary statistics for a variety of different variables. Furthermore, you've seen the value of statistical inference as a process to come to conclusions about a population by using sampling. Lastly, you've explored how to fit linear regression models and the importance of checking the conditions required so that all confidence intervals and hypothesis tests have valid interpretation. All throughout, you've learned many computational techniques and focused on writing R code that's reproducible.

We now present another set of case studies, but this time on the "effective data storytelling" done by data journalists around the world. Great data stories don't mislead the reader, but rather engulf them in understanding the importance that data plays in our lives through storytelling.

11.3.1 Bechdel test for Hollywood gender representation

We recommend you read and analyze Walt Hickey's FiveThirtyEight.com article, "The Dollar-And-Cents Case Against Hollywood's Exclusion of Women."[9] In it, Walt completed a multidecade study of how many movies pass the Bechdel test[10], an informal test of gender representation in a movie that was created by Alison Bechdel.

As you read over the article, think carefully about how Walt Hickey is using data, graphics, and analyses to tell the reader a story. In the spirit of reproducibility, FiveThirtyEight have also shared the data and R code[11] that they used for this article. You can also find the data used in many more of their articles on their GitHub[12] page.

ModernDive co-authors Chester Ismay and Albert Y. Kim along with Jennifer Chunn went one step further by creating the `fivethirtyeight` package which provides access to these datasets more easily in R. For a complete list of all 127 datasets included in the `fivethirtyeight` package, check out the package webpage at `https://fivethirtyeight-r.netlify.com/articles/fivethirtyeight.html`.

Furthermore, example "vignettes" of fully reproducible start-to-finish analyses of some of these data using `dplyr`, `ggplot2`, and other packages in the `tidyverse` are available here[13]. For example, a vignette showing how to reproduce one of the plots at the end of the article on the Bechdel test is available here[14].

11.3.2 US Births in 1999

The `US_births_1994_2003` data frame included in the `fivethirtyeight` package provides information about the number of daily births in the United States

---

[9]`http://fivethirtyeight.com/features/the-dollar-and-cents-case-against-hollywoods-exclusion-of-women/`

[10]`https://bechdeltest.com/`

[11]`https://github.com/fivethirtyeight/data/tree/master/bechdel`

[12]`https://github.com/fivethirtyeight/data`

[13]`https://fivethirtyeight-r.netlify.com/articles/`

[14]`https://fivethirtyeight-r.netlify.com/articles/bechdel.html`

between 1994 and 2003. For more information on this data frame including a link to the original article on FiveThirtyEight.com, check out the help file by running ?US_births_1994_2003 in the console.

It's always a good idea to preview your data, either by using RStudio's spreadsheet View() function or using glimpse() from the dplyr package:

```
glimpse(US_births_1994_2003)
```

```
Observations: 3,652
Variables: 6
$ year          <int> 1994, 1994, 1994, 1994, 1994, 1994, 1994, 1994, ...
$ month         <int> 1, 1, 1, 1, 1, 1, 1, 1, 1, 1, 1, 1, 1, 1, 1, 1, ...
$ date_of_month <int> 1, 2, 3, 4, 5, 6, 7, 8, 9, 10, 11, 12, 13, 14, 1...
$ date          <date> 1994-01-01, 1994-01-02, 1994-01-03, 1994-01-04,...
$ day_of_week   <ord> Sat, Sun, Mon, Tues, Wed, Thurs, Fri, Sat, Sun, ...
$ births        <int> 8096, 7772, 10142, 11248, 11053, 11406, 11251, 8...
```

We'll focus on the number of births for each date, but only for births that occurred in 1999. Recall from Section 3.2 we can do this using the filter() function from the dplyr package:

```
US_births_1999 <- US_births_1994_2003 %>%
  filter(year == 1999)
```

As discussed in Section 2.4, since date is a notion of time and thus has sequential ordering to it, a linegraph would be a more appropriate visualization to use than a scatterplot. In other words, we should use a geom_line() instead of geom_point(). Recall that such plots are called *time series* plots.

```
ggplot(US_births_1999, aes(x = date, y = births)) +
  geom_line() +
  labs(x = "Date",
       y = "Number of births",
       title = "US Births in 1999")
```

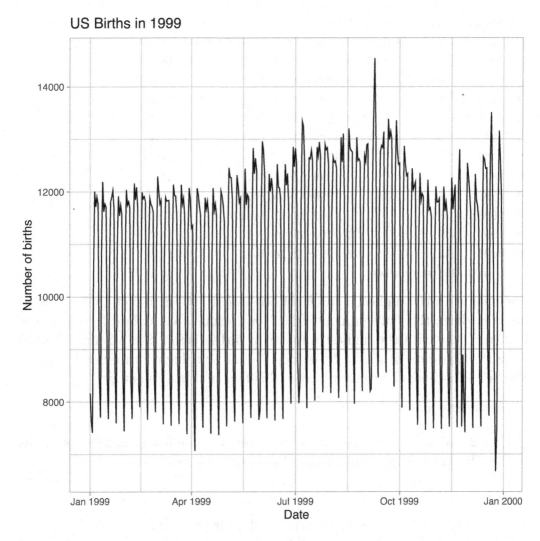

FIGURE 11.9: Number of births in the US in 1999.

We see a big dip occurring just before January 1st, 2000, most likely due to the holiday season. However, what about the large spike of over 14,000 births occurring just before October 1st, 1999? What could be the reason for this anomalously high spike?

Let's sort the rows of US_births_1999 in descending order of the number of births. Recall from Section 3.6 that we can use the arrange() function from the dplyr function to do this, making sure to sort births in descending order:

```
US_births_1999 %>%
    arrange(desc(births))
```

```
# A tibble: 365 x 6
     year month date_of_month date       day_of_week births
    <int> <int>         <int> <date>      <ord>       <int>
 1   1999     9             9 1999-09-09 Thurs       14540
 2   1999    12            21 1999-12-21 Tues        13508
 3   1999     9             8 1999-09-08 Wed         13437
 4   1999     9            21 1999-09-21 Tues        13384
 5   1999     9            28 1999-09-28 Tues        13358
 6   1999     7             7 1999-07-07 Wed         13343
 7   1999     7             8 1999-07-08 Thurs       13245
 8   1999     8            17 1999-08-17 Tues        13201
 9   1999     9            10 1999-09-10 Fri         13181
10   1999    12            28 1999-12-28 Tues        13158
# ... with 355 more rows
```

The date with the highest number of births (14,540) is in fact 1999-09-09. If
we write down this date in month/day/year format (a standard format in the
US), the date with the highest number of births is 9/9/99! All nines! Could
it be that parents deliberately induced labor at a higher rate on this date?
Maybe? Whatever the cause may be, this fact makes a fun story!

Learning check

(**LC11.2**) What date between 1994 and 2003 has the fewest number of births
in the US? What story could you tell about why this is the case?

Time to think with data and further tell your story with data! How could
statistical modeling help you here? What types of statistical inference would
be helpful? What else can you find and where can you take this analysis? What
assumptions did you make in this analysis? We leave these questions to you as
the reader to explore and examine.

Remember to get in touch with us via our contact info in the Preface. We'd
love to see what you come up with!

Please check out additional problem sets and labs at https://moderndive.com/labs
as well.

11.3.3 Scripts of R code

An R script file of all R code used in this chapter is available at https:
//www.moderndive.com/scripts/11-tell-your-story-with-data.R.

R code files saved as *.R files for all relevant chapters throughout the entire book are in the following table.

chapter	link
1	https://moderndive.com/scripts/01-getting-started.R
2	https://moderndive.com/scripts/02-visualization.R
3	https://moderndive.com/scripts/03-wrangling.R
4	https://moderndive.com/scripts/04-tidy.R
5	https://moderndive.com/scripts/05-regression.R
6	https://moderndive.com/scripts/06-multiple-regression.R
7	https://moderndive.com/scripts/07-sampling.R
8	https://moderndive.com/scripts/08-confidence-intervals.R
9	https://moderndive.com/scripts/09-hypothesis-testing.R
10	https://moderndive.com/scripts/10-inference-for-regression.R
11	https://moderndive.com/scripts/11-tell-your-story-with-data.R

Concluding remarks

Now that you've made it to this point in the book, we suspect that you know a thing or two about how to work with data in R! You've also gained a lot of knowledge about how to use simulation-based techniques for statistical inference and how these techniques help build intuition about traditional theory-based inferential methods like the *t*-test.

The hope is that you've come to appreciate the power of data in all respects, such as data wrangling, tidying datasets, data visualization, data modeling, and statistical inference. In our opinion, while each of these is important, data visualization may be the most important tool for a citizen or professional data scientist to have in their toolbox. If you can create truly beautiful graphics that display information in ways that the reader can clearly understand, you have great power to tell your tale with data. Let's hope that these skills help you tell great stories with data into the future. Thanks for coming along this journey as we dove into modern data analysis using R and the `tidyverse`!

A

Statistical Background

A.1 Basic statistical terms

Note that all the following statistical terms apply only to *numerical* variables, except the *distribution* which can exist for both numerical and categorical variables.

A.1.1 Mean

The *mean* is the most commonly reported measure of center. It is commonly called the *average* though this term can be a little ambiguous. The mean is the sum of all of the data elements divided by how many elements there are. If we have n data points, the mean is given by:

$$Mean = \frac{x_1 + x_2 + \cdots + x_n}{n}$$

A.1.2 Median

The median is calculated by first sorting a variable's data from smallest to largest. After sorting the data, the middle element in the list is the *median*. If the middle falls between two values, then the median is the mean of those two middle values.

A.1.3 Standard deviation

We will next discuss the *standard deviation* (*sd*) of a variable. The formula can be a little intimidating at first but it is important to remember that it is essentially a measure of how far we expect a given data value will be from its mean:

$$sd = \sqrt{\frac{(x_1 - Mean)^2 + (x_2 - Mean)^2 + \cdots + (x_n - Mean)^2}{n-1}}$$

A.1.4   Five-number summary

The *five-number summary* consists of five summary statistics: the minimum, the first quantile AKA 25th percentile, the second quantile AKA median or 50th percentile, the third quantile AKA 75th, and the maximum. The five-number summary of a variable is used when constructing boxplots, as seen in Section 2.7.

The quantiles are calculated as

- first quantile (Q_1): the median of the first half of the sorted data
- third quantile (Q_3): the median of the second half of the sorted data

The *interquartile range (IQR)* is defined as $Q_3 - Q_1$ and is a measure of how spread out the middle 50% of values are. The IQR corresponds to the length of the box in a boxplot.

The median and the IQR are not influenced by the presence of outliers in the ways that the mean and standard deviation are. They are, thus, recommended for skewed datasets. We say in this case that the median and IQR are more *robust to outliers.*

A.1.5   Distribution

The *distribution* of a variable shows how frequently different values of a variable occur. Looking at the visualization of a distribution can show where the values are centered, show how the values vary, and give some information about where a typical value might fall. It can also alert you to the presence of outliers.

Recall from Chapter 2 that we can visualize the distribution of a numerical variable using binning in a histogram and that we can visualize the distribution of a categorical variable using a barplot.

A.1.6   Outliers

Outliers correspond to values in the dataset that fall far outside the range of "ordinary" values. In the context of a boxplot, by default they correspond to values below $Q_1 - (1.5 \cdot IQR)$ or above $Q_3 + (1.5 \cdot IQR)$.

A.2   Normal distribution

Let's next discuss one particular kind of distribution: *normal distributions.* Such bell-shaped distributions are defined by two values: (1) the *mean μ* ("mu") which locates the center of the distribution and (2) the *standard deviation* σ ("sigma") which determines the variation of the distribution. In Figure A.1, we plot three normal distributions where:

1. The solid normal curve has mean $\mu = 5$ & standard deviation $\sigma = 2$.
2. The dotted normal curve has mean $\mu = 5$ & standard deviation $\sigma = 5$.
3. The dashed normal curve has mean $\mu = 15$ & standard deviation $\sigma = 2$.

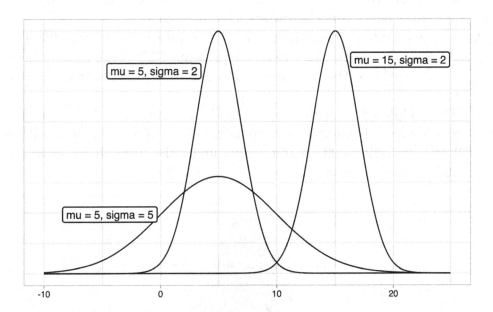

FIGURE A.1: Three normal distributions.

Notice how the solid and dotted line normal curves have the same center due to their common mean $\mu = 5$. However, the dotted line normal curve is wider due to its larger standard deviation of $\sigma = 5$. On the other hand, the solid and dashed line normal curves have the same variation due to their common standard deviation $\sigma = 2$. However, they are centered at different locations.

When the mean $\mu = 0$ and the standard deviation $\sigma = 1$, the normal distribution has a special name. It's called the *standard normal distribution* or the *z-curve*.

Furthermore, if a variable follows a normal curve, there are *three rules of thumb* we can use:

1. 68% of values will lie within $\pm$ 1 standard deviation of the mean.
2. 95% of values will lie within $\pm$ 1.96 $\approx$ 2 standard deviations of the mean.

3. 99.7% of values will lie within ± 3 standard deviations of the mean.

Let's illustrate this on a standard normal curve in Figure A.2. The dashed lines are at -3, -1.96, -1, 0, 1, 1.96, and 3. These 7 lines cut up the x-axis into 8 segments. The areas under the normal curve for each of the 8 segments are marked and add up to 100%. For example:

1. The middle two segments represent the interval -1 to 1. The shaded area above this interval represents 34% + 34% = 68% of the area under the curve. In other words, 68% of values.
2. The middle four segments represent the interval -1.96 to 1.96. The shaded area above this interval represents 13.5% + 34% + 34% + 13.5%= 95% of the area under the curve. In other words, 95% of values.
3. The middle six segments represent the interval -3 to 3. The shaded area above this interval represents 2.35% + 13.5% + 34% + 34% + 13.5% + 2.35% = 99.7% of the area under the curve. In other words, 99.7% of values.

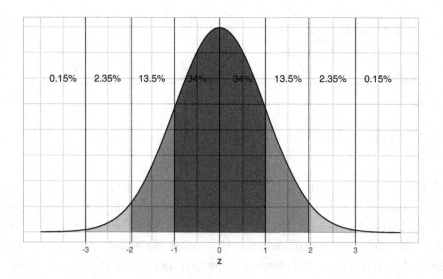

FIGURE A.2: Rules of thumb about areas under normal curves.

Learning check

Say you have a normal distribution with mean $\mu = 6$ and standard deviation $\sigma = 3$.

(LCA.1) What proportion of the area under the normal curve is less than 3? Greater than 12? Between 0 and 12?

(LCA.2) What is the 2.5th percentile of the area under the normal curve? The 95th percentile? The 100th percentile?

A.3 log10 transformations

At its simplest, log10 transformations return base 10 *logarithms*. For example, since $1000 = 10^3$, running `log10(1000)` returns `3` in R. To undo a log10 transformation, we raise 10 to this value. For example, to undo the previous log10 transformation and return the original value of 1000, we raise 10 to the power of 3 by running `10^(3) = 1000` in R.

Log transformations allow us to focus on changes in *orders of magnitude*. In other words, they allow us to focus on *multiplicative changes* instead of *additive ones*. Let's illustrate this idea in Table A.1 with examples of prices of consumer goods in 2019 US dollars.

TABLE A.1: log10 transformed prices, orders of magnitude, and examples

Price	log10(Price)	Order of magnitude	Examples
$1	0	Singles	Cups of coffee
$10	1	Tens	Books
$100	2	Hundreds	Mobile phones
$1,000	3	Thousands	High definition TVs
$10,000	4	Tens of thousands	Cars
$100,000	5	Hundreds of thousands	Luxury cars and houses
$1,000,000	6	Millions	Luxury houses

Let's make some remarks about log10 transformations based on Table A.1:

1. When purchasing a cup of coffee, we tend to think of prices ranging in single dollars, such as $2 or $3. However, when purchasing a mobile phone, we don't tend to think of their prices in units of single dollars such as $313 or $727. Instead, we tend to think of their prices in units of hundreds of dollars like $300 or $700. Thus, cups of coffee and mobile phones are of different *orders of magnitude* in price.
2. Let's say we want to know the log10 transformed value of $76. This would be hard to compute exactly without a calculator. However, since $76 is between $10 and $100 and since $\log10(10) = 1$ and $\log10(100)$

= 2, we know log10(76) will be between 1 and 2. In fact, log10(76) is 1.880814.

3. log10 transformations are *monotonic*, meaning they preserve orders. So if Price A is lower than Price B, then log10(Price A) will also be lower than log10(Price B).

4. Most importantly, increments of one in log10-scale correspond to *relative multiplicative changes* in the original scale and not *absolute additive changes*. For example, increasing a log10(Price) from 3 to 4 corresponds to a multiplicative increase by a factor of 10: $100 to $1000.

B

Versions of R Packages Used

If you are seeing different results than what is in the book, we recommend installing the exact version of the packages we used. This can be done by first installing the `remotes` package via `install.packages("remotes")`. Then, use `install_version()` replacing the `package` argument with the package name in quotes and the `version` argument with the particular version number to install.[1]

```
remotes::install_version(package = "skimr", version = "1.0.6")
```

package	version
bookdown	0.16
broom	0.5.2
dplyr	0.8.3
fivethirtyeight	0.5.0
forcats	0.4.0
gapminder	0.3.0
ggplot2	3.2.1
ggplot2movies	0.0.1
infer	0.5.1
ISLR	1.2
janitor	1.2.0
kableExtra	1.1.0
knitr	1.26
moderndive	0.4.0
mvtnorm	1.0-11
nycflights13	1.0.1
patchwork	0.0.1
purrr	0.3.3
readr	1.3.1
scales	1.1.0
skimr	1.0.6
stringr	1.4.0
tibble	2.1.3
tidyr	1.0.0
tidyverse	1.3.0
viridis	0.5.1
viridisLite	0.3.0

---

[1]As of November 2019, the `patchwork` package is not on CRAN and needs to be installed via `remotes::install_github("thomasp85/patchwork")` instead of using `install_version()`.

Bibliography

Bray, A., Ismay, C., Chasnovski, E., Baumer, B., and Cetinkaya-Rundel, M. (2019). *infer: Tidy Statistical Inference*. R package version 0.5.1.

Chihara, L. M. and Hesterberg, T. C. (2011). *Mathematical Statistics with Resampling and R*. John Wiley & Sons, Hoboken, NJ, first edition.

Diez, D. M., Barr, C. D., and Çetinkaya Rundel, M. (2014). *Introductory Statistics with Randomization and Simulation*. CreateSpace Independent Publishing Platform, Scotts Valley, CA, first edition.

Firke, S. (2019). *janitor: Simple Tools for Examining and Cleaning Dirty Data*. R package version 1.2.0.

Grolemund, G. and Wickham, H. (2017). *R for Data Science*. O'Reilly Media, Sebastopol, CA, first edition.

Ismay, C. and Kennedy, P. C. (2016). *Getting Used to R, RStudio, and R Markdown*.

James, G., Witten, D., Hastie, T., and Tibshirani, R. (2017). *An Introduction to Statistical Learning: with Applications in R*. Springer, New York, NY, first edition.

Kim, A. Y. and Ismay, C. (2019). *moderndive: Tidyverse-Friendly Introductory Linear Regression*. R package version 0.4.0.

Kim, A. Y., Ismay, C., and Chunn, J. (2019). *fivethirtyeight: Data and Code Behind the Stories and Interactives at 'FiveThirtyEight'*. R package version 0.5.0.

Quinn, M., McNamara, A., Arino de la Rubia, E., Zhu, H., and Ellis, S. (2019). *skimr: Compact and Flexible Summaries of Data*. R package version 1.0.6.

Robbins, N. (2013). *Creating More Effective Graphs*. Chart House, New York, NY, first edition.

Robinson, D. and Hayes, A. (2019). *broom: Convert Statistical Analysis Objects into Tidy Tibbles*. R package version 0.5.2.

Wickham, H. (2014). Tidy data. *Journal of Statistical Software*, Volume 59(Issue 10).

Wickham, H. (2019a). *nycflights13: Flights that Departed NYC in 2013*. R package version 1.0.1.

Wickham, H. (2019b). *tidyverse: Easily Install and Load the 'Tidyverse'*. R package version 1.3.0.

Wickham, H., Chang, W., Henry, L., Pedersen, T. L., Takahashi, K., Wilke, C., Woo, K., and Yutani, H. (2019a). *ggplot2: Create Elegant Data Visualisations Using the Grammar of Graphics*. R package version 3.2.1.

Wickham, H., François, R., Henry, L., and Müller, K. (2019b). *dplyr: A Grammar of Data Manipulation*. R package version 0.8.3.

Wickham, H. and Henry, L. (2019). *tidyr: Tidy Messy Data*. R package version 1.0.0.

Wickham, H., Hester, J., and Francois, R. (2018). *readr: Read Rectangular Text Data*. R package version 1.3.1.

Wilkinson, L. (2005). *The Grammar of Graphics (Statistics and Computing)*. Springer-Verlag, Secaucus, NJ, first edition.

Xie, Y. (2019). *bookdown: Authoring Books and Technical Documents with R Markdown*. R package version 0.16.

Index

Printed in the United States
by Baker & Taylor Publisher Services